Ceramics and Composites: Properties, Processing and Characterization

Ceramics and Composites: Properties, Processing and Characterization

Edited by **Carl Burt**

WILLFORD PRESS

New York

Published by Willford Press,
118-35 Queens Blvd., Suite 400,
Forest Hills, NY 11375, USA
www.willfordpress.com

Ceramics and Composites: Properties, Processing and Characterization
Edited by Carl Burt

International Standard Book Number: 978-1-68285-083-1 (Hardback)

Printed in the United States of America.

Contents

Permissions

List of Contributors

Preface

This book on ceramics and ceramic composites aims to provide a detailed overview of topics related to properties and processing of different ceramics and ceramic-based composites. The chapters included in this book encompass different aspects like novel forming and sintering technologies, structural characterization, thermal, mechanical and other properties, advanced ceramic materials like ceramic coatings, nanostructured ceramics, etc. Coherent flow of topics, student-friendly language and extensive use of examples make this book an invaluable source of knowledge and will provide comprehensive insight to the readers.

This book is a comprehensive compilation of works of different researchers from varied parts of the world. It includes valuable experiences of the researchers with the sole objective of providing the readers (learners) with a proper knowledge of the concerned field. This book will be beneficial in evoking inspiration and enhancing the knowledge of the interested readers.

In the end, I would like to extend my heartiest thanks to the authors who worked with great determination on their chapters. I also appreciate the publisher's support in the course of the book. I would also like to deeply acknowledge my family who stood by me as a source of inspiration during the project.

Editor

Electrical properties of $Na_2Pb_2R_2W_2Ti_4V_4O_{30}$ (R = Dy, Pr) ceramics

Piyush R. DAS[*], B. N. PARIDA, R. PADHEE, R. N. P. CHOUDHARY

Department of Physics, Institute of Technical Education & Research, Siksha 'O' Anusandahan University, Khandagiri, Bhubaneswar 751030, Odisha, India

Abstract: The polycrystalline samples of complex tungsten bronze (TB) $Na_2Pb_2R_2W_2Ti_4V_4O_{30}$ (R=Dy, Pr) compounds were prepared by solid-state reaction technique. Room- temperature preliminary structural studies confirm the formation of the compounds in the orthorhombic crystal system. Detailed studies of electrical properties of the materials using complex impedance spectroscopy technique exhibit that the impedance and related parameters are strongly dependent upon temperature and microstructure (bulk, grain boundary, etc). An observation of negative temperature coefficient of resistance (NTCR) suggests the materials have semiconducting properties. The variation of AC conductivity with temperature shows a typical Arrhenius behavior of the materials. Both the samples obey Jonscher's universal power law. The existence of hopping mechanism in the electrical transport processes in the system with non-exponential type of conductivity relaxation is confirmed by electrical modulus analysis.

Keywords: ceramics; impedance spectroscopy; electrical properties; ferroelectricity; microstructure

1 Introduction

Though a large number of ferroelectric oxides of different structural family are known today, some ferroelectrics with tungsten bronze (TB) structure have fascinated many researchers because of their interesting physical properties useful for transducer, multi-layered capacitors, microwave dielectric resonators, pyroelectric detectors, actuators, etc. The TB structure has complex crystal structure with a general chemical formula $[(A_1)_2(A_2)_4(C)_4][(B_1)_2(B_2)_8]O_{30}$, where the A-sites are usually filled by mono-trivalent cations, and the B-sites by W^{6+}, Ti^{4+}, Nb^{5+}, Ta^{5+} or V^{5+} atoms. As the C-site is generally empty, the above formula reduces to $A_6B_{10}O_{30}$. The physical properties of the materials with the TB structure can thus be tailored by the substitution of varieties of cations at different interstitial sites for various applications. Detailed literature survey on various physical properties of TB structure reveals that a lot of work has been done in the past on ferroelectric ceramics of this family [1–10]. The structure and ferroelectric properties in six-valence complex TB structure compounds such as $Na_2Pb_2Sm_2W_2Ti_4Nb_4O_{30}$ [11], $Na_2Pb_2Nd_2W_2Ti_4Nb_4O_{30}$ [12], and $Na_2Pb_2R_2W_2Ti_4V_4O_{30}$ (R = Gd, Eu) [13] have already been reported by us. We have already reported the details of thermo-gravimetry analysis (TGA), crystal structure at room temperature, microstructure, ferroelectric properties and AC conductivity of the titled compounds: $Na_2Pb_2Dy_2W_2Ti_4V_4O_{30}$ (NPDWTV) and $Na_2Pb_2Pr_2W_2Ti_4V_4O_{30}$ (NPPWTV) [14]. In the present work, we mainly report the impedance properties of these materials.

---

* Corresponding author.
E-mail: prdas63@gmail.com

2 Experiment

The polycrystalline samples of $Na_2Pb_2Dy_2W_2Ti_4V_4O_{30}$ and $Na_2Pb_2Pr_2W_2Ti_4V_4O_{30}$ were prepared relatively at low temperature using a standard mixed-oxide method with high-purity raw materials (analytical reagent grade): Na_2CO_3, PbO, Dy_2O_3, Pr_2O_3, TiO_2, V_2O_5 and WO_3. These oxides and carbonate were thoroughly mixed in dry (air) and wet (methanol) medium each for 1 h using agate mortar. The calcination temperature was optimized using TGA and repeated firing. Finally, the physical mixtures were calcined at 650 ℃ for 4 h in alumina crucibles. The formation of compounds was checked by X-ray diffraction (XRD) technique using X-ray powder diffractometer (Rigaku Miniflex, Japan, Cu Kα radiation). The calcined powders were pelletized and sintered at 675 ℃ for 4 h. The pellets were coated with high-quality silver paste and dried at 150 ℃. The impedance and related parameters were measured on the pellets using a computer-controlled Hioki-LCR Hi-TESTER (Model 3532, Japan) as a function of frequency (100 Hz–1 MHz) at different temperatures (25–500 ℃).

3 Results and discussion

3.1 Sample preparation and XRD analysis

The samples of NPDWTV and NPPWTV were prepared relatively at low temperature using a standard mixed-oxide method. In our previous work [14], the diffraction peaks of the ceramics are indexed using the standard computer program POWD [15] and the orthorhombic crystal structure is selected. The least-square lattice parameters (as reported) are: $a = 18.8696(12)$ Å, $b = 19.8915(12)$ Å, $c = 3.8018(12)$ Å for NPDWTV, and $a = 19.7863(7)$ Å, $b = 19.9355(7)$ Å, $c = 3.7269(7)$ Å for NPPWTV (estimated standard deviation in parenthesis).

3.2 Complex impedance analysis

Complex impedance spectroscopy (CIS) is a non-destructive and powerful technique to study the electrical properties of ferroelectrics and ionic conductors over a wide range of frequency and temperature. This technique also enables us to estimate the contribution of bulk, grain boundary and material electrode polarization in the impedance. For this, AC

signal is applied across the sample and the output response is then measured. The impedance and related parameters of the materials give us both real (resistive) and imaginary (reactive) components. The AC response of a material is generally expressed as follows:

complex impedance

$$Z^*(\omega) = Z' - jZ'' = R_s - \frac{j}{\omega C_s}$$

complex modulus

$$M^*(\omega) = \frac{1}{\varepsilon^*(\omega)} = M' + jM'' = j\omega C_0 Z^*$$

complex admittance

$$Y^* = Y' + jY'' = j\omega C_0 \varepsilon^* = (R_p)^{-1} + j\omega C_p$$

complex permittivity $\varepsilon^* = \varepsilon' - j\varepsilon''$

loss tangent $\tan\delta = \dfrac{\varepsilon''}{\varepsilon'} = -\dfrac{Z'}{Z''} = \dfrac{M''}{M'}$

where $\omega = 2\pi f$ is the angular frequency; C_0 is the geometrical capacitance; and $j = \sqrt{-1}$. The subscripts "p" and "s" refer to the parallel and series circuit components, respectively.

3.2.1 Impedance analysis

Figure 1 shows the variation of Z' with frequency at different temperatures. It is found that the value of Z' decreases with the increase of both frequency and temperature. The conductivity of materials increases with the rise in temperature and frequency. At high frequency, the value of Z' coincides for all temperatures, implying the possible release of space charge [16].

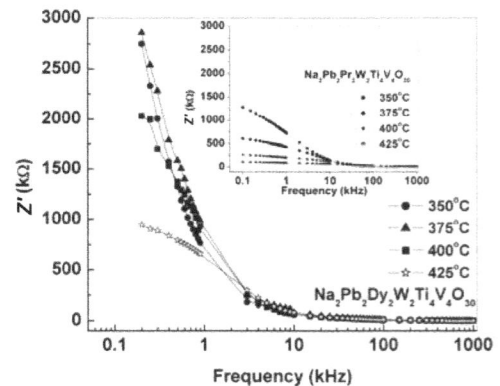

Fig. 1   Variation of Z' with frequency at different temperatures for NPDWTV and NPPWTV.

Figure 2 shows the variation of Z'' with frequency at different temperatures. The value of Z'' becomes the maximum (Z''_{max}) at higher temperatures. The maximum value of Z'' decreases with the rise in temperature. This exhibits the relaxation in the samples [17]. The increase in the broadening of peak with the rise in temperature suggests the presence of relaxation phenomenon in the materials. The relaxation process occurs due to the presence of immobile species at low temperatures and defects and vacancies at higher temperatures [18,19].

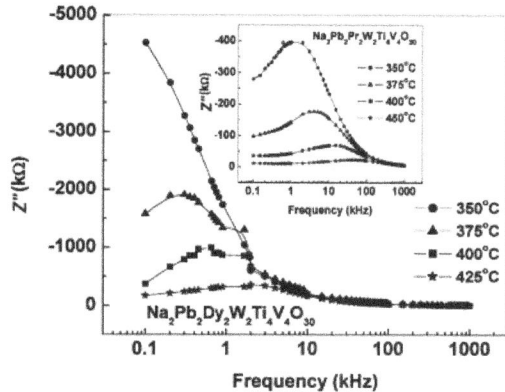

Fig. 2   Variation of Z'' with frequency at different temperatures for NPDWTV and NPPWTV.

Figures 3 and 4 show the variation of Z' with Z'' (Nyquist plot) for a wide frequency range (100 Hz– 1 MHz) at different temperatures (325–450 ℃). The depressed or deformed semicircles indicate the presence of non-Debye type of relaxation in the materials. It suggests the presence of a distribution of relaxation time instead of a single relaxation time in the materials [20,21]. The intercept of each semicircle on real Z' axis normally gives the value of bulk resistance. At higher temperatures, there is a tendency to form a second semicircle in one of the samples (NPDWTV). It clearly suggests that this compound has

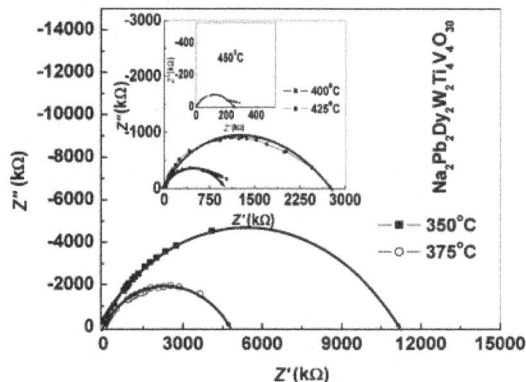

Fig. 3   Variation of Z'' with Z' at different temperatures for NPDWTV.

both grain and grain boundary contributions in impedance.

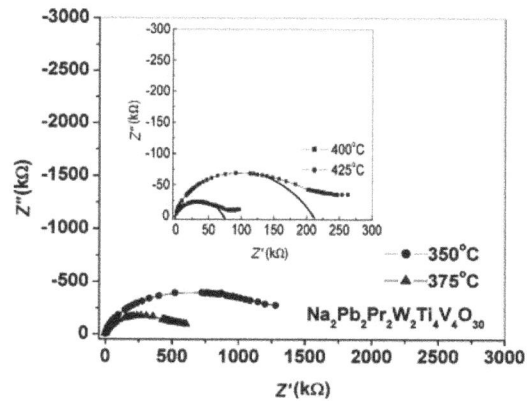

Fig. 4   Variation of Z'' with Z' at different temperatures for NPPWTV.

The semicircles of the impedance spectrum have a characteristic peak occurring at a unique relaxation frequency ($\omega_r = 2\pi f_r$). It can be expressed as

$$\omega_r RC = \omega_r \tau = 1 \quad \text{or} \quad f_r = 1/(2\pi RC)$$

where τ is the relaxation time. The relaxation time due to bulk effect (τ_b) has been calculated using the equation:

$$\omega_r \tau_b = 1 \quad \text{or} \quad \tau_b = 1/(2\pi f_r)$$

Figure 5 shows the variation of $\ln \tau_b$ with the inverse absolute temperature ($10^3/T$). It is observed that the value of τ_b decreases with the rise in temperature for the samples and its temperature dependent characteristics follows Arrhenius relation:

$$\tau_b = \tau_0 \exp(-E_a/k_B T)$$

where τ_0 is the pre-exponential factor; k_B is Boltzmann constant; and T is the absolute temperature. The value of activation energy (E_a) was found to be 1.64 eV and 1.81 eV for NPDWTV and NPPWTV,

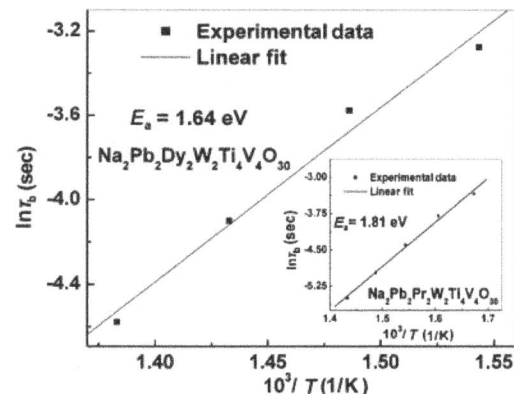

Fig. 5   Variation of relaxation time with temperature for NPDWTV and NPPWTV.

respectively [22].

3.2.2 Modulus analysis

The complex electric modulus formalism can easily distinguish between electrode polarization effect and grain boundary conduction process. It is also useful in detecting the bulk properties as apparent from relaxation time [23,24].

Figure 6 shows the variation of M'' with frequency at different temperatures. It shows that, M''_{max} shifts towards higher-frequency side with the rise in temperature. This behavior suggests that dielectric relaxation is thermally activated in which the hopping mechanism of charge carriers dominates intrinsically [17]. Asymmetric broadening of the peak indicates the spread of relaxation with different time constants, and hence the relaxation in the materials is considered as non-Debye type [20]. For NPDWTV, two peaks are observed indicating the contribution due to both bulk and grain boundary, because the capacitances of both factors are comparable. The grain boundary effect is not noticed in the impedance plot because the grain boundary resistance is small as compared to the bulk resistance. Impedance spectrum gives more emphasis to elements with the largest resistance whereas modulus spectrum enables to identify the process with the smallest capacitance.

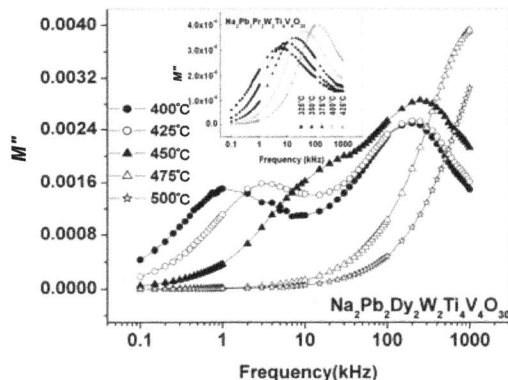

Fig. 6   Variation of M'' with frequency at different temperatures for NPDWTV and NPPWTV.

The scaling behavior of the samples is studied by plotting normalized parameters (i.e., M''/M''_{max} vs. $\log(f/f_{max})$), where f_{max} corresponds to the maximum value of M'', M''_{max}, at different temperatures (Fig. 7). The modulus scaling behavior gives an in-depth knowledge of the dielectric process occurring inside the materials. The low-frequency side of the peak represents a range of frequency in which charge carriers can perform hopping from one site to

their neighboring site. The high-frequency side of the peak represents a range of frequency in which charge carriers are spatially confined to their potential well and thus cannot make localized motion inside the well. The small region of the peak represents the transition from long-range mobility to short-range mobility with the increase in frequency [6]. At all the temperatures, the peaks coincide at $\log(f/f_{max})=1$, which indicates the temperature independent behavior of the relaxation dynamic process occurring in the material [25]. In the case of NPDWTV, two peaks are observed with different time constants, which are due to grain and grain boundary contributions.

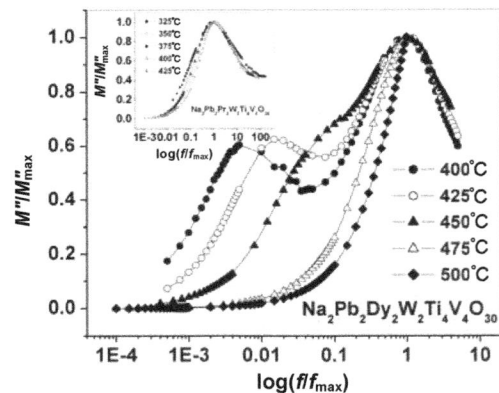

Fig. 7   Plot of M''/M''_{max} vs. $\log(f/f_{max})$ at different temperatures for NPDWTV and NPPWTV.

3.3   Electrical conductivity

3.3.1   DC conductivity

The DC conductivity of the samples has been evaluated from the impedance spectrum using a simple relation:

$$\sigma_{DC} = \frac{t}{R_b A}$$

where t and A represent the thickness and area of the samples, respectively. Figure 8 shows the temperature dependence of DC conductivity. It is observed that σ_{DC} increases with the rise in temperature which supports the negative temperature coefficient of resistance (NTCR) behavior of the samples. The nature of the curve follows Arrhenius relation [20]:

$$\sigma_{DC} = \sigma_0 e^{\frac{E_a}{k_B T}}$$

The activation energy (E_a) of NPDWTV in the temperature range of 300–375 ℃ is found to be 0.728 eV, and in the temperature range of 400–475 ℃, it is 1.974 eV. Similarly, E_a of NPPWTV in the temperature range of 225–325 ℃ is 0.441 eV, and in the temperature range of 350–425 ℃, it is 1.113 eV. These values of E_a are different from those calculated from the relaxation time plots. This implies that the charge carriers responsible for conduction and relaxation are different. Besides, the difference in activation energy in low and high temperature range supports the conduction mechanism of hopping type [11].

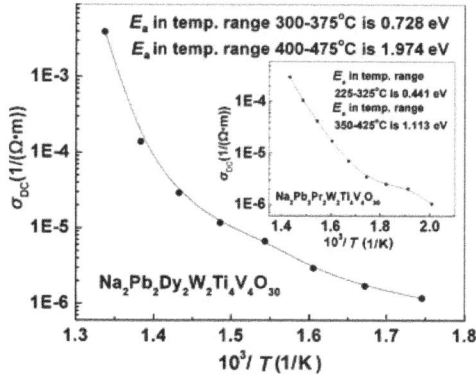

Fig. 8 Variation of DC conductivity with temperature for NPDWTV and NPPWTV.

3.3.2 AC conductivity

The frequency dependence of AC conductivity provides information regarding the nature of charge carriers. The AC electrical conductivity (σ_{AC}) is calculated using an empirical relation:

$$\sigma_{AC} = \omega \varepsilon_r \varepsilon_0 \tan \delta$$

where ε_0 is the permittivity in free space; ω is the angular frequency; and ε_r and $\tan \delta$ are the dielectric parameters. Jonscher made an attempt to explain the behavior of AC conductivity using the following law [26]:

$$\sigma_T(\omega) = \sigma(0) + \sigma_1(\omega) = \sigma_0 + a\omega^n$$

where $\sigma_T(\omega)$ is the total conductivity; $\sigma(0)$ is the frequency independent term giving DC conductivity, and $\sigma_1(\omega)$ is the pure dispersive component of AC conductivity having a characteristic of power law in terms of angular frequency ω and exponent n. The value of exponent n can have a value of $0 \leqslant n \leqslant 1$. This parameter is frequency independent but temperature and material dependent.

Figures 9(a) and 9(b) show the variation of AC conductivity with frequency (usually referred as conductivity spectrum) at different temperatures for NPDWTV and NPPWTV compounds. In these compounds, conductivity curves show dispersion in the low-frequency region. From the nature of graphs, it is

obvious that σ_{AC} decreases on the decreasing frequency and becomes nearly independent at low frequency. The extrapolation of the lower frequency gives σ_{DC}. The increasing trend of σ_{AC} with frequency (in the lower-frequency region) may be attributed to the disordering of cations between neighboring-sites and presence of space charge. In the high-frequency region, the curves approach each other for NPPWTV. For NPDWTV, the frequency independent plateau region is observed at higher temperature and higher frequency possibly due to the release of space charge. A close inspection on the conductivity plots reveals that the curves exhibit low-frequency dispersion phenomena obeying Jonscher's power law equation. According to Jonscher [18], the origin of the frequency dependence of conductivity lies in the relaxation phenomena arising due to mobile charge carriers. When a mobile charge hops to a new site from its original position, it remains in a state of displacement between two potential energy minima. Moreover, the conduction behavior of the materials obeys the power law, $\sigma(\omega) \propto \omega^n$, with a slope change governed by n in the low-temperature

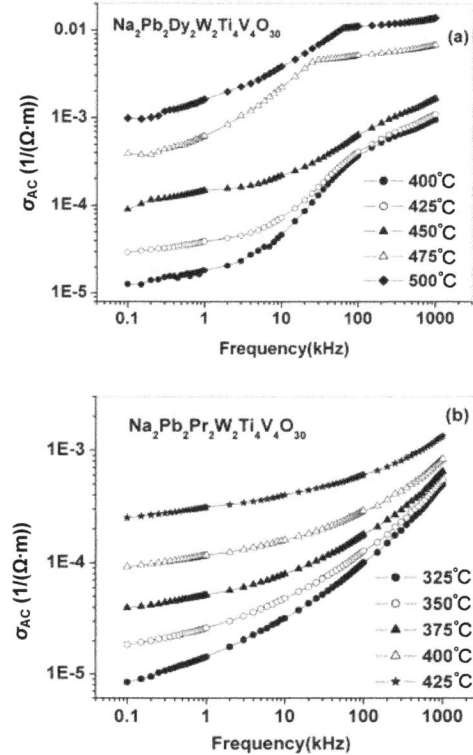

Fig. 9 (a) Variation of AC conductivity with frequency at different temperatures for NPDWTV; (b) variation of AC conductivity with frequency at different temperatures for NPPWTV.

region. According to Funke [27], the value of n has a physical meaning. The value of $n \leqslant 1$ means that the hopping motion involves a translational motion with a sudden hopping, whereas $n > 1$ means that the motion involves localized hopping without the species leaving the neighborhood. The frequency at which slope changes is known as hopping frequency of the polarons (ω_p), and is temperature dependent. The low frequency dispersion has been attributed to the AC conductivity whereas the frequency independent plateau region corresponds to the DC conductivity.

4 Conclusions

The NPDWTV and NPPWTV samples have an orthorhombic TB crystal structures. Detailed studies of electrical properties indicate that the materials exhibit: ① conduction due to the bulk material up to temperature 450 ℃ for both samples; ② electrical transport governed in both the samples at temperature $T > 475$ ℃ characterized by the appearance of two semi-circular arcs in the impedance spectrum; ③ NTCR-type behavior; and ④ temperature-dependent relaxation phenomena. The results of impedance spectrum have been used to estimate the electrical conductivity of the materials. The activation energy estimated from the conducting pattern, relaxation time pattern and modulus pattern are nearly similar. This suggests the presence of similar type of charge carriers in the materials that are responsible for both the electrical conduction and electrical relaxation phenomena in the materials. The activation energy from grain boundary conduction plot suggests the possibility of electrical conduction due to the mobility of the oxide ion. Modulus analysis indicates non-exponential type conductivity relaxation in the material. Comparing the experimental results of NPDWTV and NPPWTV with those of others (as mentioned above), it is suggested that these two materials may be used as good dielectric materials with moderate dielectric permittivity. With the low tangent loss (i.e., high quality factor) in high frequency range, the materials may be used for microwave applications, and also for pyroelectric detector.

References

[1] Singh AK, Choudhary RNP. Study of ferroelectric phase transition in $Pb_3R_3Ti_5Nb_5O_{30}$ (R = rare earth ion) ceramics. *Ferroelectrics* 2005, **325**: 7–14.

[2] Kim MS, Lee JH, Kim JJ, *et al.* Microstructure evolution and dielectric properties of $Ba_{5-x}Na_{2x}Nb_{10}O_{30}$ ceramics with different Ba–Na ratios. *J Solid State Electr* 2006, **10**: 18–23.

[3] Fang L, Zhang H, Huang TH, *et al.* Preparation, structural, and dielectric properties of $Ba_5YZnM_9O_{30}$ (M = Nb, Ta) ceramics. *J Mater Sci* 2005, **40**: 533–535.

[4] Behera B, Nayak P, Choudhary RNP. Structural, dielectric and electrical properties of $NaBa_2X_5O_{15}$ (X = Nb and Ta) ceramics. *Mater Lett* 2005, **59**: 3489–3493.

[5] Hornebecq V, Elissalde C, Reau JM, *et al.* Relaxations in new ferroelectric tantalates with tetragonal tungsten bronze structure. *Ferroelectrics* 2000, **238**: 57–63.

[6] Ganguly P, Jha AK, Deori KL. Complex impedance studies of tungsten–bronze structured $Ba_5SmTi_3Nb_7O_{30}$ ferroelectric ceramics. *Solid State Commun* 2008, **146**: 472–477.

[7] Neurgaonkar RR, Nelson JG, Oliver JR, *et al.* Ferroelectric and structural properties of the tungsten bronze system $K_2Ln^{3+}Nb_5O_{15}$, Ln = La to Lu. *Mater Res Bull* 1990, **25**: 959–970.

[8] Fang L, Zhang H, Yang JF, *et al.* Preparation, characterization and dielectric properties of $Ba_5LnZnTa_9O_{30}$ (Ln = La, Sm) ceramics. *Matet Res Bull* 2004, **39**: 677–682.

[9] Qu YQ, Li AD, Shao QY, *et al.* Structure and electrical properties of strontium barium niobate ceramics. *Mater Res Bull* 2002, **37**: 503–513.

[10] Das PR, Biswal L, Behera B, *et al.* Structural and electrical properties of $Na_2Pb_2Eu_2W_2Ti_4X_4O_{30}$ (X = Nb, Ta) ferroelectric ceramics. *Mater Res Bull* 2009, **44**: 1214–1218.

[11] Das PR, Choudhary RNP, Samantray BK. Diffuse ferroelectric phase transition in $Na_2Pb_2Sm_2W_2Ti_4Nb_4O_{30}$ ceramics. *Mater Chem Phys* 2007, **101**: 228–233.

[12] Das PR, Choudhary RNP, Samantray BK. Diffuse ferroelectric phase transition in $Na_2Pb_2Nd_2W_2Ti_4Nb_4O_{30}$ ceramic. *J Alloys Compd* 2008, **448**: 32–37.

[13] Das PR, Choudhary RNP, Samantray BK. Diffuse phase transition in $Na_2Pb_2R_2W_2Ti_4V_4O_{30}$ (R = Gd, Eu)

ferroelectric ceramics. *J Phys Chem Solids* 2007, **68**: 516–522.

[14] Das PR, Behera B, Choudhary RNP, *et al.* Ferroelectric properties of $Na_2Pb_2R_2W_2Ti_4V_4O_{30}$ (R = Dy, Pr) ceramics. *Res Lett Mater Sci* 2007, Article ID: 91796.

[15] Wu E. POWDMULT—An interactive powder diffraction data interpretation and indexing program, Version 2.5. Bedford Park, Australia: School of Physical Science, Finders University of South Australia.

[16] Plocharski J, Wieczoreck W. PEO based composite solid electrolyte containing nasicon. *Solid State Ionics* 1988, **28–30**: 979–982.

[17] Behera B, Nayak P, Choudhary RNP. Structural and impedance properties of $KBa_2V_5O_{15}$ ceramics. *Mater Res Bull* 2008, **43**: 401–410.

[18] Jonscher AK. The 'universal' dielectric response. *Nature* 1977, **267**: 673–679.

[19] Suman CK, Prasad K, Choudhary RNP. Complex impedance studies on tungsten–bronze electroceramic: $Pb_2Bi_3LaTi_5O_{18}$. *J Mater Sci* 2006, **41**: 369–375.

[20] Sen S, Choudhary RNP, Pramanik P. Structural and electrical properties of Ca^{2+}-modified PZT electroceramics. *Physica B* 2007, **387**: 56–62.

[21] Behera B, Nayak P, Choudhary RNP. Impedance spectroscopy study of $NaBa_2V_5O_{15}$ ceramic. *J Alloys Compd* 2007, **436**: 226–232.

[22] Das PS, Chakraborty PK, Behera B, *et al.* Electrical properties of $Li_2BiV_5O_{15}$ ceramics. *Physica B* 2007, **395**: 98–103.

[23] Sinclair DC, West AR. Impedance and modulus spectroscopy of semiconducting $BaTiO_3$ showing positive temperature coefficient of resistance. *J Appl Phys* 1989, **66**: 3850–3856.

[24] Hodge IM, Ingram MD, West AR. A new method for analysing the a.c. behaviour of polycrystalline solid electrolyte. *J Electroanal Chem Interfacial Electroch* 1975, **58**: 429–432.

[25] Saha S, Sihna TP. Low-temperature scaling behavior of $BaFe_{0.5}Nb_{0.5}O_3$. *Phys Rev B* 2002, **65**: 134103.

[26] Jonscher AK. *Universal Relaxation Law.* London: Chelsea Dielectrics Press, 1996.

[27] Funke K. Jump relaxation in solid electrolytes. *Prog Solid State Ch* 1993, **22**: 111–195.

2

Structural phase relations in perovskite-structured BiFeO$_3$-based multiferroic compounds

Valdirlei Fernandes FREITAS[*], Gustavo Sanguino DIAS, Otávio Algusto PROTZEK,
Diogo Zampieri MONTANHER, Igor Barbosa CATELLANI, Daniel Matos SILVA,
Luiz Fernando CÓTICA, Ivair Aparecido dos SANTOS

*UEM—Universidade Estadual de Maringá, GDDM—Grupo de Desenvolvimento de Dispositivos Multifuncionais,
Departamento de Física, Maringá, Brazil*

Abstract: In this review, the state of the art in understanding the structural phase relations in perovskite-structured BiFeO$_3$-based polycrystalline solid solutions is presented and discussed. Issues about the close relation between the structural phase and overall physical properties of the reviewed systems are pointed out and discussed. It is shown that, by adjusting the structural symmetric arrangement, the ferroelectric and magnetic properties of BiFeO$_3$-based polycrystalline solid solutions can be tuned to find specific multifunctional applications. However, an intrinsic mechanism linking structural arrangement and physical properties cannot be identified, revealing that this subject still deserves further discussion and investigation.

Keywords: bismuth ferrite; ferroelectrics; multiferroics

1   Introduction

Multifunctional compounds are materials whose two or more physical properties can be exploited simultaneously or separately for the same or different purposes. The advancement of processing multifunctional materials needs to meet a range of revolutionary technologies, such as shape memory, electrostriction, solid-state transformer and magnetorheological device [1,2]. In this context, perovskite-structured polycrystalline compounds, i.e., BiFeO$_3$–ABO$_3$-type materials can potentially be applied for constructing or developing multifunctional

smart devices. In fact, compounds with perovskite structure may have more than one applicable property (e.g., ferroelectricity, (weak, ferri) magnetism, piezoelectricity, and magnetoelectric coupling) to be classified as multifunctional. Furthermore, materials with perovskite structure that present ferroelectricity, ferroelasticity and/or some magnetic order simultaneously, are known as multiferroics; for multiferroic compounds, such properties can be intrinsically coupled, as the case of magnetoelectric materials. In these materials, an electric field can induce an electric polarization and a magnetic order simultaneously, or a magnetic field can induce a magnetization and an electric order simultaneously [3,4]. These materials can also be used in advanced electro-electronic devices, such as multiple-state non-volatile memory [5,6] and high-power solid-state transformer, by fundamentally exploiting the

---

* Corresponding author.
E-mail: freitas@dfi.uem.br

magnetoelectric coupling [7]. In recent years, some multiferroic materials have emerged as potential candidates for these specific applications, e.g., perovskite-structured polycrystalline $BiFeO_3$ (BF) and BF-based compounds.

In fact, BF is a well-known multiferroic magnetoelectric material presenting two ferroic orders, i.e., antiferromagnetic and ferroelectric states in the same phase [8,9]. BF possesses a rhombohedral distorted perovskite structure ($R3c$ space group). It is ferroelectric (Curie temperature $T_C \approx 830$ ℃) and antiferromagnetic at room temperature (with $T_N \approx 367$ ℃) [10,11], and it shows a strong magnetoeletric coupling (dE/dH as high as 3 V/(cm·Oe)) [12]. Despite being a promising candidate for multifunctional applications, polycrystalline BF shows serious problems related to electric conductivity, and a cycloidal disposition of the magnetic moments forms a typical weak-ferromagnetic arrangement that prevents practical application of BF samples. In this sense, the mixing of BF with other perovskite-structured materials (forming pseudo-binary systems), which focuses on obtaining high-resistive solid solutions with magnetic moment and magnetoelectric coupling, has been employed with good results.

In this work, a careful review concerning the state of the art about the structural phase relations in binary perovskite-structured BF-based polycrystalline compounds is presented and discussed. Furthermore, some unpublished results are added and discussed with the objective to point out the open issues relative to the intrinsic and unrevealed mechanism that links structural arrangements and ferroic properties in these systems, and show their potential for practical applications.

2 BiFeO₃-based systems

The three most investigated BF-based polycrystalline solid solutions that can be found in the literature are $(1-x)BiFeO_3-xPbZrO_3$ (BF–PZ), $(1-x)BiFeO_3-xBaTiO_3$ (BF–BT), and $(1-x)BiFeO_3-xPbTiO_3$ (BF–PT) systems. BF–PZ system is less studied among them, and its potential for multifunctional applications is only recently investigated [13]. Initially, this system was synthesized focusing on PZ's effects on the overall physical properties of BF compound. However, these studies were abandoned because of the enormous difficulty for obtaining single-phase polycrystalline samples and the strong tendency of the formation of non-perovskite phases during the synthesis process [13]. BF–BT system is the most widely studied and shows potential to be applied in ferroelectric, magnetic and piezoelectric devices. As BF–BT solid solutions are processed from $BiFeO_3$ and $BaTiO_3$ that are known to be lead-free, BF–BT solid solutions form a continuous series of compounds that also show environmental interest. Finally, BF–PT system shows the highest potential of multifunctional applications, because it presents very good ferroelectric, piezoelectric and weak ferromagnetic properties, and a strong magnetoelectric coupling. In fact, the development of BF–PT compounds can trigger the onset of a new family of multifunctional devices focusing on several advanced applications.

3 $(1-x)BiFeO_3-xPbZrO_3$ system

BF–PZ solid solutions show triclinic crystal symmetry (rhombohedral, $R3c$ space group) and perovskite structure (ABO_3 type) at room temperature [13], and low PZ concentration ($x < 0.2$). Increasing PZ concentration ($x > 0.2$), the compound acquires a pseudo-cubic symmetry, which can be represented by a rhombohedral structure whose angles between lattice parameters are close to 90°. When x is above 0.7, BF–PZ system acquires an orthorhombic symmetry ($Pba2$) [14]. Apparently, the lattice symmetry is derived from BF compound and does not change with the increase of PZ content until $x \approx 0.2$ [13]. The related lattice parameters are: for $x = 0.1$, $a \approx 5.59$ Å and $c \approx 13.9$ Å; for $x = 0.2$, $a \approx 5.63$ Å and $c \approx 13.98$ Å. The magnetic properties of BF–PZ system are also investigated by magnetic and neutron diffraction measurements. The magnetic field dependency of the magnetization indicates a typical weak-ferromagnetic behavior, revealing that the spiral spin arrangement of BF end member is effectively broken by introducing PZ into BF structure. However, the magnetic moments, obtained from the neutron diffraction experiments, indicate an antiferromagnetic/paramagnetic phase transition around 635 K for $x = 0.1$, and 500 K for $x = 0.2$, respectively. Figure 1 shows the sketch of magnetic structure of the $x = 0.1$ sample.

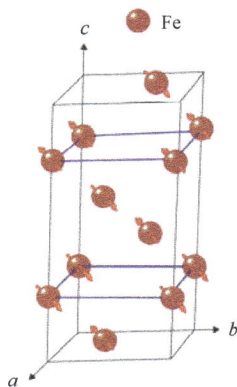

Fig. 1   Schematic magnetic structure of the $0.9BiFeO_3$–$0.1PbZrO_3$ sample at 10 K ($R3c$ space group with a G-type antiferromagnetic structure, adapted from [13]).

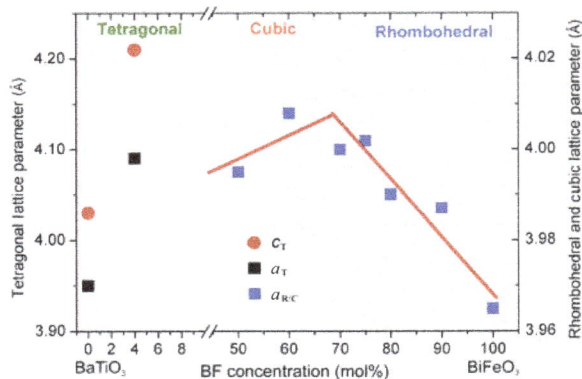

Fig. 2   Lattice parameters and space groups of BF–BT system obtained from the XRD results, as a function of BF molar concentration; the left and right scales are the values of the tetragonal (a_T, c_T) and rhombohedral/cubic ($a_{R/C}$) lattice parameters, respectively (adapted from [15]).

The magnetic phase revealed by neutron diffraction data is best described by an antiferromagnetic G-type magnetic structure model, as those reported for BF compound. Anyway, considering that few BF–PZ compositions are investigated, a close relation between the structure and magnetization has not been identified in this case, suggesting that further investigations need to be conducted to elucidate/reveal the true mechanism in this correlation.

4 $(1-x)BiFeO_3$–$xBaTiO_3$ system

X-ray diffraction (XRD) and high-resolution XRD studies show that the symmetric structural arrangements of BF–BT compounds can be controlled by changing BT content in the solid solutions. In fact, for x ranging from 0 to 0.27, BF–BT solid solutions exhibit a rhombohedral distortion ($R3c$ space group) [15,16]. However, a monoclinic phase (Cm space group) is also reported as coexisting with the $R3c$ symmetry in the polycrystalline sample of $x = 0.2$ [17]. The coexistence of cubic and rhombohedral phases is reported to occur near $x = 0.3$ [16]. For samples with x ranging from 0.27 to 0.93, a cubic phase (Fig. 2), and a tetragonal symmetric arrangement for $x > 0.92$ [15,16], are also reported.

The change in BF–BT structural symmetry, as a function of temperature, was investigated by Wang et al. [16] by performing high-resolution XRD measurements. For pure BF samples, a rhombohedral phase is observed until 825 ℃, where the symmetry changes to an orthorhombic arrangement and persists until 850 ℃, and then the sample becomes cubic. For

$x = 0.1$, a transition from rhombohedral to cubic phase is observed in the temperature range of 700–760 ℃. A similar behavior is also observed for the $x = 0.2$ sample, which reveals a rhombohedral-to-cubic phase transition near 740 ℃. For samples with $x = 0.3$, a "cubic + rhombohedral" (where the cubic phase is the majority phase) to cubic transition takes place near 680 ℃.

The dielectric, ferroelectric and magnetic properties of BF–BT solid solutions also show a strong dependence with BF–BT composition, and consequently, with the structural symmetry in each case. The electrical properties are improved with the increase of BT content, generally by increasing the dielectric response and reducing the dielectric losses [18]. In fact, Wang et al. [18] observed an increase of the dielectric constant in the $x = 0$–0.3 samples, from 30 ($x = 0$) to 500 ($x = 0.3$), at 1 MHz and room temperature.

Considering magnetic properties, it is reported a considerable change in the magnetic behavior of BF–BT solid solutions with the increase of BT content, by changing the structural symmetric arrangements of the samples [15,17,19,20]. As observed by Kumar et al. [15], BF–BT system exhibits an improvement of magnetic response for the $x = 0.1$ and 0.2 samples, showing a high hysteretic behavior. However, a decrease is observed as BF–BT symmetry becomes more and more symmetric, showing almost no hysteresis for $x = 0.75$ and so on. The $x = 0.1$, 0.2 and 0.3 samples, analyzed by Kumar et al. [15] and Shi et al. [21], exhibit an antiferromagnetic behavior with a field-induced ferromagnetic ordering. In fact, these authors reaffirmed that, through electron spin resonance measurements, the ferromagnetic behavior is shown for the $x = 0.1$, 0.2 and 0.25 samples, while for samples with $x = 0.3$ and 0.996 at room temperature, resonance peaks at different fields reveal the absence of magnetic ordering, confirming that as BT content increases, the antiferromagnetic ordering gets to be suppressed, giving order to a paramagnetic state [22].

5 $(1-x)$BiFeO$_3$-xPbTiO$_3$ system

The first published work, regarding the synthesis of BF–PT solid solutions, was reported by Fedulov et al. [23] in 1962. Later, in 1964, Fedulov et al. [24] proposed the first structural, magnetic and ferroelectric phase diagrams for these solid solutions. In fact, it reported the coexistence of rhombohedral and tetragonal symmetries at $x = 0.7$, consisting in a morphotropic phase boundary (MPB). In the early 1990s, the magnetic properties of BF–PT solid solutions were investigated once more [25]. However, it was only with the resurgence of the academic interest in the multiferroics driven by works of Spaldin et al. [12,26] at the beginning of this century, that the studies of BF–PT solid solutions gained a renewed impulse [27–30]. In this way, structural aspects of these compounds began to be investigated in order to find magnetoelectric applications, understand the magnetic [31], ferroelectric and piezoelectric behaviors [32], and even the magnetoelectric coupling [33]. Currently, studies are being conducted in order to understand the nature of the physical properties observed in these compounds in both single and polycrystalline formats. These studies are intended to discover the relationship of the physical properties of these compounds with their structure, in order to control these properties through doping/modifying the A and B perovskite sites [34,35].

In fact, BF–PT solid solutions acquire magnetoelectric, ferroelectric and piezoelectric properties from their end members, i.e., BF and PT compounds [36,37]. Furthermore, similar to the previous systems, their properties can be tuned/controlled by changing the concentration of each end member. In this way, by increasing PT concentration the ferroelectric and piezoelectric properties are enhanced, while by decreasing PT concentration the magnetic properties tend to be enhanced.

These degrees of freedom for tuning BF–PT physical properties are intrinsically linked to the structural arrangement/group symmetry, which also can be controlled by PT concentration with a still unrevealed mechanism. As BF compound shows a rhombohedral perovskite structure ($R3c$ space group) [36], when PT compound is added into BF, the $R3c$ space group is maintained at low PT concentrations, as shown in Fig. 3. However, with increasing PT concentration (always reported for $x = 0.3$, here precisely for $x = 0.31$–0.32), the structural symmetry of these solid solutions changes to a tetragonal symmetry ($P4mm$ space group), arising from PT end member.

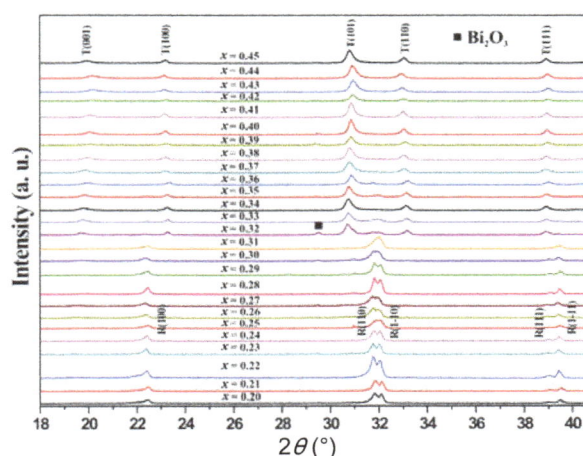

Fig. 3 XRD patterns for BF–PT polycrystalline compounds. The PT concentration is ranged from 20 mol% to 45 mol% (from bottom to top).

The structural transition is also a region of coexistence of both symmetric phases ($R3m$ and $P4mm$). This MPB region can vary with the method and parameters of synthesis (processing protocol), and occur in our samples for x ranging from 0.20 to 0.45 (Fig. 4). Systems showing MPB region have been pointed out as promising candidates for practical applications, because they present enhanced ferroelectric and piezoelectric properties as, for example, those observed in Pb(Zr,Ti)O$_3$ samples [38].

The structural, electric and magnetic properties of BF–PT system have been studied since the middle of the 20th century. The first paper reporting BF–PT phase diagram was published in 1964 [24]. In this work, the authors observed a structural phase transition from rhombohedral to tetragonal symmetry and the presence of MPB region. Recently, in 2008, a complete structural and magnetic study of BF–PT system was conducted by Zhu et al. [39]. The structural

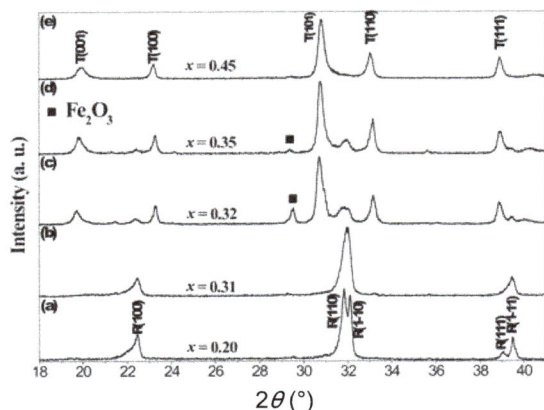

Fig. 4   XRD patterns for BF–PT solid solutions, evidencing an MPB region for x ranging from 0.20 to 0.45.

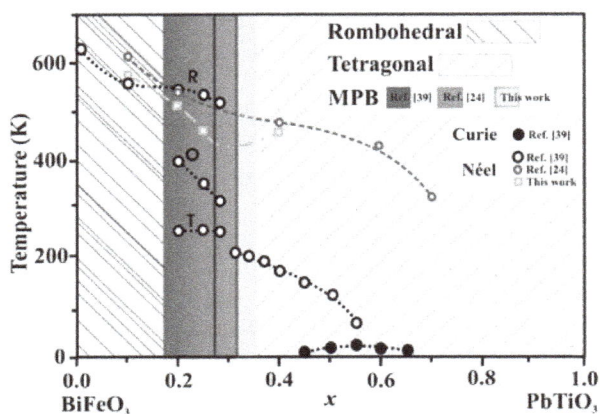

Fig. 5   Structural and magnetic phase diagrams for BF–PT system (R: rhombohedral; O: orthorhombic; T: tetragonal, adapted from [24,39]).

magnetization curves. The experimental evidence of the existence of a monoclinic phase separating or coexisting with tetragonal and rhombohedral phases means a bridge connecting the different directions of polarization in this system, as similarly described by Noheda *et al.* [38] for PZT (lead zirconate titanate) samples. In other works, it was shown that the change in the electric polarization direction does not occur abruptly, from [111] (rhombohedral symmetry) to [001] (tetragonal symmetry) direction [41]. Instead, an intermediate direction in a monoclinic arrangement was proposed (Fig. 6).

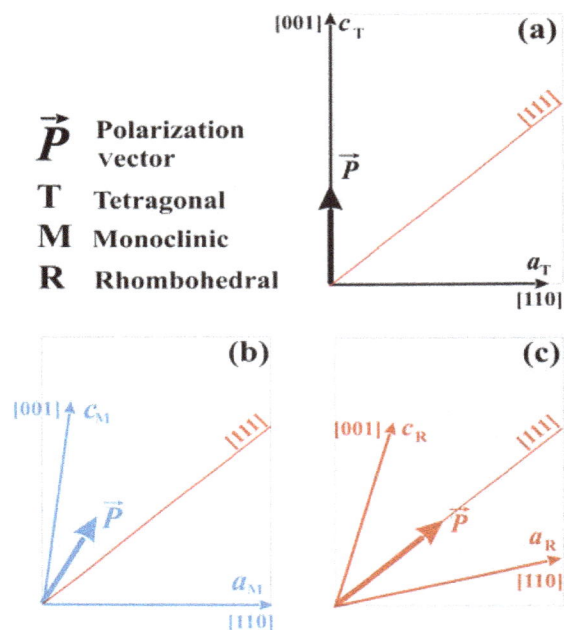

Fig. 6   Schematic illustration of (a) tetragonal, (b) monoclinic, and (c) rhombohedral distortions of the perovskite unit cell projected on the pseudo-cubic (110) plane with the respective polarization directions (adapted from [41]).

and magnetic data, comparing with current phase diagrams, is shown in Fig. 5. In fact, unlike the first authors [24], Zhu *et al.* claimed that they have found an orthorhombic phase in MPB region by deconvolutioning XRD peaks [39]. In this way, three phases are proposed for MPB region and three magnetic phase transition temperatures (antiferromagnetic–paramagnetic Néel temperatures) are identified. However, divergent Néel temperatures can be observed for high PT concentrations, probably as a consequence of the different synthesis protocols employed for samples processing. In addition, at high PT concentrations and very low temperatures, a ferromagnetic state was also proposed.

In 2009, Bhattacharjee *et al.* [40] showed evidences for a monoclinic phase (Cc space group) in MPB region of the BF–PT system by associating Rietveld analysis with some magnetic anomalies observed in

The magnetic and nuclear structure of BF–PT solid solutions were also recently investigated by neutron diffraction. In 2009, Comyn *et al.* [31] investigated the $x = 0.1$ composition by neutron diffraction and observed a perovskite structure with rhombohedral symmetry ($R3c$ space group) and an antiferromagnetic G-type magnetic arrangement. Recently, in 2011, two works by Ranjan *et al.* [42] and Comyn *et al.* [34], reported neutron diffraction investigations in BF–PT samples. In the first work [42], an interesting study about the sensible change in the structure of the $x = 0.2$ composition was conducted as a function of temperature. The authors reported a change in the chemical bonds of the compound at the magnetic Néel temperature and related this change to the force that the magnetic arrangement promotes in the structure. These observations suggest that this system exhibits some kind of coupling between spin, strain and structural degrees of freedom.

The second work [34] conducted at low temperatures ($T < 4$ K), and reported an antiferromagnetic ordering in the tetragonal arrangement of the $x = 0.3$ sample. Figure 7 shows the tetragonal structure (eight unit cells) with the antiferromagnetic G-type magnetic arrangement. In

this scheme, the ferroelectric polarization vector, pointing in [001] direction (c-direction), is approximately 50° misaligned to the antiferromagnetic magnetization vector, which points in the [111] direction of tetragonal unit cell.

6 Structure–property relations

The three BF-based compounds presented in this paper have advantages for some practical applications and disadvantage for others (see Table 1). For example, BF–PZ ceramics have high dielectric constants (at room temperature) in comparison with those found in BF–PT and BF–BT ones, being apparently more suitable for applications in areas where elevated capacitances are required, as in information or electric field storage. On the other hand, the magnetic coercive fields of BF–BT samples are smaller than those found for BF–PZ and BF–PT samples. Therefore, magnetic devices constructed with BF–BT materials will tend to require less energy for operation. However, for all other possible practical applications (magnetic, ferroelectric, piezoelectric and magnetoelectric), BF–PT samples seem to be more adequate. In fact, the results found for BF-based ceramics in the framework of multiferroic properties and applications, are listed in Table 1.

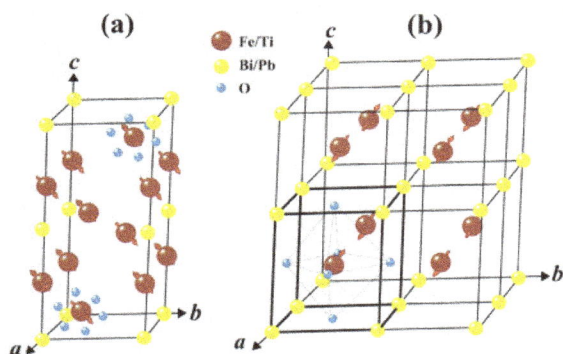

Fig. 7 Simulation of the magnetic and structural arrangements from neutron diffraction data: (a) $0.9BiFeO_3–0.1PbTiO_3$, and (b) $0.7BiFeO_3–0.3PbTiO_3$ unit cells (adapted from [22]).

Table 1   Comparison between physical properties of $(1–x)BiFeO_3–xABO_3$ solid solutions

Feature	$xPbZrO_3$	$xBaTiO_3$	$xPbTiO_3$
Structural behavior	$R3c$ ($x < 0.2$) [13]	$R3c$ or $R3c + Cm$ ($0 < x < 0.27$) [15,17]	$R3c$ ($0 < x < 0.18$) [24,39]
	P-C ($0.2 < x < 0.7$) [14]	Pm-$3m$ ($0.30 < x < 0.92$) [16]	$R3c + P4mm$ ($0.18 < x < 0.35$) [24,39]
	$Pba2$ ($x > 0.7$) [14]	$P4mm$ ($x > 0.92$) [15,16]	$P4mm$ ($x > 0.35$) [24,39]
Magnetic behavior	G-type antiferromagnetic	G-type antiferromagnetic with weak ferromagnetism	G-type antiferromagnetic with weak ferromagnetism
T_N (K)	500 ($x = 0.2$) [13]	267 ($x = 0.3$) [15]	630 ($x = 0.3$) [29]
M_r (10^{-3} emu/g)	0	1.76 ($x = 0.3$) [12]	9.91 ($x = 0.4$) [37]
H_c (Oe)	0	182 ($x = 0.3$) [12]	754 ($x = 0.4$) [37]
Electric property	Ferroelectric	Ferroelectric	Ferroelectric and piezelectric
$T_{C\text{-}FE}$ (K)	423 (MPB)	645 (MPB)	933 (MPB)
P_s ($\mu C/cm^2$)	4.0 ($x = 0.8$, 100 kHz) [14]	4.4 ($x = 0.3$, 46 Hz) [16]	22.5 ($x = 0.4$, 30 Hz) [37]
P_r ($\mu C/cm^2$)	0.7 ($x = 0.8$, 100 kHz) [14]	1.8 ($x = 0.3$, 46 Hz) [16]	17.5 ($x = 0.4$, 30 Hz) [37]
E_c (kV/cm)	Uninformed [14]	19.0 ($x = 0.3$, 46 Hz) [16]	18.1 ($x = 0.4$, 30 Hz) [37]
K' (room temp.)	800 ($x = 0.8$, 100 kHz) [14]	500 ($x = 0.3$, 1 MHz) [16]	330 ($x = 0.3$, 1 kHz) [29]
α_{33} (mV/(cm·Oe))	—	0.11 ($x = 0.2$, 300 K) [20]	0.90 ($x = 0.5$, 298 K) [33]
d_{33} (10^{-12} m/V)	—	—	49.1 ($x = 0.4$) [32]

From top to bottom: structural behavior ($R3c$ = rhombohedral, P-C = pseudo-cubic, $Pba2$ = orthorhombic, Cm = monoclinic, Pm-$3m$ = cubic, and $P4mm$ = tetragonal); magnetic behavior (T_N = antiferro–paramagnetic (Néel) transition temperature, M_r = remnant magnetization, and H_c = coercive magnetic field); electric property ($T_{C\text{-}FE}$ = ferro–paraelectric (Curie) transition temperature, P_s = polarization of saturation, P_r = remnant polarization, E_c = coercive electric field, and K' = real dielectric constant); magnetoelectric property (α_{33} = magnetoelectric coefficient); piezolectric property (d_{33} = piezoelectric coefficient).

The structural phase relations in this system appear to be coincident because all structural transitions occur with their specific modifier concentration increasing. Moreover, MPB is observed in all systems ($x \approx 0.8$ to BF–PZ, $x \approx 0.3$ to BF–BT and BF–PT), where there is an improvement of the ferroic properties. Interestingly, these MPB regions appear to be intrinsic of perovskite structures, since other known compounds, as $Pb(Zr_xTi_{1-x})O_3$ and $(1-x)Pb(Mg_{1/3}Nb_{2/3})O_3-xPbTiO_3$, also present intriguing MPBs [43].

Likewise, all revised systems present an antiferromagnetic G-type structure. However, only BF–PT and BF–BT systems exhibit a small ferromagnetic hysteresis, indicating a weak-ferromagnetic state. The weak ferromagnetism is closely linked to the structural symmetry of the compound. In fact, according to Dzyaloshinsky and Moriya's prevision [44,45], in systems with rhombohedral symmetry and antiferromagnetic order, a small canting in antiparallel magnetic dipole moments results in a localized magnetization. Among those systems mentioned in this work, BF–PT has the best magnetic parameters (T_N and M_r), which can be attributed to specific changes in the superexchange angle and to the disposition of the magnetic ions caused by specific modifications in A and B sites of the compound.

Ferroelectric properties are present in all BF–PT samples, independently of PT concentration. Like other properties, the best ferroelectric parameters are found in BF–PT system. The highest ferroelectric/paraelectric (Curie) transition temperature for the saturation and remnant polarizations, with the lowest coercive fields, are found in samples of the BF–PT system. As the ferroelectric properties are directly linked to the structural distortions, they are more pronounced in BF–PT than in other systems due to the strong and directional chemical bonds (active lone-pairs and/or hybridization [46]) between $Pb^{2+}-O^{2-}$ and $Ti^{4+}-O^{2-}$ ions.

The magnetoelectric coefficients α_{33} for BF–PT (0.90 mV/(cm·Oe)) and BF–BT (0.11 mV/(cm·Oe)) systems were obtained in the same order of magnitude. The requirements for showing magnetoeletric coupling (MC), according to Hill [47], are mainly the existence of ferroelectric response and some magnetic order. However, other requirements are also needed, as having a polar point group (1, 2, 2', m, m', 3, 3m', 4, 4m'm', m'm2', m'm'2', 6 e 6m'm') adequate for both magnetic and electric orders [47]. Another requirement,

specific for polycrystalline samples, is that they own electrical resistivities that proportionate electric polarization. At this point, an enormous difficulty arises because magnetic systems generally tend to be electrically conductive.

Thus, by meeting the requirements listed above, two of the three systems discussed in this work need to be highlighted, i.e., BF–PT and BF–BT solid solutions. In fact, both systems show high electrical resistivity, ferroelectric polarization and weak ferromagnetic order, as facilitators of the MC [47]. From another point, which highlights the relation between MC and atomic structure, is that both systems show point groups belonging to those cited above ($3m$ to BF–PT, $3m$ and/or $4mm$ to BF–PT). In fact, their high electrical resistivities are linked to the hybridization which distorts the d^n electronic sublayer and induces the magnetoelectric coupling [33]. In both cases, Ti–O bonds are hybridized [46], distorting the d^n sublayer and allowing, in some way, the coexistence of magnetic and electric (with high electrical resistivity) orders. However, Pb–O bonds are also hybridized in BF–PT system, enhancing its ferroelectric properties. In this way, the magnetoelectric coupling of BF–PT compounds (0.9 mV/(cm·Oe)) tends to be stronger than that of BF–BT compounds (0.1 mV/(cm·Oe)) (better electric and magnetic responses for BF–PT samples ($P_r = 17.5$ μC/cm^2 and $M_r = 9.91 \times 10^{-3}$ emu/g) in comparison with BF–BT ones ($P_r = 1.8$ μC/cm^2 and $M_r = 1.76 \times 10^{-3}$ emu/g)).

7   Close remarks

Perovskite-structured BF-based multiferroic compounds appear as important candidates to be applied in smart advanced multifunctional devices, as solid-state transformers, multiple state nonvolatile memories, magnetic tunable piezodevices, magnetic field sensing and actuators, and so on. In fact, a complete understanding of the close relations between structure and physical properties in these solid solutions is highly desirable. However, it seems to be clear that the physical properties of the perovskite-structured BF-based compounds are highly dependent of the structural symmetric arrangement in each case. Depending on the solid solution end member, i.e., $PbZrO_3$, $BaTiO_3$ or $PbTiO_3$, and the physical nature of the atoms in A site of the perovskite structure, the compound can assume a specific group symmetry that can favor specific physical properties, as weak ferromagnetism and piezoelectricity. In this way, new ideas and investigations, focused to identify the intrinsic mechanisms that link the structural and physical properties of multifunctional multiferroic materials, especially those composed by BF compound, are still necessary, and demonstrate that this subject is still open for discussion.

Acknowledgements

The authors would like to thank the Brazilian Agency Funding CNPq (Proc. 476964/2009-1 and 552900/2009-5), CAPES (Procad 082/2007), and Fundação Araucária de Apoio ao Desenvolvimento Científico do Paraná (Prots. 22825 and 22870) for financial support.

References

[1] Kumar A, Podraza NJ, Denev S, et al. Linear and nonlinear optical properties of multifunctional PbVO$_3$ thin films. *Appl Phys Lett* 2008, **92**: 231915.

[2] Lin YR, Sodano HA. Concept and model of a piezoelectric structural fiber for multifunctional composites. *Compos Sci Technol* 2008, **68**: 1911–1918.

[3] Ramesh R, Spaldin NA. Multiferroics: Progress and prospects in thin films. *Nat Mater* 2007, **6**: 21–29.

[4] Béa H, Paruch P. Multiferroics: A way forward along domain walls. *Nat Mater* 2009, **8**: 168–169.

[5] Bibes M, Barthélémy A. Multiferroics: Towards a magnetoelectric memory. *Nat Mater* 2008, **7**: 425–426.

[6] Scott JF. Data storage: Multiferroic memories. *Nat Mater* 2007, **6**: 256–257.

[7] Neaton JB, Ederer C, Waghmare UV, et al. First-principles study of spontaneous polarization in multiferroic BiFeO$_3$. *Phys Rev B* 2005, **71**: 014113.

[8] Ederer C, Spaldin NA. Weak ferromagnetism and magnetoelectric coupling in bismuth ferrite. *Phys Rev B* 2005, **71**: 060401.

[9] Higuchi T, Liu YS, Yao P, et al. Electronic structure of multiferroic BiFeO$_3$ by resonant soft X-ray emission spectroscopy. *Phys Rev B* 2008, **78**: 085106.

[10] Kamba S, Nuzhnyy D, Savinov M, et al. Infrared and terahertz studies of polar phonons and magnetoelectric effect in multiferroic BiFeO$_3$ ceramics. *Phys Rev B* 2007, **75**: 024403.

[11] Baetting P, Ederer C, Spaldin NA. First principles study of the multiferroics BiFeO$_3$, Bi$_2$FeCrO$_6$, and BiCrO$_3$: Structure, polarization, and magnetic ordering temperature. *Phys Rev B* 2005, **72**: 214105.

[12] Wang J, Neaton JB, Zheng H, et al. Epitaxial BiFeO$_3$ multiferroic thin film heterostructures. *Science* 2002, **299**: 1719–1722.

[13] Ivanov SA, Nordblad P, Tellgren R, et al. Influence of PbZrO$_3$ doping on the structural and magnetic properties of BiFeO$_3$. *Solid State Sci* 2008, **10**: 1875–1885.

[14] Gerson R, Chou PC, James WJ. Ferroelectric properties of PbZrO$_3$–BiFeO$_3$ solid solutions. *J Appl Phys* 1967, **38**: 55.

[15] Kumar MM, Srinath S, Kumar GS, et al. Spontaneous magnetic moment in BiFeO$_3$–BaTiO$_3$ solid solutions at low temperatures. *J Magn Magn Mater* 1998, **188**: 203–212.

[16] Wang TH, Tu CS, Ding Y, et al. Phase transition and ferroelectric properties of xBiFeO$_3$–(1–x)BaTiO$_3$ ceramics. *Curr Appl Phys* 2011, **11**: S240–S243.

[17] Gotardo RAM, Santos IA, Cótica LF, et al. Improved ferroelectric and magnetic properties of monoclinic structured 0.8BiFeO$_3$–0.2BaTiO$_3$ magnetoelectric ceramics. *Scripta Mater* 2009, **61**: 508–511.

[18] Wang TH, Ding Y, Tu CS, et al. Structure, magnetic, and dielectric properties of (1–x)BiFeO$_3$–xBaTiO$_3$ ceramics. *J Appl Phys* 2011, **109**: 07D907.

[19] Singh H, Kumar A, Yadav KL. Structural, dielectric, magnetic, magnetodielectric and impedance spectroscopic studies of multiferroic BiFeO$_3$–BaTiO$_3$ ceramics. *Mat Sci Eng B* 2011, **176**: 540–547.

[20] Kumar MM, Srinivas A, Kumar GS, et al. Investigation of the magnetoelectric effect in BiFeO$_3$–BaTiO$_3$ solid solutions. *J Phys: Condens Matter* 1999, **11**: 8131.

[21] Shi CY, Liu XZ, Hao YM, et al. Structural, magnetic and dielectric properties of Sc modified (1–y)BiFeO$_3$–yBaTiO$_3$ ceramics. *Solid State Sci* 2011, **13**: 1885–1888

[22] Kumar MM, Srinivas A, Suryanarayana SV. Structure property relations in BiFeO$_3$/BaTiO$_3$ solid solutions. *J Appl Phys* 2000, **87**: 855.

[23] Fedulov SA, Venevtsev YN, Zhdanov GS, et al. X-ray and electrical studies of the PbTiO$_3$–BiFeO$_3$

system. *Sov Phys Crys* 1962, **7**: 62–66.

[24] Fedulov SA, Ladyzhinskii PB, Pyatigorskaya IL, *et al.* Complete phase diagram of the PbTiO$_3$–BiFeO$_3$ system. *Sov Phys Cryst* 1964, **6**: 375–378.

[25] Kajima A, Kaneda T, Ito H, *et al.* Ferromagnetic amorphouslike oxide films of the Fe$_2$O$_3$–Bi$_2$O$_3$–PbTiO$_3$ system prepared by rf-reactive sputtering. *J Appl Phys* 1991, **69**: 3663.

[26] Spaldin NA, Fiebig M. The renaissance of magnetoelectric multiferroics. *Science* 2005, **309**: 391–392.

[27] Cheng JR, Cross LE. Effects of La substituent on ferroelectric rhombohedral/tetragonal morphotropic phase boundarie in $(1 - x)$(Bi,La)(Ga$_{0.05}$Fe$_{0.95}$)O$_3$–xPbTiO$_3$ piezoelectric ceramics. *J Appl Phys* 2003, **94**: 5188.

[28] Woodward DI, Reaney IM, Eitel RE, *et al.* Crystal and domain structure of the BiFeO$_3$–PbTiO$_3$ solid solution. *J Appl Phys* 2003, **94**: 3313.

[29] Cheng JR, Li N, Cross LE. Structural and dielectric properties of Ga-modified BiFeO$_3$–PbTiO$_3$ crystalline solutions. *J Appl Phys* 2003, **94**: 5153.

[30] Cheng JR, Yu SW, Chen JG, *et al.* Dielectric and magnetic enhancements in BiFeO$_3$–PbTiO$_3$ solid solutions with La doping. *Appl Phys Lett* 2006, **89**: 122911.

[31] Comyn TP, Stevenson T, Al-Jawad M, *et al.* High temperature neutron diffraction studies of 0.9BiFeO$_3$–0.1PbTiO$_3$. *J Appl Phys* 2009, **105**: 094108.

[32] Freitas VF, Santos IA, Botero É, *et al.* Piezoelectric characterization of (0.6)BiFeO$_3$–(0.4)PbTiO$_3$ multiferroic ceramics. *J Am Ceram Soc* 2011, **94**: 754–758.

[33] Singh A, Gupta A, Chatterjee R. Enhanced magnetoelectric coefficient (α) in the modified BiFeO$_3$–PbTiO$_3$ system with La substitution. *Appl Phys Lett* 2008, **93**: 022902.

[34] Comyn TP, Stevenson T, Al-Jawad M, *et al.* Antiferromagnetic order in tetragonal bismuth ferrite–lead titanate. *J Magn Magn Mater* 2011, **323**: 2533–2535.

[35] Ranjan R, Raju KA. Unconventional mechanism of stabilization of a tetragonal phase in the perovskite ferroelectric (PbTiO$_3$)$_{1-x}$(BiFeO$_3$)$_x$. *Phys Rev B* 2010, **82**: 054119.

[36] Catalan G, Scott JF. Physics and applications of bismuth ferrite. *Adv Mater* 2009, **21**: 2463–2485.

[37] Freitas VF, Cótica LF, Santos IA, *et al.* Synthesis and multiferroism in mechanically processed BiFeO$_3$–PbTiO$_3$ ceramics. *J Eur Ceram Soc* 2011, **31**: 2965–2973.

[38] Noheda B, Cox DE, Shirane G, *et al.* A monoclinic ferroelectric phase in the Pb(Zr$_{1-x}$Ti$_x$)O$_3$ solid solution. *Appl Phys Lett* 1999, **74**: 2059.

[39] Zhu WM, Guo HY, Ye ZG. Structural and magnetic characterization of multiferroic (BiFeO$_3$)$_{1-x}$(PbTiO$_3$)$_x$ solid solutions. *Phys Rev B* 2008, **78**: 014401.

[40] Bhattacharjee S, Pandey V, Kotnala RK, *et al.* Unambiguous evidence for magnetoelectric coupling of multiferroic origin in 0.73BiFeO$_3$–0.27PbTiO$_3$. *Appl Phys Lett* 2009, **94**: 012906.

[41] Noheda B, Gonzalo JA, Cross LE, *et al.* Tetragonal-to-monoclinic phase transition in a ferroelectric peorvskite: The structure of PbZr$_{0.52}$Ti$_{0.48}$O$_3$. *Phys Rev B* 2000, **61**: 8687–8695.

[42] Ranjan R, Kothai V, Senyshyn A, *et al.* Neutron diffraction study of the coupling between spin, lattice, and structural degrees of freedom in 0.8BiFeO$_3$–0.2PbTiO$_3$. *J Appl Phys* 2011, **109**: 063522.

[43] Ahart M, Somayazulu M, Cohen RE, *et al.* Origin of morphotropic phase boundaries in ferroelectrics. *Nature* 2008, **451**: 545–548.

[44] Moriya T. Anisotropic superexchange interaction and weak ferromagnetism. *Phys Rev* 1960, **120**: 91–98.

[45] Dzyaloshinsky I. A thermodinamic theory of "weak" ferromagnetism of antiferromagnetic substances. *J Phys Chem Solids* 1958, **4**: 241–255.

[46] Cohen RE. Origin of ferroeletricity in perovskite oxides. *Nature* 1992, **358**: 136–138.

[47] Hill NA. Why are there so few magnetic ferroelectrics? *J Phys Chem B* 2000, **104**: 6694–6709.

3

Hydrothermal synthesis and characterization of MnCo$_2$O$_4$ in the low-temperature hydrothermal process: Their magnetism and electrochemical properties

Lianfeng DUAN[a,b,*], Fenghui GAO[a], Limin WANG[b], Songzhe JIN[a], Hua WU[a]

[a]Key Laboratory of Advanced Structural Materials, Ministry of Education, and Department of Materials Science and Engineering, Changchun University of Technology, Changchun 130012, China
[b]State Key Laboratory of Rare Earth Resource Utilization, Changchun Institute of Applied Chemistry, Chinese Academy of Sciences, Changchun 130022, China

Abstract: MnCo$_2$O$_4$ octahedral structure with edge length about 500 nm was successfully synthesized by a simple hydrothermal route. With the use of NaOH, the chemical potential and the rate of ionic motion in the precursor solution were controlled, and the particle size was limited. The magnetization measurements revealed that the products exhibited ferrimagnetic characteristics with different saturation magnetization and coercivity at different measuring temperatures. In addition, the as-prepared MnCo$_2$O$_4$ as anodes for lithium-ion batteries (LIBs) exhibited a reversible capacity of 1180 mA·h/g and 1090 mA·h/g at current density of 0.1 C and 1 C, respectively. The excellent cyclic performance was confirmed because the value of reversible capacity for MnCo$_2$O$_4$ was 618 mA·h/g after 50 cycles at 0.1 C. Owing to the good rate performance, MnCo$_2$O$_4$ octahedral products were suggested to have a promising application as anode material for LIBs.

Keywords: MnCo$_2$O$_4$; hydrothermal synthesis; lithium-ion batteries (LIBs); anode

1 Introduction

Complex oxides (containing two or more types of cations) with spinel structure under controlled size and shape are of intense interests for both fundamental science and technological applications because of their chemical and physical properties [1–3]. Among these, transition-metal cobaltites having a spinel structure MCo$_2$O$_4$ (M = Mn, Ni, Zn) with unusual physicochemical properties are widely used in many areas such as colossal magnetoresistance (CMR),

sensors, fuel cell electrodes, electrical catalysts, microwave adsorption, etc. [4–7]. In particular, lithium-ion batteries (LIBs) have become the main power source for today's portable electronics and are being actively pursued for propelling electric vehicles in the near future. However, to match the growing demand for LIBs, higher discharge capacity is needed. Meanwhile, renewed interest in transition-metal oxides as potential anode materials for use in LIBs starts to grow [8,9]. Compared with graphite (372 mA·h/g), transition-metal cobaltites can deliver a high specific capacity. It has been reported that the mixed transition-metal oxides including NiCo$_2$O$_4$ [10], ZnCo$_2$O$_4$ [11] and MnCo$_2$O$_4$ [12] have been applied as anode materials for LIBs. Among these

---

* Corresponding author.
E-mail: duanlf@mail.ccut.edu.cn

transition-metal oxides, spinel $MnCo_2O_4$ has been studied widely due to it promising applications as a magnetic material. However, as an anode material for LIBs, $MnCo_2O_4$ has received little attention. $MnCo_2O_4$ particles have been prepared by various approaches such as Pechini method, sol–gel techniques, coprecipitation, microwave plasma synthesis, milling the oxide powders ($MnCO_3$ and Co_3O_4), spray pyrolysis of the metal nitrate aqueous solutions, heat mixture of $MnCO_3$ and CoC_2O_4, and so on [13–16]. However, the research on $MnCo_2O_4$ particles into different architectures and their structural development are comparatively limited, because the size, morphology, composition, dispersion, and surface features of the particles cannot meet the needs of further practical applications.

Up to now, reports on the morphology-controlled synthesis of $MnCo_2O_4$ through an aqueous reduction strategy under mild conditions is still lacked. The hydrothermal method demonstrates its superiority in controlling the shape of crystals [17]. In this paper, we present the synthesis of octahedral $MnCo_2O_4$ via a facile hydrothermal process. The influences of reaction time, dosage of sodium hydroxide on the morphology of the products have been investigated. To compare these results with that of a low ferrimagnetic oxide system, we have examined the magnetic behavior of $MnCo_2O_4$. The electrochemical performance of the as-prepared powders was also evaluated. By the studying of the charge–discharge properties of the submicrocrystals with octahedral structure, it can be expected that $MnCo_2O_4$ could be used as anode material for LIBs in future applications.

2   Experimental section

All chemicals of analytical grade were used without further purification. In the first step, 0.2745 g $Co(NO_3)_2$ was dissolved in 20 ml EG (ethylene glycol) under magnetic stirring at room temperature. Then 0.0527 g $KMnO_4$ and different dosages of NaOH were also added sequentially with constant magnetic stirring. The mixture was transferred into a 30 ml stainless steel autoclave, sealed, and maintained at 200 ℃ for 24 h. After completion of the reaction, the products were collected by centrifuge separation, washed several times with water and absolute ethanol, and finally dried in a vacuum oven at 40 ℃ for 6 h.

The phases were identified by means of X-ray diffraction (XRD) using a Rigaku D/max 2500pc X-ray diffractometer with Cu Kα radiation ($\lambda = 1.54156$ Å) at a scan rate of 0.04 (°)/s. The morphologies were characterized by a JEOL JSM-6700F field-emission scanning electron microscope (FESEM) operated at an acceleration voltage of 8.0 kV. Transmission electron microscopy (TEM), high-resolution TEM (HRTEM) observations, and selected-area electron diffraction (SAED) patterns were obtained by using a JEOL 2100F instrument with an emission voltage of 200 kV. Magnetic measurements were carried out using a Quantum Design superconducting quantum interference device (SQUID) magnetometer (LakeShore 7307).

The electrochemical experiments were performed via CR2025 coin-type test cells assembled in a dry argon-filled glove box with both moisture and oxygen contents below 1 ppm. The test cell consisted of a working electrode and lithium foil which were separated by a Celgard 2400 membrane. The electrolyte solution was prepared by dissolving 1 M $LiPF_6$ in EC-DMC (ethylene carbonate dimethyl carbonate) (weight-to-weight ratio = 1:1). The working electrodes were prepared by casting slurry containing 80% active material, 10% acetylene black and 10% polyvinylidene fluoride (PVDF) onto a copper foil. After vacuum drying at 80 ℃ for about 12 h, the electrode disks ($d = 12$ mm) were punched and weighed. Each electrode has approximately 1–3 mg active material. Galvanostatic charge–discharge cycling tests were performed using an LAND CT2001A multi-channel battery testing system in the voltage range between 0.5 V and 3 V at room temperature.

3   Results and discussion

The XRD pattern of the as-obtained $MnCo_2O_4$ crystals is shown in Fig. 1. All the diffraction peaks in the XRD pattern of the products can be indexed to the face-centered cubic structure of $MnCo_2O_4$ (JCPDS 23-1237). The sharpness of the XRD peaks confirms that the material should be highly crystallized $MnCo_2O_4$ without any other impurities. Further insight into the morphology and microstructure of the $MnCo_2O_4$ octahedra was gained by using FESEM and TEM. Figure 2 shows the typical FESEM and TEM images of $MnCo_2O_4$ ferrite crystals, from which we can conclude that octahedral structures are the

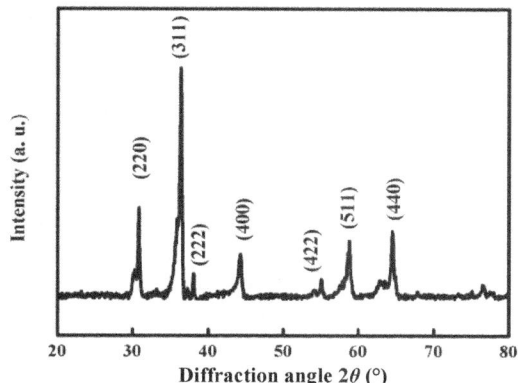

Fig. 1 XRD pattern of the as-obtained products.

exclusive products, which means that the $MnCo_2O_4$ octahedra can be prepared on a large scale. From the high-magnification FESEM and TEM images of the products (Figs. 2(b) and 2(c)), the edge length of octahedral $MnCo_2O_4$ crystals is about 500 nm. HRTEM was performed on an individual $MnCo_2O_4$ particle, as shown in Fig. 2(d). The clear and alignment of lattice fringes demonstrate that the octahedron is essentially single crystalline with no crystal defects. The lattice spacing between two adjacent fringes that can be observed corresponds to the set of ($1\overline{1}1$) planes with a lattice spacing of 0.238 nm and the set of (220) planes with a lattice spacing of 0.293 nm, respectively. In principle, a crystal growth process consists of nucleation and growth, which are affected by the intrinsic crystal structure and external conditions. It is well known that spinel $MnCo_2O_4$ is cubic in structure, and the well defined octahedral morphology is characteristic of cubic structured crystals bound by eight (111) planes. Because of the slow reaction rate under the present hydrothermal synthesis conditions and the absence of other structure-modifying ions, octahedra with entirely {111} faces are considered to be the thermodynamically favorable product structures [18].

The corresponding inverse fast-Fourier-transform (FFT) pattern and the filtered FFT of a selected area of the image are shown in the insets of Fig. 2(d). It represents a pattern of diffraction spots with hexagonal

Fig. 2 FESEM image of the as-prepared products: (a) low magnification and (b) high magnification; (c) TEM and (d) HRTEM. The insets are filtered FFT (left) and inverse FFT (right) of the HRTEM image.

symmetry. The FFT pattern illustrates a perfect single crystal nature of the $MnCo_2O_4$, where the spots can be steadily indexed to (220), (311), ($1\overline{1}1$) facets. And from the filtered FFT image which is just several atom distance wide, the atoms are aligned which means there is not any dislocation. It is further confirmed that the $MnCo_2O_4$ submicrocrystals are single crystalline.

To understand the formation mechanism of the octahedral $MnCo_2O_4$ particles, alkalinity-dependent experiments were carried out. Herein, Fig. 3 shows the SEM images of the samples obtained at various dosages of NaOH at 200 ℃ for 24 h. In fact, irregular particles are obtained when the dosage of NaOH is 0.4 g. With the increase in NaOH concentration, the morphology of the products evolves into a perfect octahedral structure with few irregular particles. When the amount of NaOH is increased to 1 g, it is interesting to note that the morphology of particles presents an octahedral shape (Fig. 2). The higher chemical potential is mainly determined by the concentration of NaOH. The pH value is supposed to exert an impact on both the rates of crystal nucleation and crystal growth [19]. Peng and Peng [20,21] have elucidated the influence of chemical potential on the shape evolution. The hydroxide with high concentrations is easily adsorbed onto (111) facets, which are then stabilized and grow slowly; hence, the growth of other facets will gradually diminish [22]. The alkalinity affects the balance between the chemical potential and the rate of ionic motion in the precursor solution, and a high concentration of NaOH accelerates the nucleation [23]. Especially, nucleation and aggregation growth in the EG solution are kinetically slower than those in the aqueous solution due to fewer surface hydroxyls and greater viscosity, thus allowing the particles to rotate adequately to find the low-energy configuration interface and form assemblies [24]. With the low concentration of hydroxide, $MnCo_2O_4$ can no longer be formed. In this case, $Co(OH)_3$ and $Mn(OH)_2$ deposit first, then complex compounds with Co^{3+} and Mn^{2+} are formed. So, with the increase of hydroxide we speculate that the redox reaction between MnO_4^-

Fig. 3   FESEM images of the samples prepared with different dosages of NaOH: (a) low magnification and (b) high magnification of 0.4 g; (c) low magnification and (d) high magnification of 0.8 g.

and Co^{2+} in alkaline solution is very important. The chemical reaction in the hydrothermal process can be written as follows:

$$4MnO_4^- + 8Co^{2+} + 12OH^- \rightarrow 4MnCo_2O_4 + 3O_2 \uparrow + 6H_2O$$

From the formula, the final product is strongly dependent on the mole ratio of MnO_4^- and Co^{2+}, which should be 1:2 for the formation of $MnCo_2O_4$. From the above experimental results, the hole is formed, where the gases produced in reaction solution with the dosage of 1 g NaOH destroy the interior of loose aggregates from the chemical formula. It is further verified for the rough surface of particles. With increasing in the reaction time, Ostwald ripening process should gradually replace the aggregation-based crystal growth due to the increasing size difference between the center and outer parts of aggregates, which could recover the hole under the crystal growth (Fig. 4). And it seems that the concentration of NaOH plays a key role in the microstructures of the final products. Both high concentration of OH^- ions and high chemical potential in solution favor the growth of octahedral structures over other possible crystal forms.

In order to further understand the growth evolution of the $MnCo_2O_4$ microcrystals, the effect of the reaction time (3 h, 6 h, 12 h and 18 h) was systematically investigated. The FESEM images of the samples are shown in Fig. 4 as a function of reaction time. At the initial step of the reaction (Fig. 4(a)), a few small $MnCo_2O_4$ nucleuses are formed due to redox reaction between MnO_4^-, Co^{2+} and OH^-. As thermodynamic growth occurs, there is physical adsorption of the diffusing nanoparticles once they are in contact with one another. When the reaction time is increased to 6 h, the aggregated particles with irregular shape are formed, as shown in Fig. 4(b). When it is extended to 12 h, the embryo octahedra already generate, because Ostwald ripening process would gradually replace the aggregation-based crystal growth in order for the increasing size. If the hydroxy groups are adsorbed on some areas of its surface, the growth rate of crystal in certain directions will be confined. Due to the gases produced, the hole is formed at the particle surface from Fig. 4(c). After 18 h, the relatively octahedral microcrystal is synthesized, and

Fig. 4 FESEM images of the samples prepared under different reaction time: (a) 2 h; (b) 6 h; (c) 12 h; (d) 18 h.

the edge length of MnCo$_2$O$_4$ octahedra is 300 nm. At last, the typical octahedral morphology character of the MnCo$_2$O$_4$ is observed at 24 h.

The magnetic properties of MnCo$_2$O$_4$ microcrystals, which were tested at 300 K and 2 K by SQUID are shown in Fig. 5. The saturation magnetization (M_s) and coercivity (H_c) values of MnCo$_2$O$_4$ nanooctahedra are 0.649 emu/g and 0.657 emu/g, 64.68 Oe and 168.56 Oe at 300 K and 2 K, respectively. In addition, the H_c of the samples at 2 K are larger than those at 300 K due to the reduced influence of thermal fluctuation on the rotation of magnetic dipoles. Such a rearrangement would give rise to a small distortion of the structure and affect the local ordering of the ions in the octahedral sites. Further detailed investigation is required to understand the unusual magnetic hysteresis behavior of MnCo$_2$O$_4$.

The as-prepared MnCo$_2$O$_4$ particles are then tested as anodes for LIBs. The typical first charge–discharge profiles of anode material of MnCo$_2$O$_4$ particles at a current density of 0.1 C are shown in Fig. 6(a). The initial discharge is 1180 mA·h/g, which is almost four times higher than that of the common carbonaceous

materials. The discharge voltage plateau at about 0.7 V is quite stable. The electrochemical mechanism of the MnCo$_2$O$_4$ electrode obeys to the displacive redox mechanism confirmed by many researchers [25–27].

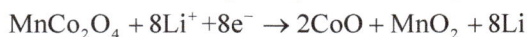

$$MnCo_2O_4 + 8Li^+ + 8e^- \rightarrow 2CoO + MnO_2 + 8Li$$

$$2CoO + MnO_2 + 8Li \rightarrow MnCo_2O_4 + 8Li^+ + 8e^-$$

Figure 6(b) shows the charge–discharge capacities versus cycle number and efficiency for the MnCo$_2$O$_4$ at current densities of 0.1 C and 1 C between 0.50 V and 3.00 V (vs. Li/Li$^+$) at room temperature. Discharge capacities obtained for the second cycle at 0.1 C is 1060 mA·h/g. After 50 cycles, polyhedron structure remains about 618 mA·h/g, which is much larger than the other structure particles. Even at a high current density of 1.0 C, a discharge capacity of 380 mA·h/g is retained after 50 cycles. Although this value is much lower than those of the earlier cycles, it is still higher than the other transition-metal oxides, such as Co$_3$O$_4$, CoFe$_2$O$_4$ and MnFe$_2$O$_4$ [28]. The improved electrochemical performance makes such MnCo$_2$O$_4$ different structure promising as anode material for next-generation LIBs.

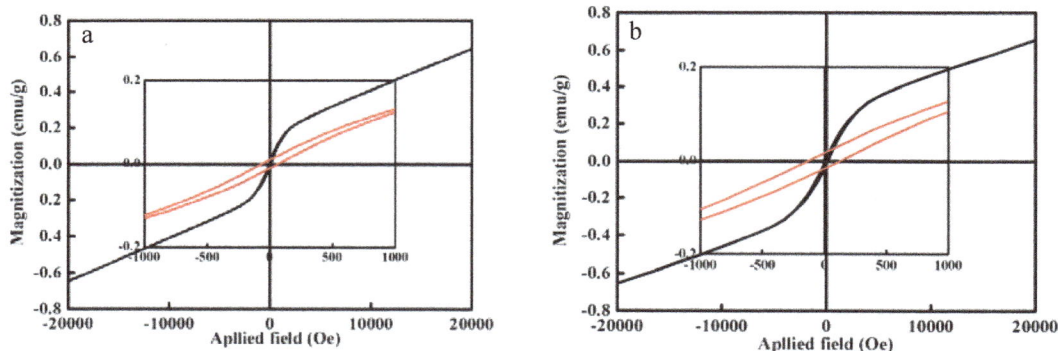

Fig. 5   Magnetization curves measured at different temperatures: (a) 300 K; (b) 2 K.

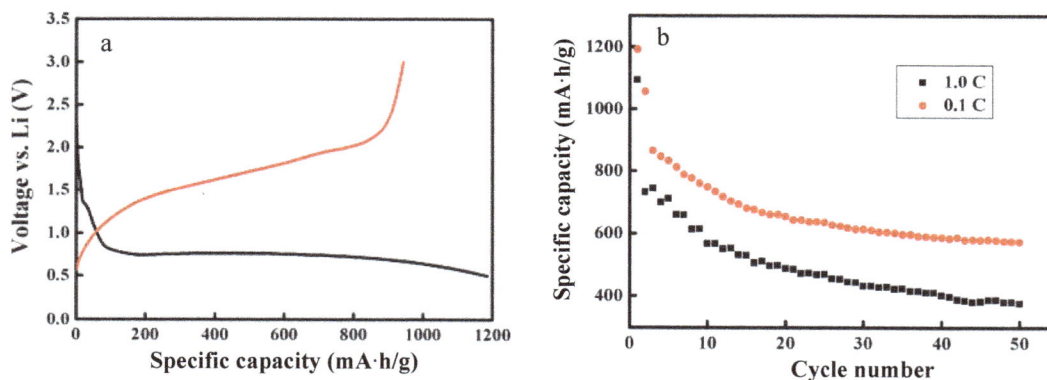

Fig. 6   (a) Charge–discharge plots and (b) cyclic performance of the MnCo$_2$O$_4$ electrode at a current density of 0.1 C and 1.0 C between 0.50 V and 3.00 V (vs. Li/Li$^+$) at room temperature.

4   Conclusions

In summary, $MnCo_2O_4$ octahedra have been successfully prepared via a facile hydrothermal synthetic approach. It has been found that the concentration of hydroxide ions plays a key role in control of size, morphology and structure of the products, which affect the balance between the chemical potential and the rate of ionic motion in the precursor solution, and the hydroxide with high concentration could be adsorbed onto (111) facets easily. It has been observed that the H_c of the samples at 2 K are much larger than that at room temperature by magnetization curve. In addition, the as-prepared $MnCo_2O_4$ as anodes for LIBs exhibits a reversible capacity of 1180 mA·h/g at a current density of 0.1 C, retaining 618 mA·h/g after 50 cycles. The results show that the as-prepared $MnCo_2O_4$ octahedral microcrystal is a kind of promising anode material for LIBs.

Acknowledgements

This work was supported by the Natural Science Foundation of Jilin Province (201215118) and the Special Funds of Changchun University of Technology.

References

[1]   Hinklin TR, Azurdia J, Kim M, et al. Finding spinel in all the wrong places. Adv Mater 2008, 20: 1373–1375.

[2]   Yoon TJ, Kim JS, Kim BG, et al. Multifunctional nanoparticles possessing a "magnetic motor effect" for drug or gene delivery. Angew Chem Int Edit 2005, 44: 1068–1071.

[3]   Marco JF, Gancedo JR, Gracia M, et al. Cation distribution and magnetic structure of the ferrimagnetic spinel NiCo2O4. J Mater Chem 2001, 11: 3087–3093.

[4]   Rios E, Gautier J-L, Poillerat G, et al. Mixed valency spinel oxides of transition metals and electrocatalysis: Case of the $Mn_xCo_{3-x}O_4$ system. Electrochim Acta 1998, 44: 1491–1497.

[5]   Nissinen T, Kiros Y, Gasik M, et al. Comparison of preparation routes of spinel catalyst for alkaline fuel cells. Mater Res Bull 2004, 39: 1195–1208.

[6]   Zhu J, Gao Q. Mesoporous MCO2O4 (M = Cu, Mn and Ni) spinels: Structural replication, characterization and catalytic application in CO oxidation. Microporous Mesoporous Mater 2009, 124: 144–152.

[7]   Zhao D-L, Lv Q, Shen Z-M. Fabrication and microwave absorbing properties of Ni–Zn spinel ferrites. J Alloys Compd 2009, 480: 634–638.

[8]   Grugeon S, Laruelle S, Dupont L, et al. An update on the reactivity of nanoparticles Co-based compounds towards Li. Solid State Sci 2003, 5: 895–904.

[9]   Larcher D, Bonnin D, Cortes R, et al. Combined XRD, EXAFS, and Mössbauer studies of the reduction by lithium of α-Fe2O3 with various particle sizes. J Electrochem Soc 2003, 150: A1643–A1650.

[10]   Alcántara R, Jaraba M, Lavela P, et al. NiCo2O4 spinel: First report on a transition metal oxide for the negative electrode of sodium-ion batteries. Chem Mater 2002, 14: 2847–2848.

[11]   Sharma Y, Sharma N, Subba Rao GV, et al. Nanophase ZnCo2O4 as a high performance anode material for Li-ion batteries. Adv Funct Mater 2007, 17: 2855–2861.

[12]   Zhou L, Zhao D, Lou XW. Double-shelled CoMn2O4 hollow microcubes as high-capacity anodes for lithium-ion batteries. Adv Mater 2012, 24: 745–748.

[13]   Borges FMM, Melo DMA, Câmara MSA, et al. Magnetic behavior of nanocrystalline MnCo2O4 spinels. J Magn Magn Mater 2006, 302: 273–277.

[14]   Lavela P, Tirado JL, Vidal-Abarca C. Sol–gel preparation of cobalt manganese mixed oxides for their use as electrode materials in lithium cells. Electrochim Acta 2007, 52: 7986–7995.

[15]   Rios E, Poillerat G, Koenig JF, et al. Preparation and characterization of thin Co3O4 and MnCo2O4 films prepared on glass/SnO2:F by spray pyrolysis at 150 ℃ for the oxygen electrode. Thin Solid Films 1995, 264: 18–24.

[16]   Choi J-J, Ryu J, Hahn B-D, et al. Dense spinel MnCo2O4 film coating by aerosol deposition on ferritic steel alloy for protection of chromic evaporation and low-conductivity scale formation. J Mater Sci 2009, 44: 843–848.

[17] Duan L, Jia S, Cheng R, *et al*. Synthesis and characterization of Co sub-micro chains by solvothermal route: Process design, magnetism and excellent thermal stability. *Chem Eng J* 2011, **173**: 233–240.

[18] Wang YQ, Cheng RM, Wen Z, *et al*. Synthesis and characterization of single-crystalline $MnFe_2O_4$ ferrite nanocrystals and their possible application in water treatment. *Eur J Inorg Chem* 2011, **2011**: 2942–2947.

[19] Zhou X-M, Wei X-W. Single crystalline $FeNi_3$ dendrites: Large scale synthesis, formation mechanism, and magnetic properties. *Cryst Growth Des* 2009, **9**: 7–12.

[20] Peng ZA, Peng X. Mechanisms of the shape evolution of CdSe nanocrystals. *J Am Chem Soc* 2001, **123**: 1389–1395.

[21] Peng ZA, Peng X. Nearly monodisperse and shape-controlled CdSe nanocrystals via alternative routes: Nucleation and growth. *J Am Chem Soc* 2002, **124**: 3343–3247.

[22] Zhao L, Zhang H, Tang J, *et al*. Fabrication and characterization of uniform Fe_3O_4 octahedral micro-crystals. *Mater Lett* 2009, **63**: 307–309.

[23] Hou X, Feng J, Xu X, *et al*. Synthesis and characterizations of spinel $MnFe_2O_4$ nanorod by seed–hydrothermal route. *J Alloys Compd* 2010, **491**: 258–263.

[24] Duan L, Jia S, Cheng R, *et al*. Synthesis and characterization of Co sub-micro chains by solvothermal route: Process design, magnetism and excellent thermal stability. *Chem Eng J* 2011, **173**: 233–240.

[25] Ai L-H, Jiang J. Rapid synthesis of nanocrystalline Co_3O_4 by a microwave-assisted combustion method. *Powder Technol* 2009, **195**: 11–14.

[26] Li ZH, Zhao TP, Zhan XY, *et al*. High capacity three-dimensional ordered macroporous $CoFe_2O_4$ as anode material for lithium ion batteries. *Electrochim Acta* 2010, **55**: 4594–4598.

[27] Kang Y-M, Song M-S, Kim J-H, *et al*. A study on the charge–discharge mechanism of Co_3O_4 as an anode for the Li ion secondary battery. *Electrochim Acta* 2005, **50**: 3667–3673.

[28] Lavela P, Ortiz GF, Tirado JL. High-performance transition metal mixed oxides in conversion electrodes: A combined spectroscopic and electrochemical study. *J Phys Chem C* 2007, **111**: 14238–14246.

Structural, magnetic and dielectric properties of nano-crystalline Ni-doped BiFeO$_3$ ceramics formulated by self-propagating high-temperature synthesis

Yogesh A. CHAUDHARI[a,b], Chandrashekhar M. MAHAJAN[c],
Prashant P. JAGTAP[a], Subhash T. BENDRE[a,*]

[a]*Department of Physics, School of Physical Sciences, North Maharashtra University, Jalgaon 425001, India*
[b]*Department of Engineering Sciences and Humanities (DESH), SRTTC-FOE, Kamshet, Pune 410405, India*
[c]*Department of Engineering Sciences and Humanities (DESH), Vishwakarma Institute of Technology (VIT), Pune 411037, India*

Abstract: Ni-doped BiFeO$_3$ powders with the composition BiFe$_{1-x}$Ni$_x$O$_3$ (x = 0.05, 0.1 and 0.15) were prepared by a self-propagating high-temperature synthesis (SHS), using metal nitrates as oxidizers and glycine as fuel. The X-ray diffraction (XRD) patterns depict that Ni-doped BiFeO$_3$ ceramics crystallize in a rhombhohedral phase. The scanning electron micrographs of Ni-doped BiFeO$_3$ ceramics show a dense morphology with interconnected structure. It is found that, the room-temperature magnetization measurements in Ni-incorporated BiFeO$_3$ ceramics give rise to nonzero magnetization. The magnetization of Ni-doped BiFeO$_3$ ceramics is significantly enhanced when Ni doping concentration reaches to $x = 0.1$ at 5 K. The variations of dielectric constant with temperature in BiFe$_{0.95}$Ni$_{0.05}$O$_3$, BiFe$_{0.9}$Ni$_{0.1}$O$_3$ and BiFe$_{0.85}$Ni$_{0.15}$O$_3$ samples exhibit clear dielectric anomalies approximately around 450 ℃, 425 ℃ and 410 ℃ respectively, which correspond to antiferromagnetic to paramagnetic phase transition of the parent compound BiFeO$_3$.

Keywords: Ni-doped BiFeO$_3$; self-propagating high-temperature synthesis (SHS); X-ray diffraction (XRD); magnetic properties; dielectric properties

1 Introduction

Multiferroic materials exhibit electric and magnetic natures which result in a mutual existence of ferroelectricity and ferromagnetism in a single phase [1]. Because of room-temperature coupling between ferroelectric and magnetic order parameters, it brings forth a novel phenomenon known as magnetoelectric effect (ME), in which polarization can be tuned by magnetic field and vice versa. This coupling provides an additional opportunity for the design of magnetoelectric and spintronic devices [2–4]. Multiferroic materials have gained tremendous attention on account of their potential applications in various fields, such as bubble memory device, microwave, satellite communication, audio-video, digital recording [4,5], sensor, multiple state memory element, electro-ferromagnetic resonance

---

* Corresponding author.
E-mail: bendrest@gmail.com

device [6], thin film capacitor, non-volatile memory [7], optoelectronics, solar energy device [4], high-density ferroelectric magnetic random access memory [8], and permanent magnet [9]. A multifunctional $BiFeO_3$ compound demonstrates a magnetoelectric coupling having Curie temperature $T_C \approx 1100$ K and G-type antiferromagnetism temperature $T_N \approx 640$ K [10,11].

However, $BiFeO_3$ has serious problems as a ferroelectric material because of its quite large leakage current density at room temperature, which is mainly attributed to the oxygen vacancy and oxidation state of Fe. Therefore, the higher conductive nature of $BiFeO_3$ makes it hard to get excellent ferroelectric property. To overcome this problem, various approaches have been proposed, such as reduction in oxygen vacancies, domination of the ohmic conduction, and intergrain depletion in grain boundary limited conduction. The efforts have been made to reduce the leakage current density by either introducing dopants or using different fabrication methods [12–16]. At present, many researchers are engaged in the enrichment of multiferroic properties of $BiFeO_3$-relevant materials, using different trivalent dopants such as La [17], Mn [18], Sm [19] and Ti [20]. There are many reports on the ferroelectric and magnetic behaviors of $BiFeO_3$. Xu et al. [21] observed a room-temperature saturated magnetic hysteresis loop in Zn-doped $BiFeO_3$ ceramic by rapid sintering method. Chaudhari et al. [3] observed a superparamagnetic nature at 5 K and weak ferromagnetic behavior in $BiFe_{1-x}Zn_xO_3$ ($0.05 \leqslant x \leqslant 0.15$) ceramic by solution combustion method (SCM). The multiferroic $Bi_{1-x}Ca_xFeO_3$ ceramic presents enhanced magnetic property which suppresses spin modulated structure [22]. Recently, Wang et al. [23] reported the enhanced magnetic property of Ni-substituted $BiFeO_3$ at doping concentration of 0.5% by hydrothermal method.

Amid this vision, the present paper investigates the structural, magnetic and dielectric properties of $BiFeO_3$ doped by Ni at Fe site.

2   Experiment

2. 1   Formulation of $BiFe_{1-x}Ni_xO_3$ (x=0.05, 0.1 and 0.15)

The starting precursors used to execute self-propagating high-temperature synthesis (SHS) reaction were $Bi(NO_3)_3 \cdot 5H_2O$, $Fe(NO_3)_3 \cdot 9H_2O$ and $Ni(NO_3)_2 \cdot 6H_2O$ acting as oxidizers, and glycine (NH_2CH_2COOH) used as fuel. Figure 1 shows the flowchart of SHS process, in which the oxidizer/fuel ratio was figured on the basis of oxidizing valencies of the metal nitrates and reducing valency of the fuel [24]. The above mentioned metal nitrates and glycine in stoichiometric proportions were totally dissolved in distilled water. Afterwards, the mixtures were heated in a Pyrex dish till the excess of free water was evaporated and spontaneous ignition occurred, and finally the powders were obtained. These powders with different doping concentration of Ni in $BiFeO_3$ were calcined at 650 ℃ for 4 h. In addition, these powders were pelletized through the addition of polyvinyl alcohol (PVA) as binder. The pellets of $BiFe_{0.95}Ni_{0.05}O_3$, $BiFe_{0.9}Ni_{0.1}O_3$ and $BiFe_{0.85}Ni_{0.15}O_3$ samples were sintered for 30 min at higher temperatures such as 670 ℃, 675 ℃ and 680 ℃, respectively. Finally, these pellets were conveyed for characterization and measurement.

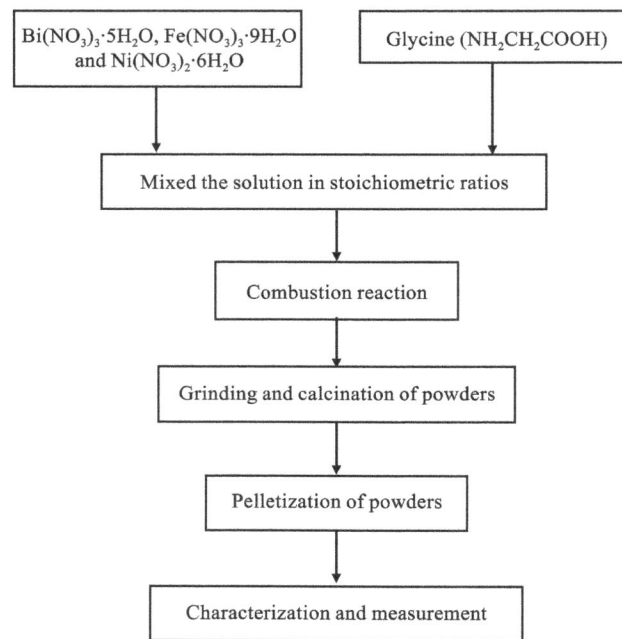

Fig. 1   Flowchart of SHS process.

2. 2   Characterization

The phase identification of the sintered pellets was performed on an X-ray diffractometer (Philips X'Pert PRO) with Cu Kα radiation in the 2θ range of 20°–60°. For ferroelectric and dielectric measurements, the two opposite faces of the sintered pellets were polished with silver paste, because the silver layer served as electrode. The ferroelectric measurement was performed at room temperature using a ferroelectric tester (Precision Premier II, Radiant Technologies, USA). Dielectric constant as a function of temperature in the range of 30–500 ℃ at certain fixed frequencies of 10 kHz and 1 MHz was carried out using an impedance analyzer (Agilent HP 4192A).

3 Results and discussion

3.1 Structural study

Figure 2 demonstrates the room-temperature X-ray diffraction (XRD) patterns of $BiFe_{0.95}Ni_{0.05}O_3$, $BiFe_{0.9}Ni_{0.1}O_3$ and $BiFe_{0.85}Ni_{0.15}O_3$ ceramics. The XRD patterns depict that, $BiFe_{1-x}Ni_xO_3$ samples crystallize in a rhombhohedral perovskite phase in the doping range of $0.05 \leqslant x \leqslant 0.15$. Moreover, an additional impurity phase corresponding to $Bi_{12}NiO_{19}$ has been spotted around 30° in the 2θ range. Typically, it is very difficult to prepare a single-phase $BiFeO_3$, as the product is frequently contaminated with some secondary phases like Bi_2O_3, $Bi_2Fe_4O_9$ and $Bi_{12}(Bi_{0.5}Fe_{0.5})O_{19.5}$ [21,25]. The XRD results are in well accord with the reported results by Wang *et al.* [23].

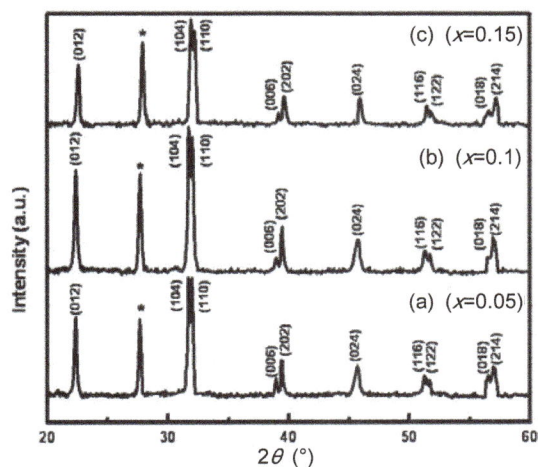

Fig. 2 XRD patterns of (a) $BiFe_{0.95}Ni_{0.05}O_3$ sintered at 670 ℃, (b) $BiFe_{0.9}Ni_{0.1}O_3$ sintered at 675 ℃, and (c) $BiFe_{0.85}Ni_{0.15}O_3$ sintered at 680 ℃ for 30 min respectively, obtained by SCM (* symbolizes secondary phases).

3.2 Surface morphology

The surface morphology of Ni-doped $BiFeO_3$ ceramics demonstrates dense morphology with interconnected structure shown in Fig. 3.

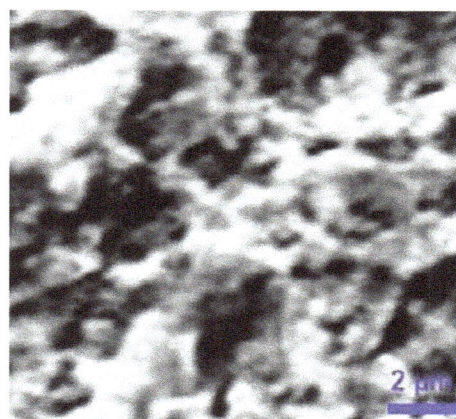

(a) $BiFe_{0.95}Ni_{0.05}O_3$

(b) $BiFe_{0.9}Ni_{0.1}O_3$

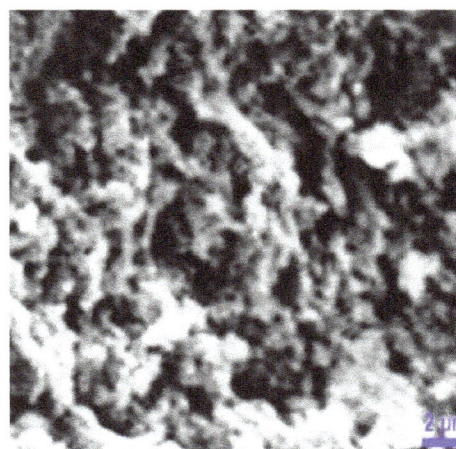

(c) $BiFe_{0.85}Ni_{0.15}O_3$

Fig. 3 scanning electron micrographs of (a) $BiFe_{0.95}Ni_{0.05}O_3$, (b) $BiFe_{0.9}Ni_{0.1}O_3$, and (c) $BiFe_{0.85}Ni_{0.15}O_3$ ceramics.

3.3 Magnetic hysteresis (M–H) loops

Figure 4 represents the room-temperature magnetic

hysteresis (M–H) loops of $BiFe_{0.95}Ni_{0.05}O_3$, $BiFe_{0.9}Ni_{0.1}O_3$ and $BiFe_{0.8}Ni_{0.15}O_3$ ceramics. From the magnetization curves A, B and C, it is assured that, the nonzero remnant magnetization (M_r) and coercive field (H_c) are observed in Ni-doped $BiFeO_3$. It may also be noted that, with varying Ni concentration, substitution-improved magnetic property is observed in $BiFeO_3$. The insets of Fig. 4 present the

room-temperature M–H loops of $BiFe_{0.95}Ni_{0.05}O_3$, $BiFe_{0.9}Ni_{0.1}O_3$ and $BiFe_{0.8}Ni_{0.15}O_3$ samples at higher field up to 15 000 Oe.

Figure 5 presents the M–H loops of $BiFe_{0.95}Ni_{0.05}O_3$, $BiFe_{0.9}Ni_{0.1}O_3$ and $BiFe_{0.8}Ni_{0.15}O_3$ ceramic samples at 5 K. It can be seen that, with increasing Ni doping concentration from $x = 0.05$ to $x = 0.15$, the loops are saturated with the saturation magnetizations (M_s) equal

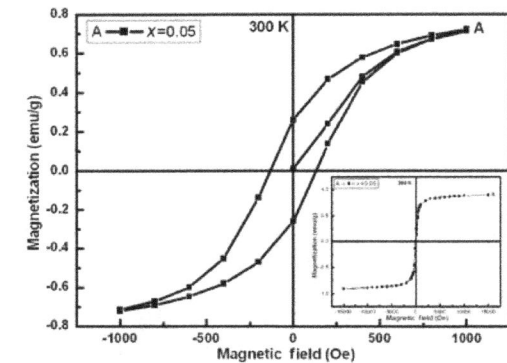

(a) $BiFe_{0.95}Ni_{0.05}O_3$

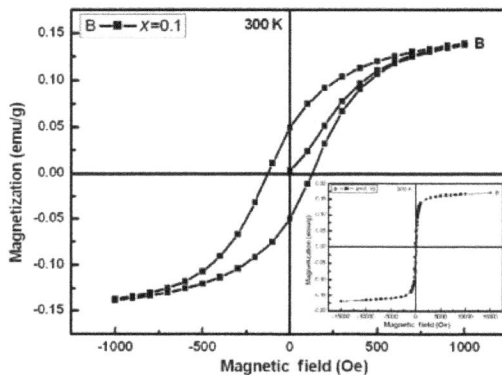

(b) $BiFe_{0.9}Ni_{0.1}O_3$

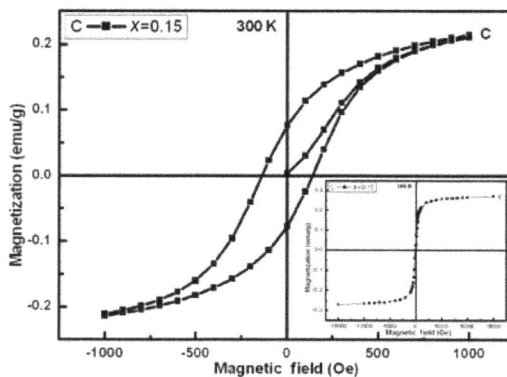

(c) $BiFe_{0.85}Ni_{0.15}O_3$

Fig. 4 Room-temperature M–H loops under the applied magnetic field of 1000 Oe for (a) $BiFe_{0.95}Ni_{0.05}O_3$, (b) $BiFe_{0.9}Ni_{0.1}O_3$, and (c) $BiFe_{0.85}Ni_{0.15}O_3$ samples. The insets show the higher-field M–H data at 15 000 Oe.

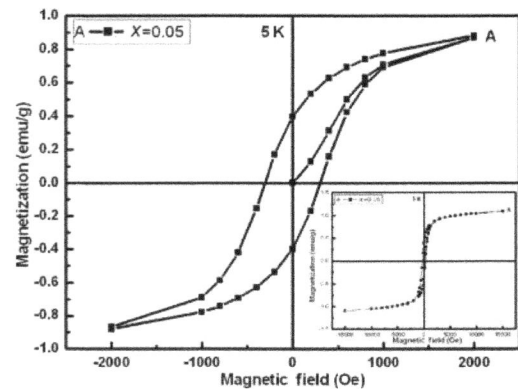

(a) $BiFe_{0.95}Ni_{0.05}O_3$

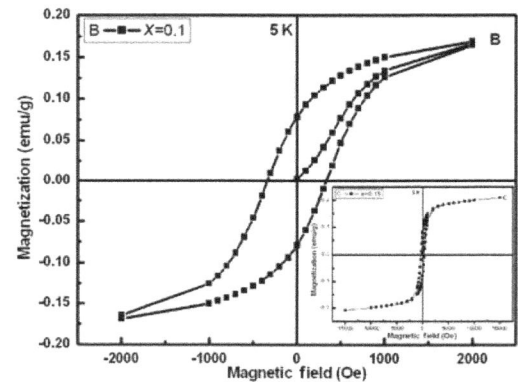

(b) $BiFe_{0.9}Ni_{0.1}O_3$

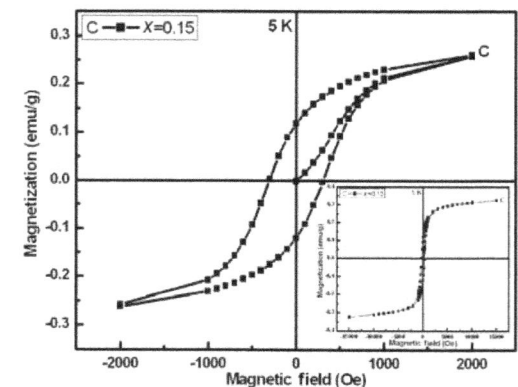

(c) $BiFe_{0.85}Ni_{0.15}O_3$

Fig. 5 M–H loops at 5 K under the applied field of 2000 Oe for (a) $BiFe_{0.95}Ni_{0.05}O_3$, (b) $BiFe_{0.9}Ni_{0.1}O_3$, and (c) $BiFe_{0.85}Ni_{0.15}O_3$ samples. The insets show the higher-field M–H data at 15 000 Oe.

to 0.88 emu/g, 0.16 emu/g and 0.26 emu/g, because Ni doping at the Fe site is responsible for the collapse of the space-modulated spin structure in $BiFeO_3$. The insets of Fig. 5 present the M–H loops of $BiFe_{0.95}Ni_{0.05}O_3$, $BiFe_{0.9}Ni_{0.1}O_3$ and $BiFe_{0.8}Ni_{0.15}O_3$ samples at 5 K and higher field up to 15 000 Oe.

3.4 Dielectric properties

Figure 6 shows the temperature-dependent variation of dielectric constant for $BiFe_{0.95}Ni_{0.05}O_3$, $BiFe_{0.9}Ni_{0.1}O_3$ and $BiFe_{0.85}Ni_{0.15}O_3$ ceramics at 10 kHz and 1 MHz.

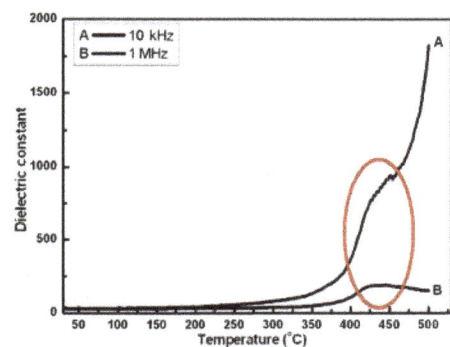

(a) $BiFe_{0.95}Ni_{0.05}O_3$

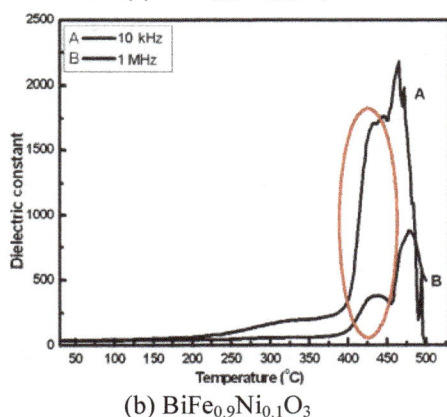

(b) $BiFe_{0.9}Ni_{0.1}O_3$

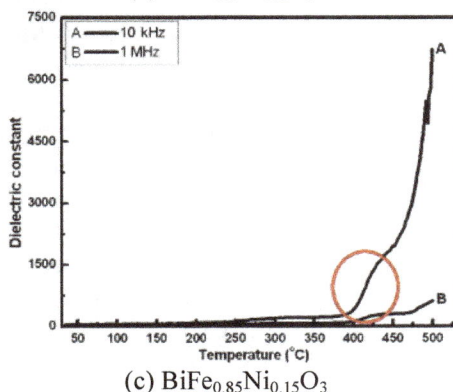

(c) $BiFe_{0.85}Ni_{0.15}O_3$

Fig. 6   Dielectric constant versus temperature at 10 kHz and 1 MHz for (a) $BiFe_{0.95}Ni_{0.05}O_3$, (b) $BiFe_{0.9}Ni_{0.1}O_3$, and (c) $BiFe_{0.85}Ni_{0.15}O_3$ samples in the temperature range of 30–500 ℃.

The dielectric constant shows a continuous increase with temperature for $BiFe_{1-x}Ni_xO_3$ ($x = 0.05$, 0.1 and 0.15) ceramics. Apparent dielectric anomalies have been detected in the three ceramics around 450 ℃, 425 ℃ and 410 ℃, respectively. These anomalies seem to be pertained with antiferromagnetic to paramagnetic phase transformation in $BiFeO_3$. From Fig. 6, we observe that, the anomaly shifts towards the direction of lower temperature with increasing the doping range of Ni in $BiFeO_3$. The similar results were reported by Kumar and Yadav [18]. The anomaly proves a possible coupling between the electric and magnetic dipole moments of $BiFeO_3$, which is associated with the antiferromagnetic Neel temperature (T_N) of bulk $BiFeO_3$ [26].

4 Conclusions

SHS-synthesized $BiFe_{1-x}Ni_xO_3$ ($x = 0.05$, 0.1 and 0.15) ceramic samples crystallize in a rhombhohedral phase. Magnetization measurement of Ni-substituted $BiFeO_3$ shows the appearance of nonzero magnetization at room temperature, whereas the M–H loops are saturated at 5 K. Dielectric constant measurements with temperature in $BiFe_{0.95}Ni_{0.05}O_3$, $BiFe_{0.9}Ni_{0.1}O_3$ and $BiFe_{0.85}Ni_{0.15}O_3$ samples exhibit anomalies around 450 ℃, 425 ℃ and 410 ℃, respectively, which prove the antiferromagnetic to paramagnetic phase transition. This transition temperature (T_N) also manifests a possible coupling between electric and magnetic dipoles of $BiFeO_3$.

Acknowledgements

This study was supported by UGC-SAP, DRS Phase II of India, and the author Y. A. Chaudhari is very much thankful for the funding agency.

References

[1] Cheong SW, Mostovoy M. Multiferroics: A magnetic

twist for ferroelectricity. *Nat Mater* 2007, **6**: 13–20.

[2] Kumar N, Panwar N, Gahtori B, *et al*. Structural, dielectric and magnetic properties of Pr substituted $Bi_{1-x}Pr_xFeO_3$ ($0 \leqslant x \leqslant 0.15$) multiferroic compounds. *J Alloys Compd* 2010, **501**: L29–L32.

[3] Chaudhari YA, Singh A, Abuassaj EM, *et al*. Multiferroic properties in $BiFe_{1-x}Zn_xO_3$ ($x = 0.1$–0.2) ceramics by solution combustion method (SCM). *J Alloys Compd* 2012, **518**: 51–57.

[4] Qin W, Guo YP, Guo B, *et al*. Dielectric and optical properties of $BiFeO_3$–$(Na_{0.5}Bi_{0.5})TiO_3$ thin films deposited on Si substrate using $LaNiO_3$ as buffer layer for photovoltaic devices. *J Alloys Compd* 2012, **513**: 154–158.

[5] Farhadi S, Rashidi N. Microwave-induced solid-state decomposition of the $Bi[Fe(CN)_6]\cdot5H_2O$ precursor: A novel route for the rapid and facile synthesis of pure and single-phase $BiFeO_3$ nanopowder. *J Alloys Compd* 2010, **503**: 439–444.

[6] Shami MY, Awan MS, Anis-ur-Rehman M. Phase pure synthesis of $BiFeO_3$ nanopowders using diverse precursor via co-precipitation method. *J Alloys Compd* 2011, **509**: 10139–10144.

[7] Garcia FG, Riccardi CS, Simões AZ. Lanthanum doped $BiFeO_3$ powders: Syntheses and characterization. *J Alloys Compd* 2010, **501**: 25–29.

[8] Wang YY. A giant polarization value in bismuth ferrite thin films. *J Alloys Compd* 2011, **509**: L362–L364.

[9] Azam A, Jawad A, Ahmed AS, *et al*. Structural, optical and transport properties of Al^{3+} doped $BiFeO_3$ nanopowder synthesized by solution combustion method. *J Alloys Compd* 2011, **509**: 2909–2913.

[10] Minh NV, Quan NG. Structural, optical and electromagnetic properties of $Bi_{1-x}Ho_xFeO_3$ multiferroic materials. *J Alloys Compd* 2011, **509**: 2663–2666.

[11] Kothari D, Reddy VR, Gupta A, *et al*. Eu doping in multiferroic $BiFeO_3$ ceramics studied by Mossbauer and EXAFS spectroscopy. *J Phys: Condens Mat* 2010, **22**: 356001.

[12] Dho J, Qi X, Kim H, *et al*. Large electric polarization and exchange bias in multiferroic $BiFeO_3$. *Adv Mater* 2006, **18**: 1445–1448.

[13] Qi XD, Dho J, Tomov R, *et al*. Greatly reduced leakage current and conduction mechanism in aliovalent-ion-doped $BiFeO_3$. *Appl Phys Lett* 2005, **86**: 062903.

[14] Wang C, Takahashi M, Fujino H, *et al*. Leakage current of multiferroic $(Bi_{0.6}Tb_{0.3}La_{0.1})FeO_3$ thin films grown at various oxygen pressures by pulsed laser deposition and annealing effect. *J Appl Phys* 2006, **99**: 054104.

[15] Xiao XH, Zhu J, Li YR, *et al*. Greatly reduced leakage current in $BiFeO_3$ thin film by oxygen ion implantation. *J Phys D: Appl Phys* 2007, **40**: 5775–5778.

[16] Pabst GW, Martin LW, Chu YH, *et al*. Leakage mechanisms in $BiFeO_3$ thin films. *Appl Phys Lett* 2007, **90**: 072902.

[17] Jiang QH, Nan CW, Shen ZJ. Synthesis and properties of multiferroic La-modified $BiFeO_3$ ceramics. *J Am Ceram Soc* 2006, **89**: 2123–2127.

[18] Kumar M, Yadav KL. Rapid liquid phase sintered Mn doped $BiFeO_3$ ceramics with enhanced polarization and weak magnetization. *Appl Phys Lett* 2007, **91**: 242901.

[19] Nalwa KS, Garg A, Upadhyaya A. Effect of samarium doping on the properties of solid-state synthesized multiferroic bismuth ferrite. *Mater Lett* 2008, **62**: 878–881.

[20] Kumar M, Yadav KL. Study of room temperature magnetoelectric coupling in Ti substituted bismuth ferrite system. *J Appl Phys* 2006, **100**: 074111.

[21] Xu QY, Zai HF, Wu D, *et al*. The magnetic properties of $BiFeO_3$ and $Bi(Fe_{0.95}Zn_{0.05})O_3$. *J Alloys Compd* 2009, **485**: 13–16.

[22] Chen SY, Wang LY, Xuan HC, *et al*. Multiferroic properties and converse magnetoelectric effect in $Bi_{1-x}Ca_xFeO_3$ ceramics. *J Alloys Compd* 2010, **506**: 537–540.

[23] Wang Y, Xu G, Yang L, *et al*. Enhancement of ferromagnetic properties in Ni-doped $BiFeO_3$. *Mater Sci-Poland* 2009, **27**: 219–224.

[24] Saha S, Ghanawat SJ, Purohit RD. Solution combustion synthesis of nano particle $La_{0.9}Sr_{0.1}MnO_3$ powder by a unique oxidant–fuel combination and its characterization. *J Mater Sci* 2006, **41**: 1939–1943.

[25] Dutta DP, Jayakumar OD, Tyagi AK, *et al*. Effect of doping on the morphology and multiferroic properties of $BiFeO_3$ nanorods. *Nanoscale* 2010, **2**: 1149–1154.

[26] Jia DC, Xu JH, Ke H, *et al*. Structure and multiferroic properties of $BiFeO_3$ powders. *J Eur Ceram Soc* 2009, **29**: 3099–3103.

Microstructure and piezoelectric properties of K$_{5.70}$Li$_{4.07}$Nb$_{10.23}$O$_{30}$-added K$_{0.5}$Na$_{0.5}$NbO$_3$ ceramics

Xuming PANG^{b}, Jinhao QIU^{a,*}, Kongjun ZHU^{a}

^{a}State Key Laboratory of Mechanics and Control of Mechanical Structures, Nanjing University of Aeronautics and Astronautics, Nanjing 210016, China
^{b}Department of Mechanical Engineering, Nanjing Tech University, Nanjing 210009, China

Abstract: Lead-free piezoelectric ceramics K$_{0.5}$Na$_{0.5}$NbO$_3$–$x$mol%K$_{5.70}$Li$_{4.07}$Nb$_{10.23}$O$_{30}$ (x = 0–2.5, KNN–$x$mol%KLN) were prepared by conventional sintering technique. The phase structure and electrical properties of KNN ceramics were investigated as a function of KLN concentration. The results showed that small amount of KLN introduced into the lattice formed a single phase perovskite structure. The KLN modification lowered the phase transition temperature of orthorhombic–tetragonal ($T_{\text{O–T}}$) and increased the Curie temperature (T_{C}). Some abnormal coarse grains were formed in a matrix when the content of KLN was relatively low (1 mol%). However, normally grown grains were only observed when the sintering aid content was increased to 2 mol%. Proper content of KLN decreased the amount of defects, thus the remanent polarization increased and the coercive field decreased markedly, and the sinterability of the KNN ceramics was simultaneously improved with significant increase of piezoelectric properties.

Keywords: ceramics; sintering aid; phase transformation; electrical properties

1 Introduction

Lead-based ferroelectric materials, such as Pb(Zr,Ti)O$_3$ (PZT), Pb(Mg$_{1/3}$Nb$_{2/3}$)O$_3$–PbTiO$_3$ (PMN–PT) and Pb(Zn$_{1/3}$Nb$_{2/3}$)O$_3$–PbTiO$_3$ (PZN–PT), show excellent piezoelectric properties [1–3] and have been adopted for many applications. However, the development of harmless lead-free piezoceramics has gained a great deal of attention because of the near future restriction on the lead-based materials due to the environmental issues.

In the past several years, much attention has been paid to the alkaline niobate-based materials, and especially to the potassium sodium niobate K$_{0.5}$Na$_{0.5}$NbO$_3$ (KNN) family. KNN is one of the most promising candidates for lead-free piezoelectric ceramics because of its high Curie temperature (about 420 °C) and large electromechanical coupling factors [4,5]. However, the difficulty in sintering KNN under the atmospheric conditions is a serious drawback, and various techniques such as hot pressing and spark plasma sintering have been utilized in order to improve the sinterability of KNN ceramics [6,7]. Since these techniques are found unsuitable for use in industrial production, many sintering aids are researched by several researchers in order to sinter KNN under atmospheric conditions, such as K$_{5.4}$Cu$_{1.3}$Ta$_{10}$O$_{29}$, CuO and MnO$_2$ [8–15]. Nevertheless, the piezoelectric properties of the KNN system are degraded in these cases although these sintering aids can improve the sinterability of the KNN ceramics. Therefore, the novel

* Corresponding author.
E-mail: qiu@nuaa.edu.cn

sintering aids, which improve both the sintering behavior and piezoelectric properties, are key research.

Because $K_{5.70}Li_{4.07}Nb_{10.23}O_{30}$ (KLN) as aid has not been studied, the sintering behaviors and piezoelectric properties of KNN ceramics with KLN added are investigated by conventional sintering technique.

2 Experimental

A conventional ceramic fabrication technique was used to prepare $K_{0.5}Na_{0.5}NbO_3$–$x$mol%$K_{5.70}Li_{4.07}Nb_{10.23}O_{30}$ ($0 \leqslant x \leqslant 2.5$, KNN–$x$mol%KLN) ceramics using analytical-grade metal oxides or carbonate powders: K_2CO_3 (99%), Na_2CO_3 (99.8%), Li_2CO_3 (98%) and Nb_2O_5 (99.5%). The KNN and KLN powders were first synthesized at 900 ℃ for 5 h by a solid-state reaction method. After the calcination, KNN and KLN powders were weighted according to the formula of KNN–$x$mol%KLN and ball milled for 12 h. The resulting mixture was further mixed with polyvinyl alcohol binder solution thoroughly and then pressed into disk samples. The disk samples were sintered at 1100 ℃ for 2 h in air.

Density of the samples was determined by the Archimedes method. The crystalline phase was analyzed using an X-ray diffractometer (D8 Advance). The microstructure was observed by a scanning electron microscope (JSM-5610LV/Noran-Vantage). Dielectric properties as functions of temperature and frequency were measured by an impedance analyzer (HP4294A). Polarization vs. electric field hysteresis loops were measured using a ferroelectric test system (TF Analyzer 2000). Silver electrodes were fired on the top and bottom surfaces of the sintered samples. The ceramics were poled under a DC field of 2 kV/mm at 110 ℃ in a silicon oil bath for 30 min. The piezoelectric constant d_{33} was measured using a quasistatic piezoelectric constant testing meter (ZJ-3A, Institute of Acoustics, Chinese Academy of Sciences, Beijing, China).

3 Results and discussion

Figure 1 shows the X-ray diffraction (XRD) patterns of KNN–$x$mol%KLN ceramics. All of the ceramics exhibit single phase perovskite structure which indicates that excess Li^+ and K^+ may incorporate into the lattice. There are two peaks at about 45° which change obviously with different KLN contents. The

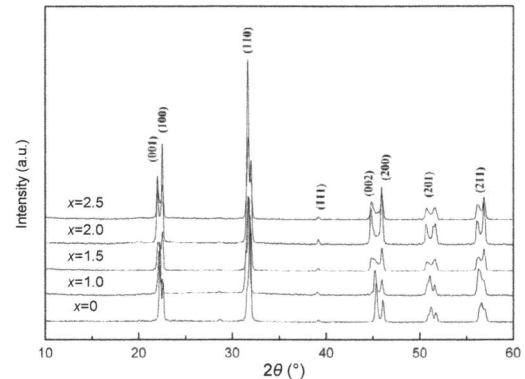

Fig. 1 XRD patterns of KNN–$x$mol%KLN ceramics with different KLN contents.

crystal structure can be distinguished from the relative intensity of these two peaks. For orthorhombic structure, the left peak at about 45° has higher intensity than that of the right peak, and it was indexed as (202) and (020), respectively. For tetragonal structure, the left peak at about 45° has lower intensity than that of the right peak, and it was indexed as (002) and (200), respectively. The phase structure of KNN–1mol%KLN is the orthorhombic structure while KNN–1.5mol%KLN exhibits tetragonal structure. Orthorhombic and tetragonal phases coexist in the ceramics when x is in the range of $1 < x < 1.5$.

The lattice parameters of KNN–$x$mol%KLN ceramics are calculated by fitting the diffraction peak profile, as shown in Fig. 2. Clearly, there is a transition zone between the orthorhombic and tetragonal phases in the range of $1 < x < 1.5$. When x is larger than 1.5, the materials become pure tetragonal phase. The tetragonality c/a is ~1.011 for KNN–1.5mol%KLN and increases to ~1.013 for the composition with $x = 2.5$.

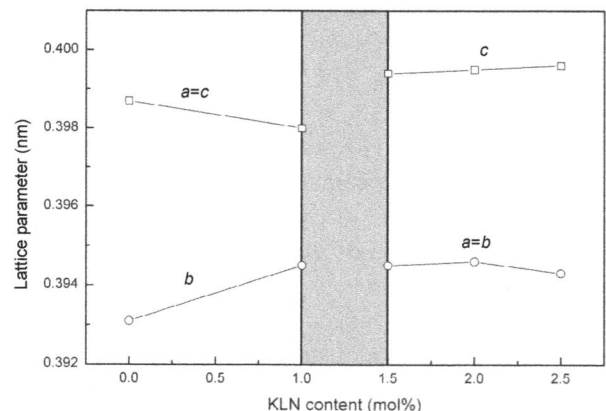

Fig. 2 Lattice parameters of KNN–$x$mol%KLN ceramics as a function of the KLN contents.

An increase in tetragonality usually corresponds to a rise in the Curie temperature for a couple of perovskite solid solution ceramics, such as Pb(Zr,Ti)O$_3$ [16].

Figure 3 shows the microstructures of KNN–$x$mol%KLN ceramics sintered at 1100 °C. For the pure KNN ceramic (i.e., $x=0$), the grains have a diameter in the range of 10 μm, and small amount of pores are observed (Figs. 3(a) and 3(b)). By the increasing x to 1, the grains become larger and more nonuniform (Figs. 3(c) and 3(d)). The ceramics are denser and almost no pore is observed. These results clearly show that the addition of KLN can improve the sintering performance of the ceramics. The grain growth is inhibited and average grain size is decreased with increasing x to 1, while the amount of pores decreases. This result explains that the grains of the KNN–1mol%KLN sample grow sufficiently.

For the KNN–1mol%KLN ceramic sintered at 1100 °C, the average grain size of matrix grains in Figs. 3(c) and 3(d) is 3 μm. However, some coarse grains (or areas), which are indexed by grains 1 and grains 2, are also observed. In particular, the extremely large grains with diameter up to 20–30 μm as indexed by grains 1, can be clearly seen in Fig. 3(d). Obviously, this is a kind of abnormal grain growth (AGG) behavior, whose characteristic is the formation of some exceptionally large grains in a matrix of fine grains [17–19]. However, it seems that the AGG in the present material is different from the classical AGG behavior reported by other authors [20,21], because the abnormal grains in this study are much more like finer matrix grains aggregated together probably due to the formation of a liquid phase [22,23]. Because K$^+$ incorporates into the lattice which gives rise to the relatively higher content of K in KNN–$x$mol%KLN samples, the solidus temperature of KNN decreases with increasing the content of KLN from phase diagram of the KNbO$_3$–NaNbO$_3$ system [24]. On the other hand, the amount and fluidity of liquid phase increase with increasing KLN at 1100 °C. When the content of KLN increases to 1 mol%, the numbers of large grains apparently increase. Meanwhile, two kinds of large grains can be classified. The one is abnormal large grains indexed by grains 1, while the other one is normal large grains indexed by grains 2 in Figs. 3(c) and 3(d). Furthermore, by increasing x to 2, only normal grain growth (NGG) behavior takes place and no abnormal large grain is observed as shown in Figs. 3(g)–3(j).

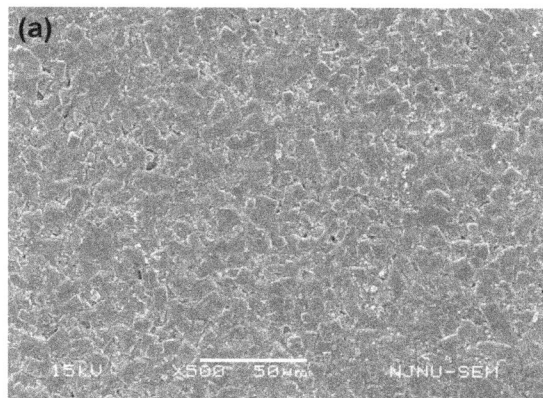

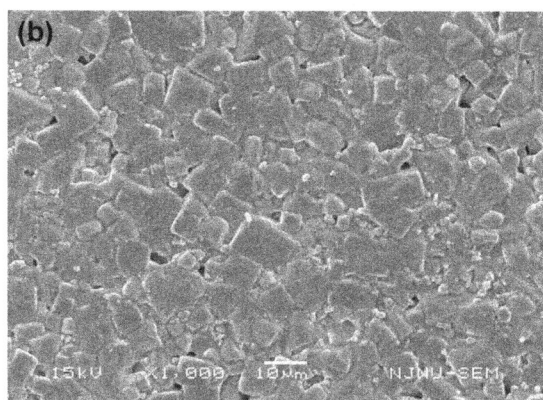

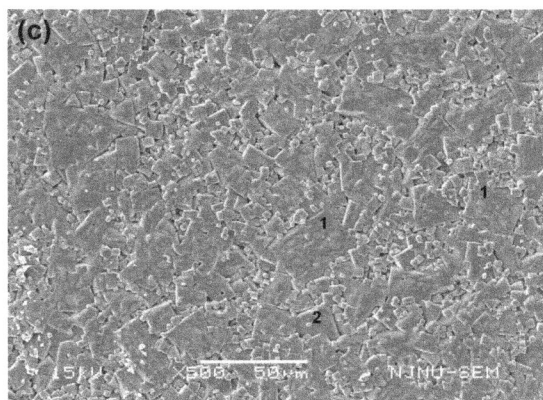

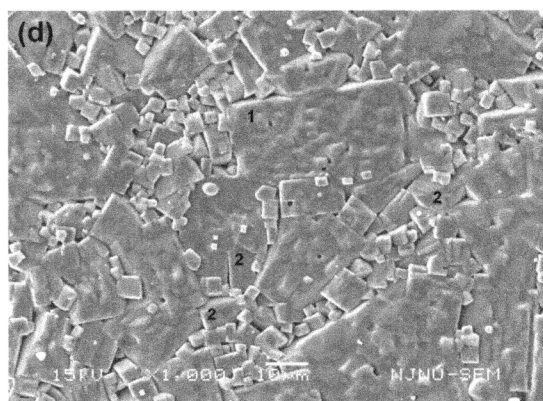

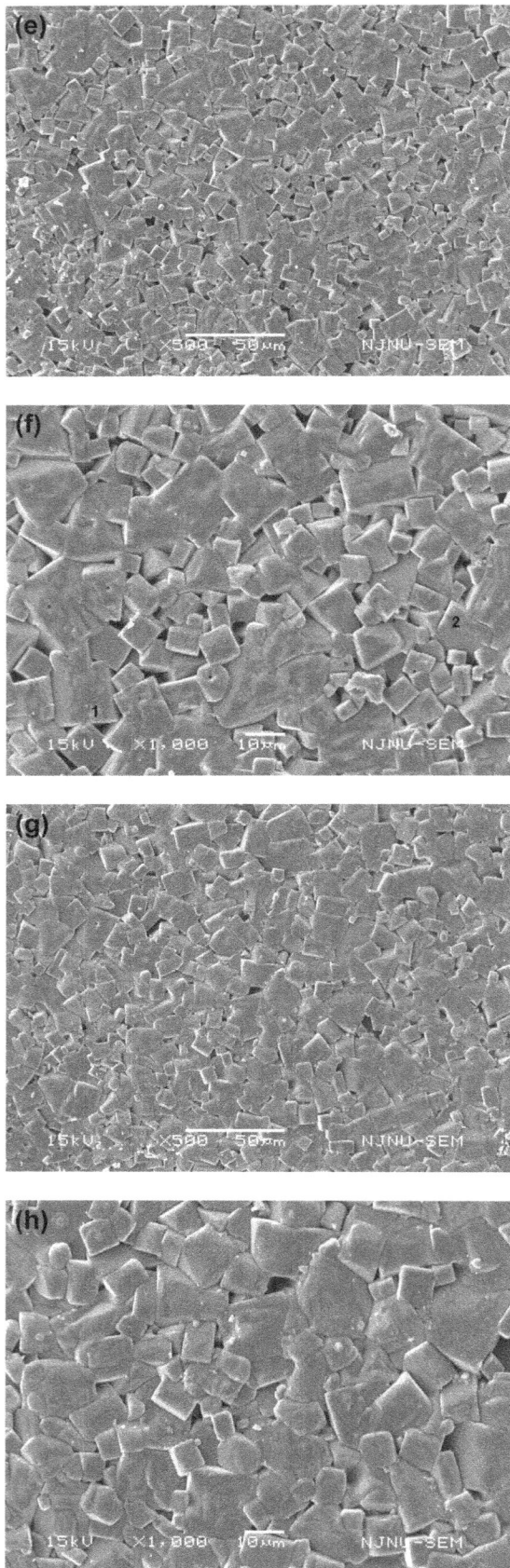

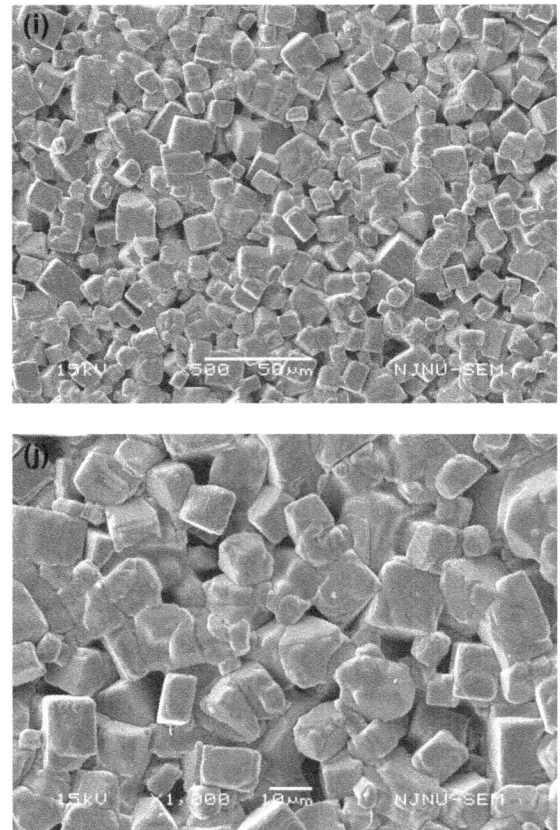

Fig. 3   SEM micrographs of KNN–$x$mol%KLN ceramics with various KLN contents: (a) and (b) $x=$ 0; (c) and (d) $x=1$; (e) and (f) $x=1.5$; (g) and (h) $x=$ 2; (i) and (j) $x=2.5$.

On the basis of the microstructure evolution and our previous study on sintering aid, it is thought that AGG and NGG are related to the presence of a liquid phase. The formation mechanisms of a liquid phase in KNN and KNN-based ceramics have been just discussed. A small amount of liquid phases may form at first in some local areas probably when the content of KLN is 1 mol%, and the liquid phase amount in different local areas may be nonuniform owing to low fluidity and volatilization of alkali metal ions [25]. Based on the powder sintering theory [26], the grain growth is controlled by dissolution and precipitation mechanisms for the more amount of liquid phase, while it is controlled by diffusion for the less one. Therefore, classical large grains are observed in the local areas where the amount of liquid phase is more. Because of less liquid phase amount in the other areas, the small grains can not dissolve and become self-organized to be aligned into clusters, as can be seen in Fig. 3(d) where the small grains obey a discipline of controlled alignment by diffusion. Generally, a liquid phase contributes to sintering by accelerating particle redistribution because of the enhanced high atom mobility. Besides, similar to the organic additives that accelerate transformation as the surfactants, liquid phase is supposed to act as a kind of surfactant during sintering at high temperature [22]. Also, the volatilization of alkali metal oxides in KNN-based ceramics might play a specific role in the microstructure formation [25]. Thus, the self-assembly of aggregation can be aided through a combined action of small grain surfactant interactions. Another positive effect of a liquid phase on the sintering is enhancement of the final sample density via a high capillary force. Figure 4 is a schematic showing how the AGG structure of a coarse grain is formed. As shown in Fig. 4, when the less liquid phase appears, the small grains distribute randomly and some small holes are even not eliminated. As the sintering time increases, the surfactant-mediated interactions accelerate the formation of self-assembled clusters by small grains. However, when the sintering time further increases, several groups of clusters with small disorientations are supposed to be bounded together to form coarse clusters. According to the classical grain growth [26], the grain growth rate is strongly dependent on the differences in grain radius and disorientation angles with the surrounding grains. For the small grains in the inner cluster structure, the grain boundary driving force is too small to move the boundary migration due to their small disorientations. Therefore, the small grains in the inner cluster structure have almost no growth, as shown in Fig. 3(d). Because of the more amount and higher fluidity of liquid phase, the matrix is easily filled and the proportion of NGG is improved

with increasing KLN as shown in Figs. 3(e) and 3(f). When the liquid phase amount exceeds a certain value, the grain growth is controlled by dissolution and precipitation mechanisms, so AGG completely disappears.

Figure 5 shows the density and property variations of KNN–$x$mol%KLN ceramics sintered at 1100 ℃ for 2 h. As shown in Fig. 5, it has been observed that the KNN ceramic without KLN has a lower bulk density. The density of the KNN–$x$mol%KLN samples increases when the content of KLN increases from 0 to 1 mol%, and then slightly decreases when the content of KLN is above 1 mol%. Proper amount of KLN modification can increase the piezoelectric properties markedly. The d_{33} and k_p for KNN–1mol%KLN are 121 pC/N and 0.39, respectively. The significant enhancement in the piezoelectric properties results from the increase of bulk density and the phase structure of KNN–1mol%KLN which is around the polymorphic phase transition of the orthorhombic and tetragonal phases, as shown in Fig. 1. The polymorphic phase transition causes the higher piezoelectric activity owing to the more possible polarization states resulting from the coexistence of the orthorhombic and tetragonal phases. For $x > 1$, d_{33} and k_p decrease probably because lower density with the increasing content of KLN. The Q_m value of KNN–$x$mol%KLN ceramics is enhanced with increasing KLN within the compositional range of KLN from 0 to 1 mol%. The highest Q_m value of 68 is achieved in the KNN–1mol%KLN ceramic. Q_m might be related to domain motion difficulty. The decrease in Q_m value is caused by the low density and the excess addition of KLN. As shown in Fig. 5, ε_r increases with increasing

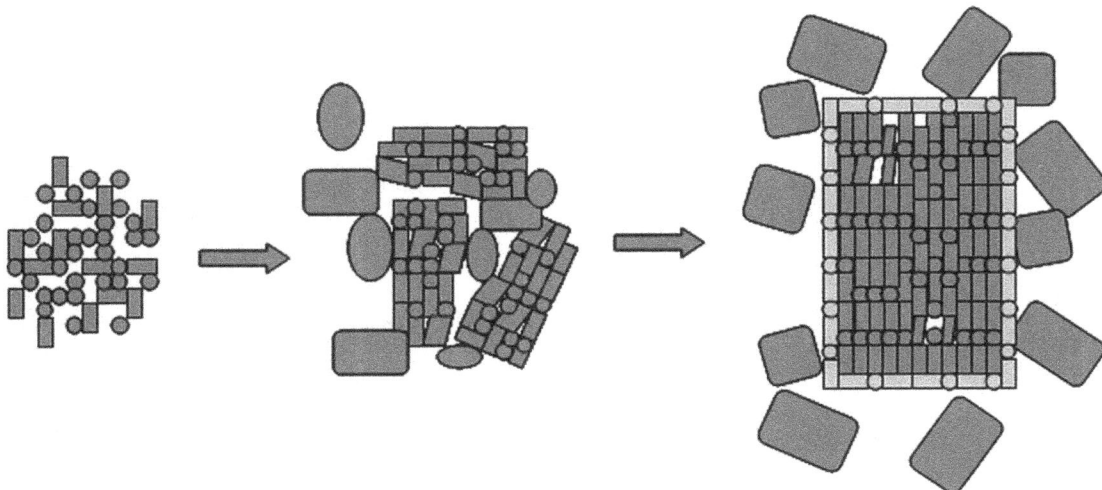

Fig. 4   Schematic diagram showing the formation procedure of AGG.

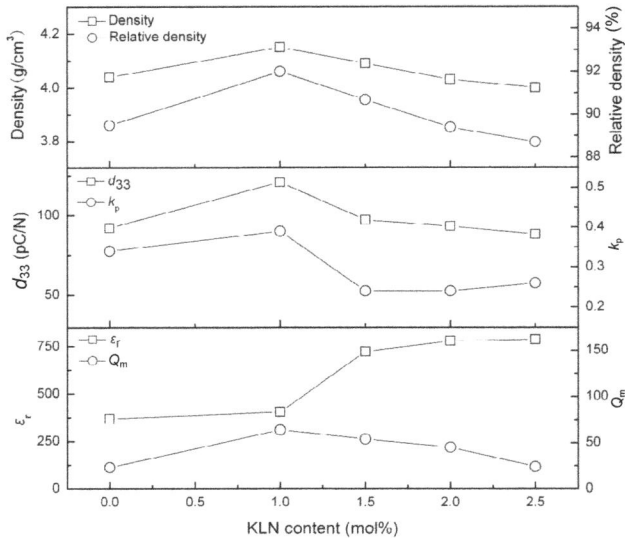

Fig. 5 Density, d_{33}, k_p, ε_r and Q_m values of the KNN–$x$mol%KLN ceramics with $0 \leqslant x \leqslant 2.5$.

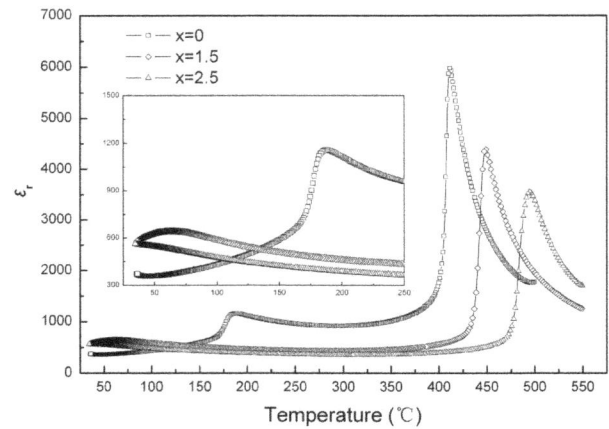

Fig. 6 Temperature dependence of dielectric constant ε_r for KNN–$x$mol%KLN ceramics.

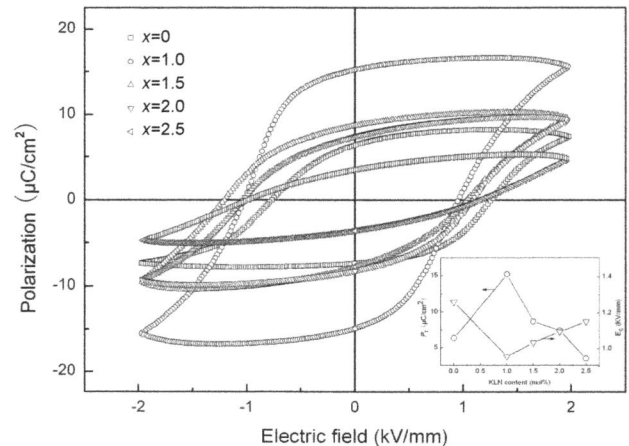

Fig. 7 P–E hysteresis loops of KNN–$x$mol%KLN ceramics.

KLN. These results confirm that the properties of KNN–1mol%KLN ceramic become optimum. The sample with $x = 1.0$ exhibits good properties: d_{33}, k_p, ε_r and Q_m show their peak values of 121 pC/N, 39%, 404 and 64, respectively.

Figure 6 shows the temperature dependence of dielectric constant ε_r (at 1 kHz) for KNN–$x$mol%KLN ($x = 0$, 1.5, 2.5) ceramics, and the inset shows the low temperature ferroelectric–ferroelectric transition (T_{O-T}). With the increase of KLN, the T_C increases and the maximum dielectric constant at T_C decreases. The ferroelectric–ferroelectric phase transition shifts to lower temperature, and the peak is slightly broadened which shows a diffusive nature with the increase of KLN. T_{O-T} is near room temperature at $x = 1.5$. When the content of KLN increases, Li$^+$ and K$^+$ could be incorporated into the lattice and compensate the volatilization of K$^+$ [25], so T_C increases. According to previous results [27,28], the incorporation of Li$^+$ will also cause the increase of T_C and decrease of T_{O-T}.

Figure 7 shows the P–E hysteresis loops of KNN–$x$mol%KLN ceramics. The P_r and E_C values for pure KNN ceramic are 6.36 μC/cm^2 and 1.26 kV/mm, respectively. Proper amount of KLN modification increases the remnant polarization and decreases the coercive field. It should be noted that the remnant polarization increases by 140% but the coercive field decreases by 24% when $x = 1$ ($P_r = 15.21$ μC/cm^2, $E_C = 0.957$ kV/mm). There are some mechanisms which are related to the increase of polarization and decrease of coercive field: the incorporation of Li$^+$ and the

decrease of defects. Li$^+$ has smaller ionic radius than K$^+$ and will cause the tilting of the Nb–O octahedron which increases the remnant polarization. The decrease of coercive field is mainly attributed to the decrease of the amount of defects. It is well-known that K$_2$O is volatile and K$^+$ vacancies will be left after sintering. The K$^+$ in KLN can compensate these vacancies and the amount of defects decreases. After $x \geqslant 1$, P_r decreases and E_C increases due to low density and poor microstructure as shown in Figs. 3(e)–3(j).

4 Conclusions

KNN–$x$mol%KLN ceramics were prepared by the solid-state reaction method. The results show that the small amount of KLN ($0 \leqslant x \leqslant 2.5$) incorporates into the lattice and forms the single phase perovskite structure. It exhibits the polymorphic phase transition from orthorhombic structure to tetragonal structure when the content of KLN increases from 1 mol% to 1.5 mol%. The KNN–1mol%KLN ceramic shows AGG behavior. However, when the amount of KLN is increased above 1.5 mol%, abnormally grown grains disappear and relatively uniform microstructure with normally grown grains is formed. The AGG and NGG behaviors should be related to the volatilization of alkali metal oxides and the presence of liquid phase amount under different amount of KLN. KLN modification lowers T_{O-T} and increases T_C with increase of x. KNN–1mol%KLN ceramic shows a very high remnant polarization and low coercive field, and its piezoelectric properties are also the best among them.

Acknowledgements

This work was supported by the Natural Science Foundation of Jiangsu Province, China (BK20130791), Projects of International Cooperation and Exchanges NSFC (51161120326), Aeronautical Science Fund (20131552025), the NUAA Fundamental Research Funds (NS2013008), and a project funded by the Priority Academic Program Development of Jiangsu Higher Education Institutions (PAPD).

References

[1] Jaffe B, Roth RS, Marzullo S. Piezoelectric properties of lead zirconate–lead titanate solid–solution ceramics. *J Appl Phys* 1954, **25**: 809–810.

[2] Park S-E, Shrout TR. Ultrahigh strain and piezoelectric behavior in relaxor based ferroelectric single crystals. *J Appl Phys* 1997, **82**: 1804–1811.

[3] Fan H, Kim H-E. Effect of lead content on the structure and electrical properties of Pb((Zn$_{1/3}$Nb$_{2/3}$)$_{0.5}$(Zr$_{0.47}$Ti$_{0.53}$)$_{0.5}$)O$_3$ ceramics. *J Am Ceram Soc* 2001, **84**: 636–638.

[4] Egerton L, Dillom DM. Piezoelectric and dielectric properties of ceramics in the system potassium–sodium niobate. *J Am Ceram Soc* 1959, **42**: 438–442.

[5] Zuo R, Fang X, Ye C. Phase structures and electrical properties of new lead-free (Na$_{0.5}$K$_{0.5}$)NbO$_3$–(Bi$_{0.5}$Na$_{0.5}$)TiO$_3$ ceramics. *Appl Phys Lett* 2007, **90**: 092904.

[6] Haertling GH. Properties of hot-pressed ferroelectric alkali niobate ceramics. *J Am Ceram Soc* 1967, **50**: 329–330.

[7] Wang R, Xie R, Sekiya T, et al. Piezoelectric properties of spark-plasma-sintered (Na$_{0.5}$K$_{0.5}$)NbO$_3$–PbTiO$_3$ ceramics. *Jpn J Appl Phys* 2002, **41**: 7119–7122.

[8] Matsubara M, Kikuta K, Hirano S. Piezoelectric properties of (K$_{0.5}$Na$_{0.5}$)(Nb$_{1-x}$Ta$_x$)O$_3$–K$_{5.4}$CuTa$_{10}$O$_{29}$ ceramics. *J Appl Phys* 2005, **97**: 114105.

[9] Takao H, Saito Y, Aoki Y, et al. Microstructural evolution of crystalline-oriented (K$_{0.5}$Na$_{0.5}$)NbO$_3$ piezoelectric ceramics with a sintering aid of CuO. *J Am Ceram Soc* 2006, **89**: 1951–1956.

[10] Yang M-R, Hong C-S, Tsai C-C, et al. Effect of sintering temperature on the piezoelectric and ferroelectric characteristics of CuO doped 0.95(Na$_{0.5}$K$_{0.5}$)NbO$_3$–0.05LiTaO$_3$ ceramics. *J Alloys Compd* 2009, **488**: 169–173.

[11] Park S-H, Ahn C-W, Nahm S, et al. Microstructure and piezoelectric properties of ZnO-added (Na$_{0.5}$K$_{0.5}$)NbO$_3$ ceramics. *Jpn J Appl Phys* 2004, **43**: L1072–L1074.

[12] Kosec M, Kolar D. On activated sintering and electrical properties of NaKNbO$_3$. *Mater Res Bull* 1975, **10**: 335–339.

[13] Tashiro S, Nagamatsu H, Nagata K. Sinterability and piezoelectric properties of KNbO$_3$ ceramics after substituting Pb and Na for K. *Jpn J Appl Phys* 2002, **41**: 7113–7118.

[14] Guo Y, Kakimoto K, Ohsato H. Dielectric and piezoelectric properties of lead-free (Na$_{0.5}$K$_{0.5}$)NbO$_3$–SrTiO$_3$ ceramics. *Solid State Commun* 2004, **129**: 279–284.

[15] Kakimoto K, Masuda I, Ohsato H. Ferroelectric and piezoelectric properties of KNbO$_3$ ceramics containing small amounts of LaFeO$_3$. *Jpn J Appl Phys* 2003, **42**: 6102–6105.

[16] Choi SW, Shrout TR, Jang SJ, et al. Morphotropic phase boundary in Pb(Mg$_{1/3}$Nb$_{2/3}$)O$_3$–PbTiO$_3$ system. *Mater Lett* 1989, **8**: 253–255.

[17] Kim M-S, Lee D-S, Park E-C, et al. Effect of Na$_2$O additions on the sinterability and piezoelectric properties of lead-free 95(Na$_{0.5}$K$_{0.5}$)NbO$_3$–5LiTaO$_3$ ceramics. *J Eur Ceram Soc* 2007, **27**: 4121–4124.

[18] Choi S-Y, Kang S-JL. Sintering kinetics by structural transition at grain boundaries in barium titanate. *Acta Mater* 2004, **52**: 2937–2943.

[19] Kim M-S, Jeong S-J, Song J-S. Microstructures and

piezoelectric properties in the Li_2O-excess $0.95(Na_{0.5}K_{0.5})NbO_3$–$0.05LiTaO_3$ ceramics. *J Am Ceram Soc* 2007, **90**: 3338–3340.

[20] Park CW, Yoon DY. Abnormal grain growth in alumina with anorthite liquid and the effect of MgO addition. *J Am Ceram Soc* 2002, **85**: 1585–1593.

[21] Kim M-S, Fisher JG, Kang S-JL, *et al.* Grain growth control and solid-state crystal growth by Li_2O/PbO addition and dislocation introduction in the PMN–35PT system. *J Am Ceram Soc* 2006, **89**: 1237–1243.

[22] Li J-F, Wang K, Zhang B-P, *et al.* Ferroelectric and piezoelectric properties of fine-grained $Na_{0.5}K_{0.5}NbO_3$ lead-free piezoelectric ceramics prepared by spark plasma sintering. *J Am Ceram Soc* 2006, **89**: 706–709.

[23] Zhen Y, Li J-F. Abnormal grain growth and new core–shell structure in $(K,Na)NbO_3$-based lead-free piezoelectric ceramics. *J Am Ceram Soc* 2007, **90**: 3496–3502.

[24] Ringgaard E, Wurlitzer T. Lead-free piezoceramics based on alkali niobates. *J Eur Ceram Soc* 2005, **25**: 2701–2706.

[25] Zhao P, Zhang B-P, Li J-F. High piezoelectric d_{33} coefficient in Li-modified lead-free $(Na,K)NbO_3$ ceramics sintered at optimal temperature. *Appl Phys Lett* 2007, **90**: 242909.

[26] Guo SJ. *Powder Sintering Theory*. Beijing: Metallurgical Industry Press, 1998.

[27] Guo Y, Kakimoto K, Ohsato H. Phase transitional behavior and piezoelectric properties of $(Na_{0.5}K_{0.5})NbO_3$–$LiNbO_3$ ceramics. *Appl Phys Lett* 2004, **85**: 4121–4123.

[28] Hollenstein E, Davis M, Damjanovic D, *et al.* Piezoelectric properties of Li- and Ta-modified $(K_{0.5}Na_{0.5})NbO_3$ ceramics. *Appl Phys Lett* 2005, **87**: 182905.

6

Property mapping of polycrystalline diamond coatings over large area

Awadesh Kumar MALLIK[a,*], Sandip BYSAKH[a], Monjoy SREEMANY[a],
Sudakshina ROY[a], Jiten GHOSH[a], Soumyendu ROY[b],
Joana Catarina MENDES[c], Jose GRACIO[d], Someswar DATTA[a]

[a]CSIR -Central Glass & Ceramic Research Institute, Kolkata 700032, West Bengal, India
[b]Department of Physics, Indian Institute of Technology Bombay, Powai, Mumbai 400076, Maharashtra, India
[c]Instituto de Telecomunicações, Campus Universitário de Santiago, 3810-193, Portugal
[d]Nanotechnology Research Division, Centre for Mechanical Technology and Automation,
University of Aveiro, 3810-193, Portugal

Abstract: Large-area polycrystalline diamond (PCD) coatings are important for fields such as thermal management, optical windows, tribological moving mechanical assemblies, harsh chemical environments, biological sensors, etc. Microwave plasma chemical vapor deposition (MPCVD) is a standard technique to grow high-quality PCD films over large area due to the absence of contact between the reactive species and the filament or the chamber wall. However, the existence of temperature gradients during growth may compromise the desired uniformity of the final diamond coatings. In the present work, a thick PCD coating was deposited on a 100-mm silicon substrate inside a 915-MHz reactor; the temperature gradient resulted in a non-uniform diamond coating. An attempt was made to relate the local temperature variation during deposition and the different properties of the final coating. It was found that there was large instability inside the system, in terms of substrate temperature (as high as $\Delta T = 212$ ℃), that resulted in a large dispersion of the diamond coating's final properties: residual stress (-15.8 GPa to $+6.2$ GPa), surface morphology (octahedral pyramids with (111) planes to cubo-octahedrals with (100) flat top surfaces), thickness (190 μm to 245 μm), columnar growth of diamond (with appearance of variety of nanostructures), nucleation side hardness (17 GPa to 48 GPa), quality (Raman peak FWHM varying from 5.1 cm^{-1} to 12.4 cm^{-1} with occasional splitting). This random variation in properties over large-area PCD coating may hamper reproducible diamond growth for any meaningful technological application.

Keywords: plasma-enhanced chemical vapor deposition (CVD); diamond film; mechanical properties; nanostructures

1 Introduction

Large-area polycrystalline diamond (PCD) coatings

* Corresponding author.
E-mail: amallik@cgcri.res.in

have a wide range of technological applications, from micro-electromechanical systems to gyrotron windows for mm-wave transmissions and electrochemical sensors under harsh environments. Due to their high thermal conductivity and biocompatibility, PCD coatings are also used as heat sink substrates and bio-sensors. For all these applications, the uniformity of the coatings is a major requirement, i.e., the point variation of the properties of the coatings should be as low as possible. For example, the loss tangent of large-area diamond gyrotron windows [1] and strip particle detectors [2,3] should be as uniform as possible. Large-area PCD coatings have various industrial applications; as an example, we can mention "Diafilm", a freestanding diamond disc manufactured by DeBeers Industrial Diamond Division since the early 1990s. The optical, thermal and dielectric properties of the discs are very uniform [4], and researchers all over the world use them as windows in high-power electron tubes.

Several authors have evaluated the growth [5–14] and nucleation [15] sides of large-area PCD coatings deposited by chemical vapor deposition (CVD). Sussmann et al. [4] mapped the tanδ loss over a 120-mm-diameter freestanding PCD disc. Yokota et al. [7] reported that {100}-faceted crystals can be deposited with high uniformity on a 150-mm-diameter silicon substrate using 915-MHz and 60-kW ASTeX reactor when the parameter α (the ratio of growth velocities along {100} and {111} directions, $\alpha = v_{100}/v_{111}$) is 1. Schelz et al. [8] evaluated the radial variation of the Raman peak shift, roughness, nucleation density and thickness of PCD films deposited on a 75-mm-diameter substrate in a 915-MHz reactor. Vikharev et al. [9] deposited PCD films on 60–90-mm-diameter substrates with large growth rate (9 μm/h) in a 10-kW and 30-GHz system, but did not comment the uniformity of the properties. Zuo et al. [11] outlined the thickness uniformity of a PCD film deposited on a 75-mm-diameter substrate in a 2.45-GHz system, but did not analyze other properties. Tsai and Kuo [13] studied the properties of PCD films deposited on a 4-inch-diameter silicon substrate with addition of nitrogen, and reported the center-to-edge variation in microstructure, texture, thickness, and Raman and X-ray diffraction (XRD) peak patterns. Meykens et al. [16] evaluated the optical absorption coefficient of diamond infrared (IR) windows by quasi-parallel collinear photothermal deflection (PTD) of a 10.6-mm CO_2 laser [17]. Low- and high-power mm-wave diamond gyrotron windows have thoroughly been investigated [18,19], but no report about the variation of properties over the large-area coatings was made. 2-inch-diameter PCD optical lenses were fabricated in large quantities with consistent quality and characteristics over the whole

area [20]. The flexural strength of the nucleation and growth sides was evaluated by ring-on-ring method for 50 17–19-mm-diameter PCD disks deposited by microwave plasma CVD (MPCVD), with thicknesses ranging from 0.1 mm to 0.7 mm, for potential missile dome application in supersonic jets [21]. All the cited studies reported the investigation of large-area PCD coatings; however, to date, there is no systematic study of point-to-point variation of different physical properties like microstructure, thickness, crystallinity, quality, etc.

Substrate temperature is a key parameter in the deposition of diamond by CVD [22–30]. CVD diamond grows between 700 ℃ and 1100 ℃ at sub-atmospheric pressure of 20–130 Torr; to deposit high-quality diamond with well-faceted morphology, the substrate temperature has to be on the higher side of 900 ℃, whereas substrate temperature lower than 700 ℃ produces cauliflower nanocrystalline morphology [31–34]. The growth rate initially increases with temperature, and after attaining maximum slowly decreases down.

Zimmer et al. [35] reported a lower-than-20 ℃ variation of the substrate surface temperature during deposition of diamond on a 300-mm-diameter silicon wafer by hot filament CVD (HFCVD) under nominal substrate temperature of 895 ℃. They reported left-to-right, front-to-back and circular temperature variations, with corresponding non-uniformity in thickness, bowing, warping and Raman spectra of the diamond coating. However, even though the thermal management of the wafer substrate plays a major role in maintaining the diamond coating uniformity over large area, thorough temperature distribution studies are still lacking for MPCVD reactors. Authors have already addressed the issue of thermal management for large-area PCD deposition in their previously published work [36], but that study did not have thorough physical characterizations of the grown PCD coatings, which are necessary to better understand the influence of substrate temperature variation on the properties of the grown coatings. In addition to the systematic characterization of large-area PCD coating, the influence of CH_4/H_2 ratio on the uniformity of such coating is now fully understood through the present work, where a 100-mm-diameter silicon wafer is coated with diamond inside a 915-MHz low-power MPCVD reactor. Special care is taken to minimize the temperature gradient across the substrate during growth [36]. Nevertheless, the properties of the PCD coating (quality, morphology, thickness, stress, crystallinity and hardness) differ from place to place, reflecting non-uniformity in the plasma that causes a temperature gradient on the silicon substrate during the diamond deposition. Diamond nanostructures also appear in some places of the freestanding diamond wafer. To get a further insight on these effects, a thick PCD wafer is grown with a high CH_4/H_2 ratio; the sample is fully characterized by scanning electron microscopy (SEM), Raman, XRD and high-resolution transmission electron microscopy (HRTEM). The results are then compared with samples deposited with a lower CH_4/H_2 ratio [36] and suggest that the increase in the methane content not only decreases the overall diamond quality (as will be seen from the Raman spectra) but also increases the temperature distribution gradient, leading to a film with a large dispersion of the characteristics.

2 Materials and methods

2.1 Substrate seeding and diamond growth

PCD was deposited on a (100) 100-mm-diameter and 6-mm-thick p-type single crystal silicon wafer with a 915-MHz and 9-kW MPCVD reactor (DIAMOTEK 1800, Lambda Technologies Inc. USA) [37] (Fig. 1(a)). The silicon substrate was previously seeded with a mixture of detonation nano-diamond (DND) suspension [38] and methanol of 1:3 ratio for 15 min. The substrate was placed inside the chamber with a stage height of −0.1 cm. The chamber was initially evacuated down to 10^{-3} Torr, and the microwave power and pressure were slowly increased till the final growth conditions were reached (9 kW and 110 Torr). Diamond growth proceeded for 95 h with a hydrogen flow rate of 1000 sccm and an initial methane flow rate of 40 sccm. After 4 h, this value was decreased to 35 sccm; after 8.5 h, the stage height was changed to +0.1 cm.

2.2 Thermal management of the wafer substrate

To minimize the temperature gradient during diamond growth, the silicon wafer was placed on top of a quartz plate. The diameter of the quartz plate was smaller than the diameter of the silicon substrate, and cold water was circulated through the molybdenum substrate

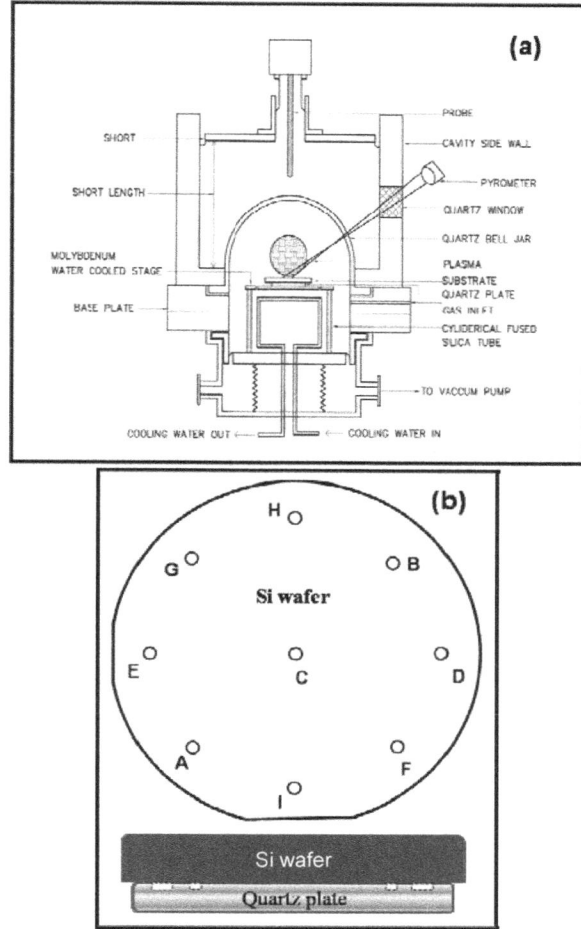

Fig. 1 (a) Schematic of the MPCVD reactor. (b) (100) 100-mm-diameter silicon wafer sitting on a structured quartz plate, depicting the positions of the nine different "zones" studied in the present work.

holder stage, as described in Fig. 1(b) [36]. At the beginning of the deposition, the hemispherical plasma ball was centered in the middle of the substrate.

The substrate temperature was measured with a double-wavelength optical pyrometer (Williamson, USA, Model: PRO 82-40-C). The pyrometer was focused at the center of the substrate throughout the 95-h long run. The temperature was monitored from time to time at nine different spots (a few millimeters in area), identified in Fig. 1(b). These spots were located 40 mm away from the center of the substrate and were separated from each other by other 40 mm. For each reading, the maximum temperature difference between the spots was determined as

$$\Delta T = T_{max} - T_{min} \tag{1}$$

where T_{max} and T_{min} are the maximum and

minimum temperature readings, respectively.

2. 3   Physical characterization of the PCD coating

The silicon substrate was etched with a solution of hydrofluoric acid, nitric acid and acetic acid in 1:1:1 ratio; the freestanding PCD coating was then broken into pieces to collect specimens A to H from the sites described in Fig. 1(b). An extra sample M was collected in the middle region of the 100-mm-diameter area. The samples A to H were characterized in detail. The roughness of the nucleation and growth sides of the samples was measured by coherence scanning interferometry (CSI, Contour GT-K, Bruker Nano GmbH, Germany). Each scan was performed on a surface area of 0.8 mm × 0.6 mm, generating millions of data points. A 10× objective lens was used, with back scan and length parameter of 35 μm and 70 μm respectively for the rough side, and 15 μm and 20 μm for the nucleation side. The micro and nanostructural features of the samples were evaluated by SEM (LEO 430i STEROSCAN, UK), and the thickness was measured by cross-section SEM. The nucleation surface morphology of the MPCVD diamond films was characterized by field-emission SEM (FESEM, Supra 35VP, Carl Zeiss, Germany). The structural characterization of the samples was done with XRD technique. The XRD patterns of the samples were recorded in an X'pert Pro MPD diffractometer (PANalytical) with an X'Celerator operating at 40 kV and 30 mA using Cu Kα radiation with step size of 0.05° (2θ) and step time of 30 s from 20° to 90°. The weight percentages of the crystalline phases were estimated using Rietveld analysis of the XRD line profiles [39,40] by X'pert high score plus software (PANalytical)[①]. In addition, the quality of the diamond films was examined by Raman spectroscopy (HR-800, HORIBA JOBIN YVON, Japan) with a 514-nm Ar$^+$ laser with 10-mW laser power and 50-μm spot size. The micro hardness of the nucleation side was measured by applying 1-kgf load for 60 s of duration using a Vickers indenter (model Leica VMHT– microsystem GmbH, Australia).

---

① http://www.PANalytical.com (last accessed on November 14, 2013).

3   Results and discussion

3. 1   Substrate temperature

Figure 2(a) shows the temperature distribution pattern and ΔT values recorded at the different surface during the first 24 h of diamond deposition. After 4 h, ΔT rises up to 160 ℃ and the methane flow rate is decreased from 40 sccm to 35 sccm; after 5.5 h, ΔT decreases to 114 ℃, reaching a minimum value of 100 ℃ after 7 h of diamond deposition. After 8.5 h, the stage height is increased from −0.1 cm to +0.1 cm and the effect is immediate, and ΔT decreases down to 90 ℃, stabilizing at 71 ℃ after 24 h of deposition. With this new set of parameters (35-sccm methane flow rate and +0.1-cm stage height), the coating is grown unattended over four days. At the end of the fourth day, the temperature distribution has become very non-uniform, with ΔT as high as 212 ℃. Throughout the initial 24 h of deposition, the temperature at the center of the wafer is always the highest, in and around 1050 ℃, but after 95 h, the center temperature rises to 1127 ℃ and

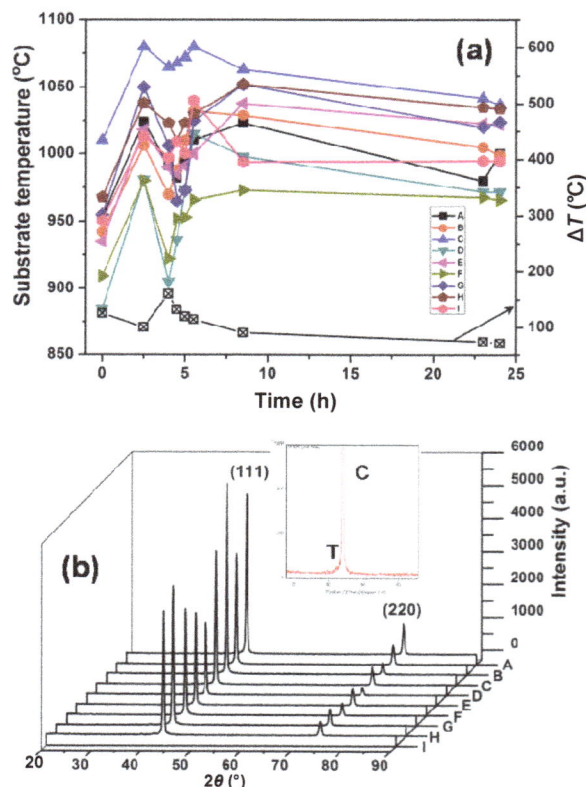

Fig. 2   (a) Substrate temperature distribution during the first 24 h of deposition. (b) XRD patterns on the growth side (zoom-in: presence of trace amounts of trigonal phase).

the points marked as H, G and E—the opposite ends of the wafer cut marks (I and F)—become hotter than the center of the wafer; the temperature of region E is as high as 1180 ℃, higher than the initial value of 1022 ℃. After 95 h of deposition, points H and G have become red hot and the deposition is stopped to prevent the cracking of the wafer. This temperature gradient over 95 h of deposition is clearly reflected in the characteristics of the PCD coating.

3. 2　Growth rate

The samples were mounted vertically on the holder and the cross sections were observed under SEM. The thicknesses are listed in Table 1. The differences on the thickness, and consequently on the local growth rate, reflect the temperature gradient across the 100-mm wafer during diamond growth. Samples I and F are the thinnest (190 μm thick and 192 μm thick, respectively). All other thicknesses are above 200 μm; samples B, C and D are about 210 μm thick, samples A and E have comparable thicknesses of 219 μm and 220 μm, and samples G and H are widely different with thicknesses of 233 μm and 245 μm respectively. The growth rate across the 100-mm wafer ranges from 2 μm/h for sample I up to 2.57 μm/h for sample H.

Different parameters, like pressure, substrate temperature, methane percentage, hydrogen flow rate, etc., are known to affect the growth rate [41]. Axial and radial distributions of gas species and temperatures inside an MPCVD reactor were elaborately discussed by Ma in his thesis [42]; he showed that there is a wide variation in the species concentration and gas temperature away from the discharge core and the substrate. However, even though the hemispherical plasma symmetrically covers the whole 100-mm-diameter substrate surface, the temperature distribution is not radial. In fact, experimental data shows evidence of a strong temperature gradient across the substrate, probably due to the asymmetry that the wafer cuts introduce in the circular shape of the substrate (Table 1).

The temperature distribution also changes as diamond growth proceeds. At the beginning of deposition, the temperature of the substrate center is 1010 ℃, increasing to 1063 ℃ after the methane gas flow rate is reduced to 35 sccm and the stage position adjusts from −0.1 cm to +0.1 cm (Fig. 2(a)). After 24 h of deposition, the temperature at point C decreases further to 1037 ℃ and ΔT reaches its minimum recorded value (71 ℃). The temperatures at points G and H are 1024 ℃ and 1034 ℃, respectively, still lower than that at point C. However, after 95 h of deposition, this area has become red hot.

This temperature evolution suggests the stability of the plasma at the core of the discharge. However, as the coatings become thicker and thicker with time, the heat distribution becomes more and more non-uniform. The values of thickness and growth rate at the center of the wafer (214 μm and 2.25 μm/h, respectively) are found to be essentially the average values of the two extreme thicknesses and growth rates at points H and I. These points correspond to two opposite positions of the substrate: point I is near the substrate cut mark and is at lower temperature values throughout the four-day long run (mostly below 1000 ℃) in comparison to point H (always above 1000 ℃). It can be concluded that, in spite of the quartz plate that improves the temperature homogeneity of the substrate, the

Table 1　Substrate temperature, mechanical properties and SEM analysis results (G: growth side; N: nucleation side)

Sample	Substrate temperature after 24 h (℃)	Mechanical property				SEM			
		Hardness N (GPa)	Roughness		Top surface microstructure	Grain size		Thickness (μm)	Growth rate (μm/h)
			G (μm)	N (nm)		G (μm)	N (nm)		
A	1000	19.50 ± 2.79	5.86	23.4	(111)+octahedral	6.23 ± 0.71	63	219	2.30
B	999	17.07 ± 1.14	6.87	16.3	(110)+cubo-octahedral	9.76 ± 1.69	49	210	2.21
C	1037	18.22 ± 1.27	5.84	17.1	(110)+(100)+cubo-octahedral	10.52 ± 1.68	282	214	2.25
D	972	35.00 ± 3.83	7.39	12.9	(111)+octahedral	21.43 ± 1.86	—	211	2.22
E	1022	30.03 ± 6.05	6.45	11.7	(100)+(110)+cubo-octahedral	18.58 ± 0.86	1450	220	2.31
F	966	24.21 ± 3.70	3.59	11.4	(111)+octahedral	10.02 ± 0.44	280	192	2.02
G	1024	29.90 ± 7.85	7.52	9.54	(100)+(110)+cubo-octahedral	21.29 ± 1.76	68	233	2.45
H	1034	48.07 ± 3.16	5.56	18.8	(110)+(100)+cubo-octahedral	10.82 ± 2.15	—	245	2.57
I	995	18.72 ± 3.42	6.15	27.1	(111) + octahedral	10.19 ± 1.17	—	190	2.00

existence of a discontinuity in the shape of the wafer (the wafer cut close to point I) induces the loss of the plasma radial symmetry, thereby causing a temperature gradient through the substrate that is reflected, for instance, in the difference of the growth rates that are measured across the diamond coating. The temperature gradient can be reduced to some extent by adjusting the process parameters (such as methane flow rate and stage height), but cannot be completely avoided, and deteriorates the uniformity of the diamond film deposited on the wafer. As growth proceeds and the diamond film thickens, the temperature gradient increases, suggesting that the plasma radial symmetry is further reduced as growth proceeds.

3. 3   Surface roughness

The roughness of the nucleation side of the diamond coating (Table 1) is comparable to the roughness of the mirror-polished silicon surface and varies from 9.5 nm to 27 nm. The roughness of the growth surface is much higher, ranging from 3.5 μm to 7.5 μm, depending on the point of measurement. Figure 3 shows typical 0.8 mm × 0.6 mm CSI scans of the growth and nucleation sides and bare silicon substrate.

3. 4   Surface morphology

3. 4. 1   Growth side

Figure 4 shows SEM micrographs of the growth

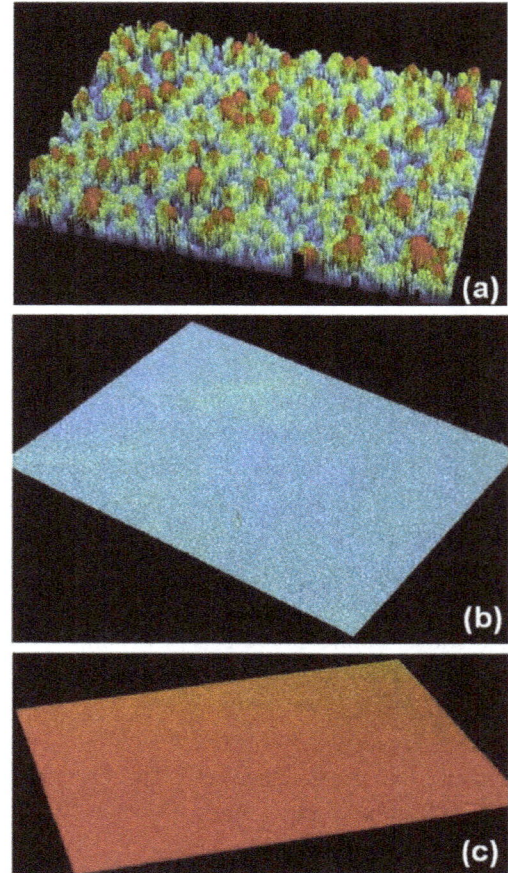

Fig. 3   0.8 mm × 0.6 mm CSI scans of the (a) grown and (b) nucleation sides of the PCD coating and (c) bare silicon substrate.

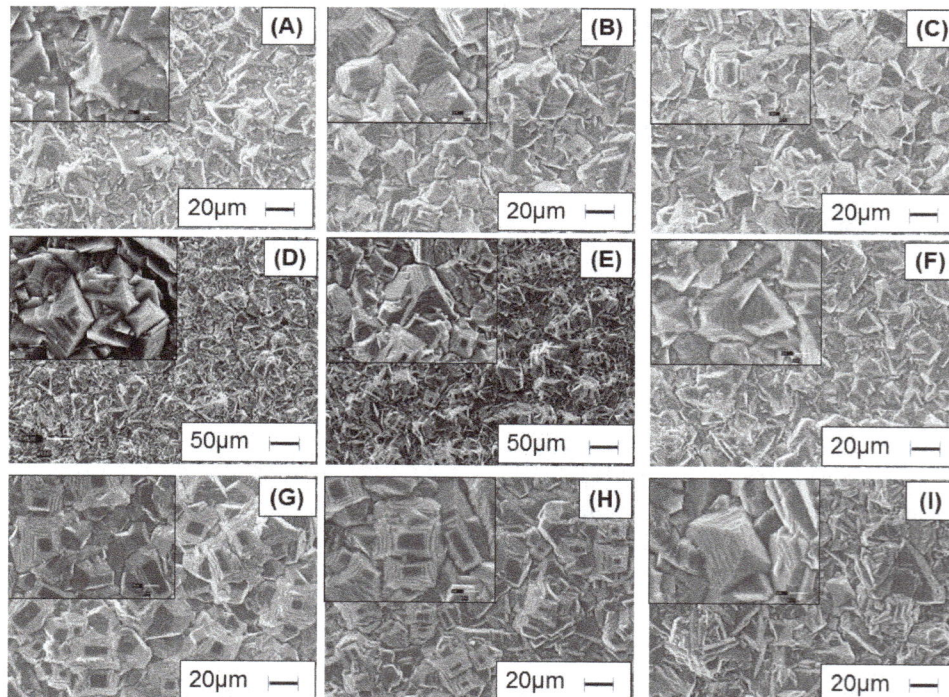

Fig. 4   SEM images of growth surface points as described in Fig. 1(b).

surface of the different regions of the wafer (labeling corresponds to positions depicted in Fig. 1(b)). The microstructure can be classified in terms of crystal size, shape and noticeable crystal facets on the top surface of the diamond coating. The grain sizes, calculated following ASTM line intercept method (ASTM Standard E 112-88), vary from 6 µm to 21 µm (Table 1). Based on the grain size, the nine surface regions can be divided into three groups, namely 10-µm-sized region, about-20-µm-sized region and region A with the smallest 6-µm grain size.

One can notice an expected tendency: larger grains correspond to higher roughness values due to the presence of bigger diamond octahedral pyramids in the microstructure. For instance, the average grain size is approximately three times larger than the roughness of regions G, E and D. However, this tendency is not a rule, since the Ra roughness and grain size of region A, for instance, are of the same magnitude.

The octahedral pyramids found in region A suggest the parameter α near or equal to 3. Similar near-perfect octahedral diamond pyramids can be seen in the regions D and F and also to some extent in region I. In regions B, C, E, G and H, cubo-octahedral pyramids with distinguishable flat top surfaces are clearly visible. These top surfaces are rectangular, which imply the appearance of (110) planes. The presence of (100) planes would translate in perfect square top surfaces. XRD peaks (Fig. 2(b)) reveal the presence of (220) planes at 74.3°, but there is no diffraction peak corresponding to (100) plane. The reason can be explained on the basis of XRD "extinction rules". It is known that for successful diffraction to occur in the diamond cubic crystal structure, the Miller indices h, k and l should meet the following criteria: (i) all odd, (ii) all even or (iii) ($h+k+l$) divisible by 4 [43]. Following such argument, it can be concluded from the XRD data that (100), (200) and (300) peaks will be missing, although one could have detected the (400) plane peak at 2θ value of 119°; however, in the present study, the scan was performed from 20° to 90° due to physical limitations of the goniometers. The XRD peaks from most regions indicate the presence of 100% intensity peaks corresponding to (111) planes in comparison to 17% relative intensity peaks for (220) set of planes; these planes can be correlated to the appearance of cubo-octahedral surface morphological features revealed under SEM.

3.4.2 Nucleation side

While the growth side is populated with cubo-octahedral pyramids, the nucleation side consists of agglomerated colonies/grains (Fig. 5). These individual colonies are separated by distinguishable

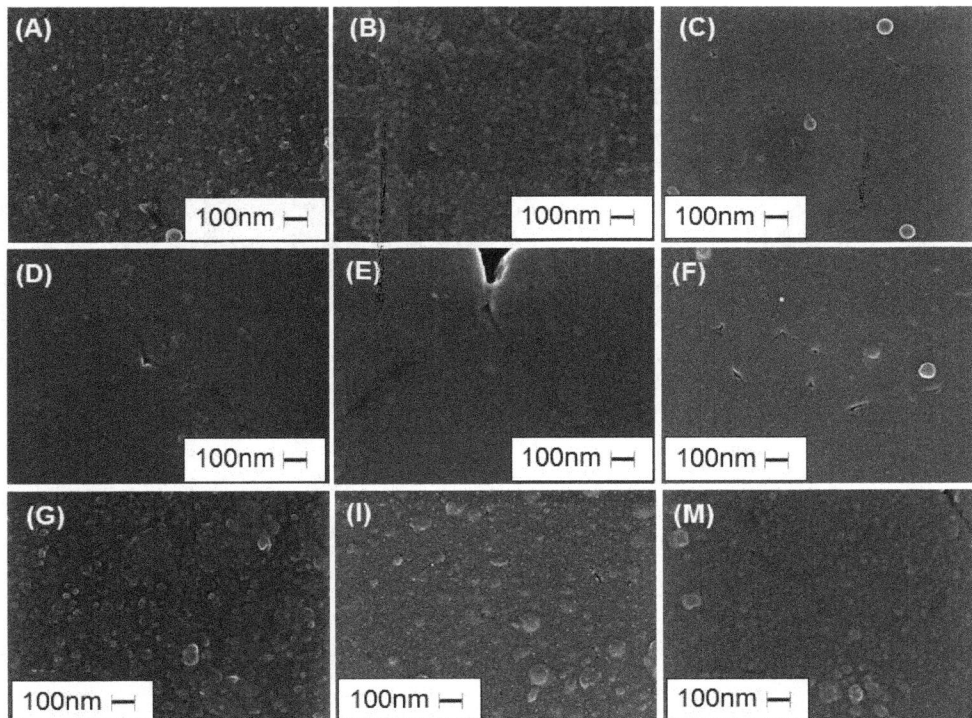

Fig. 5 FESEM images of nucleation surface points as described in Fig. 1(b).

grain boundaries (Figs. 5(C)–5(F) and 5(I)). The size of the colonies varies from 0.28 μm (points C and F) to 1.45 μm (point E), depending on the location of the substrate.

Higher-magnification images of these colonies reveal nanocrystalline agglomerations of diamond crystals (more clearly in Figs. 5(A), 5(B) and 5(G)). Typical sizes of these diamond nanocrystals vary from 49 nm (point B) to 68 nm (point G). It is interesting to notice that the initial temperature at these regions is in-between 940 ℃ (Fig. 5(B)) and 955 ℃ (Fig. 5(G)). These agglomerations are visible neither in spot C (higher initial temperature) nor in spots D, E or F (lower initial temperatures).

Energy dispersive X-ray analysis (EDAX) signals of the boundaries and the colonial features are shown in Figs. 6(a) and 6(b), respectively. The scans taken on the colonial features (Fig. 6(b)) produce only carbon peaks, whereas signals from the boundaries (Fig. 6(a)) show silicon and to some extent oxygen peaks, in addition to the normal carbon peaks. Silicon is present as SiC and SiO_2 compounds.

3. 4. 3 Diamond nanostructure

The cross sections of the different surface points (not shown) reveal the traditional columnar growth of diamond crystals along the z-direction from the substrate side. However, sample collected from point B has a whitish 40–50-μm-thick layer at the middle of the cross section (Fig. 7(a)). This white layer runs parallel to the substrate and through the entire length of the sample, while the characteristic columnar diamond growth is apparent in the growth direction (z). From the growth rate at this point (2.21 μm/h), it can be estimated that this layer is deposited during 20 h, after the initial 55 h of deposition.

The double-arrows (Fig. 7(a)) indicate that the microscopic columnar growth is kept at the nanostructured whitish layer of sample B. However, when this layer is magnified, nanostructures become evident: the region is filled with elongated nano-features which are either oriented randomly (Fig. 7(b) nanoworms) or in particular directions (Fig. 7(c) nanoplates). The high aspect ratio plates are 30–80 nm thick and 0.5–2 μm long. They run parallel along two directions separated by an angle slightly greater than 70° (Fig. 7(c)). Literature suggests that the angle between (311) and (100) planes is equal to 72.4°, so we raise the hypothesis that these plates correspond to (311) and (100) diamond planes formed during CVD growth, forming square and rectangular grids as shown in Fig. 7(c). Such diamond nanogrids can have futuristic application of "nano-shelf" for storing different materials.

MPCVD of diamond nanoplate arrays on (111) planes was reported in a recent paper in the view of field-emission applications [44]. The authors attributed the high aspect ratio growth to the extension of

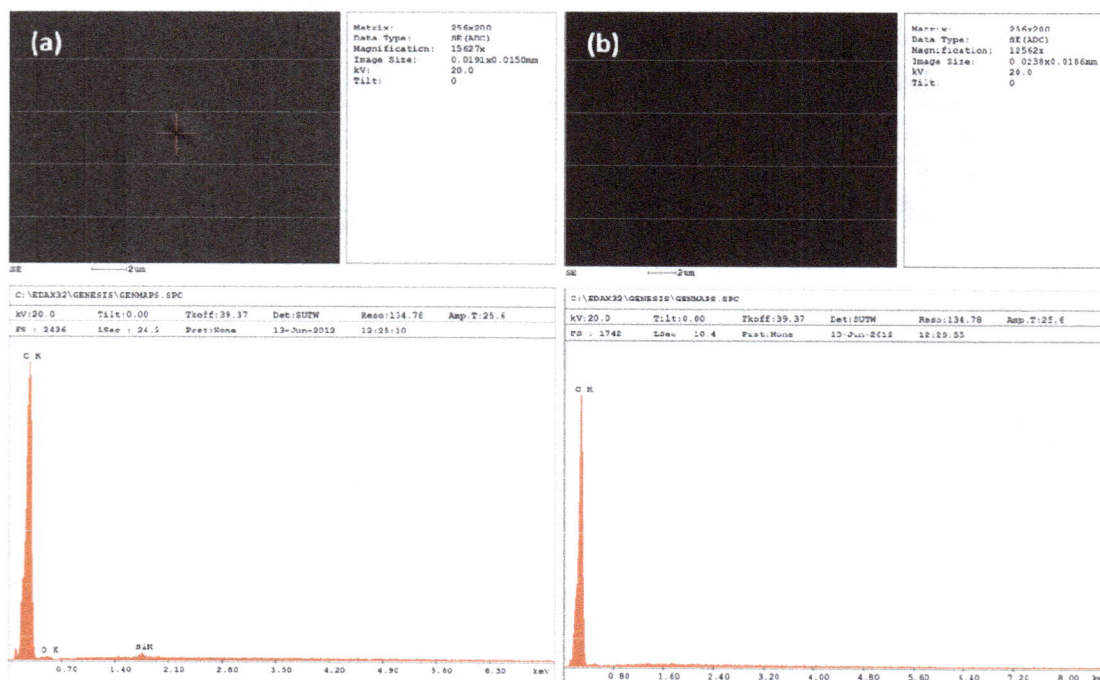

Fig. 6   EDAX signals from (a) the boundaries and (b) the colonial features of the freestanding coating.

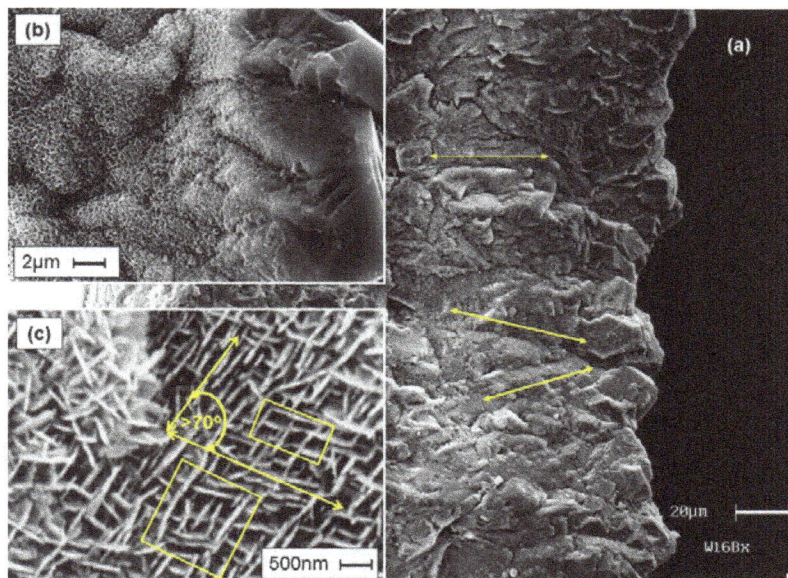

Fig. 7   (a) Cross section of the white layer of sample B showing typical microcrystalline-columnar growth; (b) higher-magnification SEM image showing nanoworms; (c) higher-magnification SEM image revealing a mesh of nanoplates forming square and rectangular grids, with plates running parallel along two directions separated by angle greater than 70°.

penetration {111} twins from underlying diamond grains. But such structures are selectively grown in 20-nm-thick layers. There have been several reports about reactive ion etching (RIE) fabrication techniques of diamond nanorods/nanowires/whiskers of 50–400 nm thick and several microns long [45–49]. Moreover, it has been shown that CVD diamond can be grown into solid or hollow fiber-, wire- or tube-like structures by using suitable templates[2]. One, two and three dimensional nanodiamond structures have been discussed in detail by Shenderova and McGuire [50], who reported the occurrence of diamond crystallites embedded in amorphous carbon clusters, nanocrystalline diamond (NCD)–carbon nanotube (CNT) composites grown by CVD, diamond filament structures obtained by laser ablation of pellets of pressed ultra-nanocrystalline diamond (UNCD) particle, etc. But none of the reported nanostructures resemble the presently obtained diamond nanoplates along the cross section of the freestanding PCD. The closest structure was deposited by MPCVD on a Ni-coated HFCVD-grown PCD substrate [51]. The nanoplates grown by Chen and Chang are single crystalline and 30–70 nm thick. The plates are triangular or parallelogram-like with top and bottom surfaces parallel to (111) planes, whereas the edges are along the <110> directions. There is no more report on such nanoplates maybe because of their limited application/repeatability. The thicknesses of the nanoplates obtained in the present study are of similar dimension, but they are not (111) planes as shown by Ref. [51]. Very recently, the CVD diamond group at Bristol University published some SEM micrographs of diamond oddities and unusual structures (rod, plate, star or spaghetti like carbon/diamond structures) without confirming their repeatability and standard method of deposition[3].

Similar whitish layers with different nanocrystalline features are also present in samples A, C and G (Fig. 8). Other than well-aligned nanoplates, the whitish region of sample A consists of a unique nanoporous structure (Fig. 8(a))—here termed as nanosieves—typical sintered ceramic structures. The pore size varies from 30 nm to 130 nm. There are reports of PCD sintered at 5–10 GPa and 1300–1800 ℃ during 5–60 min [52,53]. The appearance of this nanosieve structure suggests that under MPCVD growth environment, there may exist a pressure–temperature regime where diamond particles are melted/sintered, forming a granular network with open pores. In fact, the pressure rise due

② http://www.chm.bris.ac.uk/pt/diamond/semwires.htm (last accessed on November 14, 2013).

③ http://www.chm.bris.ac.uk/pt/diamond/semoddities.htm (last accessed on November 14, 2013).

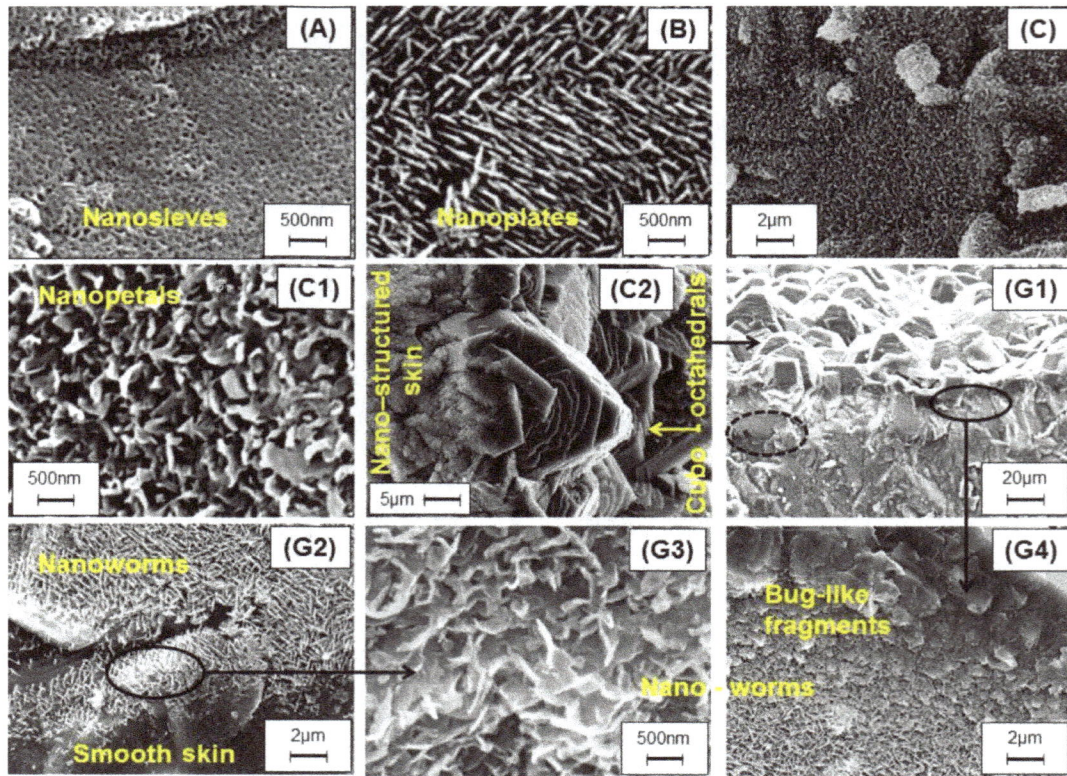

Fig. 8   SEM images of nanostructural features found in samples A, B, C and G.

to nanosize may push the meta-stable CVD growth environment into the stable diamond phase region in nanocarbon-phase diagram [54,55]. In addition, gas temperature may reach 1400–3300 ℃ under typical MPCVD pressure of 105 Torr ($\approx$ 14 kPa) [41]. It is possible that, under the present growth conditions, the nanosize-induced pressure rises up to several GPa; this effect, together with the high plasma temperature, may have induced sintering of the diamond grains/crystals from secondary nucleation sites, forming the nanosieve network.

A similar sintered/melted porous microstructure is found in the cross section of sample H (not shown); however, the grains here are elongated and bent with less porosity.

Wavy nanocrystalline features (nanoworms) are found at some locations of the cross section of sample G (Fig. 8(G3)). On some regions of sample C, the diamond crystals form curved nanopetals (Fig. 8(C1)). These nanopetals are folded or bent in nature but are not fused together, like in the case of sample G (Fig. 8(G3)) and H (not shown); instead, this nanostructure resembles bent onion structures. The cross section of sample C also has regions with grids of nanoplates or randomly dispersed nanoworms.

Figures 8(C2), 8(C1) and 8(G4) show the transition region between the whitish layer and the as-grown diamond surface. The top surface of sample C (Fig. 8(C2)) consists of typical diamond cubo-octahedral pyramids with (111) planar side surfaces and flat top surfaces (either (100) or (110), as found in Fig. 4(C)). The base size of the pyramids is around 30 μm; 8 μm down from the pyramid base, the texture changes from smooth to rough. A similar transition from microcrystalline grains into nanostructures is also observed below the growth surface of sample G (Fig. 8(G1)). The surface is densely populated with diamond cubo-octahedral pyramids (Fig. 4(G)), with different top and bottom surface textures. A whitish layer, 10–20 μm below the top surface, is also present; upon higher magnification (Fig. 8(G4)), the interface between this whitish layer and the remaining film reveals a gradual skin transformation: diamond crystals, with initial smooth texture, seem to be broken down into smaller fragments. These bug-like fragments are less than 1 μm long, and as the distance from the top surface increases, they disappear with emergence of elongated worms/plates, initially running randomly with respect to each other but gradually forming a mesh/grid-like nanostructure (Fig. 7(c)) throughout the outer skin of the typical diamond columns. The other end of this whitish layer (Fig. 8(G2)) also contains similar elongated plates or worms. A higher magnification of the encircled portion of Fig. 8(G2) (Fig. 8(G3)) reveals somewhat molten nanoworm structures randomly oriented.

3.5 Quality

Figure 9 shows the Raman spectra collected from both sides of the nine different surface regions. Since the spot size of the 514.5-nm laser was 50 μm, it can be assumed that the Raman spectra are generated from several grains. Both signals corresponding to diamond (I_d) and graphite-like-carbon (I_{glc}) are present in all spectra.

The FWHM (full width at half maximum) of the diamond peaks (Table 2) vary between 5 cm^{-1} (point G) and 12.39 cm^{-1} (point A) on the growth surface, and between 9.33 cm^{-1} (points A, B, C and E) and 11.79 cm^{-1} (point G) on the nucleation surface. Generally speaking, the broadening of the peaks is higher on the nucleation surface, since it contains more grain boundaries and defect sites. The dispersion is higher on the growth surface, reflecting once again the non-homogeneity of the growth side. Raman peak broadening can be due to two reasons: (i) homogeneous lifetime broadening and (ii) size effect of crystal. Scattering of phonons at grain boundaries and defect sites shortens their lifetime, increasing the FWHM. The Raman peaks are symmetrical, so peak broadening due to phonon confinement in small domain size might not have occurred in the present study.

The quality factor, a semi-quantitative estimation of the diamond phase ratio

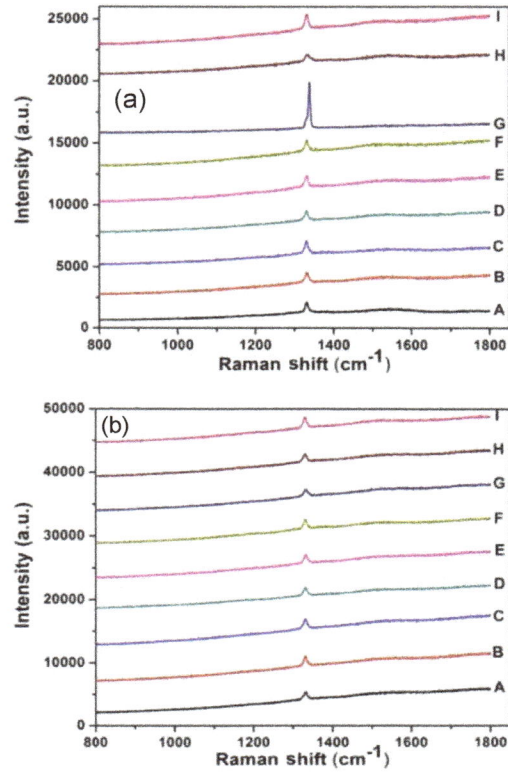

Fig. 9 Raman spectra taken on the (a) growth and (b) nucleation sides of the different samples.

$$Q = \frac{I_d}{I_d + I_{glc}} \times 100\% \qquad (2)$$

was calculated for the growth and nucleation surfaces, considering the 1332 cm^{-1} diamond peak and 1500–1600 cm^{-1} graphite-like carbon band (Table 2). For the nucleation side, Q varies between 50% and 51%. The Q values for the growth surface vary between 50% (point H) and 79% (point G). This high dispersion is a consequence of the poor homogeneity of the thick

Table 2 Raman and XRD results

Sample	1332 cm^{-1} peak shift (cm^{-1})		FWHM (cm^{-1})		Q (%)		Stress (GPa)		Crystalline phases (%)		Cell parameter (Å)	Unit cell volume (Å^3)	Average crystallite size (nm)	Average lattice strain (%)
	G	N	G	N	G	N	G	N	Cubic	Trigonal				
A	−1.37	0.58	12.39	9.33	57	50	0.7	−0.33	90.9	9.1	$a=b=c=3.56742$	45.400	40.38	0.046
B	0.58	−0.39	9.40	9.33	53	51	−0.3	0.22	90.1	9.9	$a=b=c=3.56749$	45.403	41.64	0.041
C	−2.36	0.09	8.36	9.33	56	51	1.33	−0.05	99.6	0.4	$a=b=c=3.57043$	45.515	39.23	0.039
D	−0.88	−0.88	7.68	10.81	52	51	0.5	0.50	98.3	1.7	$a=b=c=3.56628$	45.357	41.74	0.043
E	−0.88	0.09	7.91	9.33	54	50	0.5	−0.05	99.3	0.7	$a=b=c=3.56725$	45.394	41.25	0.039
F	−0.88	1.08	8.17	9.82	52	51	0.5	−0.60	96.1	3.9	$a=b=c=3.56547$	45.326	41.87	0.040
G	5.99	0.09	5.07	11.79	79	50	−3.3	−0.05	98.0	2.0	$a=b=c=3.56910$	45.465	41.55	0.040
H	1.57	−1.37	11.84	10.32	50	50	−0.9	0.78	82.2	17.8	$a=b=c=3.57003$	45.500	39.20	0.038
I	−0.88	−1.37	7.38	9.82	53	51	0.5	0.78	99.0	1.0	$a=b=c=3.56377$	45.261	39.62	0.041

PCD coating due to the large temperature gradient during deposition. The Q values are in general very low, due in one hand to the high CH_4/H_2 ratio used (3.5%) and on the other hand to the high Raman scattering efficiency of non-diamond phase [56].

3. 6 Stress

CVD diamond film can have extrinsic and intrinsic stresses. The extrinsic stress arises from thermal expansion mismatch and lattice parameter mismatch between the substrate and the diamond coating. In this study, the silicon substrate was etched, so the extrinsic stress component was removed.

The intrinsic internal stress arises from defects present within the PCD coating and leads to a diamond peak shift. The intrinsic stress (hydrostatic stress symmetrically distorts the diamond lattice and hence also its optical modes phonon) can be compressive leading to up-lift of the peak shift, or tensile leading to down-shift of the diamond peak in the Raman spectrum. Further, due to the presence of biaxial stress, the diamond peak can split into two or three maxima, depending on the degeneracy of the phonon [57,58]. Factors like temperature, grain domain size, hydrostatic stress, have a strong influence on the peak shifting [59–61].

The residual stress (γ) is calculated in GPa using

$$\gamma = -0.567 \times \Delta v \qquad (3)$$

where Δv is the peak shift in cm^{-1} [62] (Table 2). The minus and plus signs are assigned to compressive and tensile stresses, respectively.

On the growth surface, stress varies from highly compressive (3.3 GPa at point G) to highly tensile (1.33 GPa at point C). The residual stress on the nucleation side is considerably lower and varies from tensile (0.78 GPa at points H and I) to compressive (0.6 GPa at point F). The Raman spectra taken on the growth surface of samples B, C, E, H and G show a diamond peak splitting, indicating an inhomogeneous stress distribution.

The stress level variation is wider on the growth side. Tensile stress arises from voids, dislocations and grain boundaries in the diamond crystal lattice, whereas impurities or hydrogen clusters introduce compressive stress in the PCD. Growth defects (twins, dislocations, stacking faults, etc.) are present more on the growth surface, producing tensile stress. The possibility of contamination is lower for the growth surface, and this is reflected in less points under compression on this surface; on the contrary, the nucleation side has many non-carbon phases such as silicon and oxygen (maybe in the form of amorphous silica, as evidenced from the EDAX signals, Fig. 6), which are responsible for the compressive stress on the nucleation side after the etching of the silicon substrate. In spite of the large stress variation on the growth surface, no observable trend could be established between these values and the temperature distribution pattern.

3. 7 Hardness

The hardness of CVD diamond films is beyond 60 GPa [63], high enough to cause damage to the indenter tip used for Vickers microhardness testing. CSI surface scans (Fig. 3(a)) reveal a very high surface roughness ($Ra = 6.87\,\mu m$) due to the presence of very big diamond cubo-octahedral pyramids on the growth surface, which produce peak ($Rp = 23.9\,\mu m$) to valley ($Rv = -30.7\,\mu m$) distance as high as 56 μm. These high values prevent reliable microhardness measurements. On the contrary, the nucleation side (Fig. 3(b)) has a very low surface roughness ($Ra = 16$ nm) with peak ($Rp = 0.67\,\mu m$) to valley ($Rv = -0.35\,\mu m$) distance as low as 1 μm; this result is supported by FESEM images (Fig. 5) of the smooth nucleation side that reveal the existence of diamond colonies with nanocrystalline agglomeration (expected to be softer than the growth surface).

Taking into account the experimental limitations, the hardness measurements with Vickers indenter were performed only on the nucleation side (Table 1). It can be observed that, although the hardness values are very high in comparison to any other known hard ceramic materials, they are much lower than the theoretical value of natural diamond (100 GPa), varying from 17 GPa (point B) to 48 GPa (point H). The reason behind this fact may be the existence of agglomerates of diamond nanocrystals (Fig. 5); these agglomerates are 0.3–1.5 μm in size, and are formed by 50–70-nm-sized particles and separated by an intergranular region which contains silicon and oxygen. The diameter of the indentation marks varies from 6 μm to 10 μm; if the indenter probe penetrates these agglomerates, the local hardness value will be lowered. However, this local lowering cannot account for the large difference in the measurements (the lowest and highest measured values differ by a factor of almost 3). This large variation can be explained on the basis of the different percentages of crystalline phases present, as measured by XRD peak broadening. Rietveld analysis of the XRD profiles (inset of Fig. 2(b)) of the samples reveals that all the samples contain variable percentages of cubic carbon phase having space group Fd-3ms (227) as a major phase and trigonal carbon phase having space group R-3mH (166) as a minor or trace phase (Table 2). The values of crystal size, average crystal strain, cell parameters and cell volume are also shown in Table 2. In particular, sample H has a considerable amount of trigonal phase (17.8%), which may be responsible for the high hardness value (48 GPa) measured at that region.

4 Conclusions

A thick PCD coating was deposited on a 100-mm-diameter silicon wafer during four days. The thick PCD coating was cut into smaller pieces, further investigated through SEM, Raman, and Vickers indentation. The temperature of the substrate during deposition was not uniform; in fact, the existence of a temperature gradient, that was reduced to some extent by adjusting the process parameters (such as methane flow rate and stage height), deteriorated the uniformity of the diamond film deposited.

The temperature gradient was kept stable after 24 h of deposition ($\Delta T = 71\ ℃$) but further increased with increasing deposition time, reaching the value $\Delta T = 212\ ℃$ at the end of deposition; the growth rate, and hence film thickness, followed this gradient; lower substrate temperatures corresponded to perfect octahedrals, and higher substrate temperatures corresponded to cubo-octahedral morphology; other properties varied in a somewhat random way. Raman spectra of the growth surface also differed considerably; the FWHM varied between $7.32\ cm^{-1}$ and $5.07\ cm^{-1}$. There was no correlation with temperature distribution, morphology or growth rate patterns. Raman spectra of nucleation surface had a little variance of $\pm 1\ cm^{-1}$ with a high average value of $10\ cm^{-1}$. Most of the surface points were under tension, while the others were under compressive residual stress; most of the nucleation side points were under compression with little variance and without any apparent reason. No correlation could be established between the presence of trigonal phase and the corresponding hardness values among different surface points.

Colonies or micron-sized agglomerations of 50–

70 nm diamond nanocrystals were identified on the nucleation surface. Different nanostructures, random in nature, were also found on the cross section of five samples. They consisted of elongated nanoflake-like structures, sometimes randomly dispersed (nanoworms), sometimes running parallel in definite directions (nanoplates), sometimes bent and flat (nanopetals), and sometimes fused/sintered together to form nanoporous microstructure (nanosieves). It seemed apparent that regions deposited at lower growth rates didn't develop nanostructures, whereas higher growth rate regions had varied nanodiamond features across the thickness.

As a general conclusion, it could be said that there is no general trend in the physical properties of the PCD coating, although the deposition was carried out uninterruptedly in a controlled way, and special care was taken in order to minimize the temperature gradient during deposition; only growth rates and octahedral morphologies could be correlated with substrate surface temperatures; other physical parameters varied in a random way.

Acknowledgements

CSIR, India provided financial support for the network project "Very High Power MW Tubes: Design and Development Capabilities (MTDDC)" (CSIR Grant Nos. NWP 0024 and PSC0101) under the 11th and 12th five-year plan periods. Dr. J. C. Mendes is thankful to the Instituto de Telecomunicações for providing the necessary working conditions. Lambda Technologies Inc. USA was helpful in carrying out the CVD experimental runs.

References

[1] Thumm M. MPACVD-diamond windows for high-power and long-pulse millimeter wave transmission. *Diam Relat Mater* 2001, **10**: 1692–1699.
[2] Oh A. Particle detection with CVD diamond. Ph.D. thesis. Univ Hamburg Germany, Inst Experim Physics, 1999.

[3] Meier D. CVD diamond sensors for particle detection and tracking. Geneva(Switzerland): CERN, 1999.

[4] Sussmann RS, Brandon JR, Coe SE, et al. CVD diamond: A new engineering material for thermal, dielectric and optical applications. Ind Diamond Rev 1998, 58: 69–77.

[5] Füner M, Wild C, Koidl P. Novel microwave plasma reactor for diamond synthesis. Appl Phys Lett 1998, 72: 1149–1151.

[6] Silva F, Hassouni K, Bonnin X, et al. Microwave engineering of plasma-assisted CVD reactors for diamond deposition. J Phys: Condens Matter 2009, 21: 364202.

[7] Yokota Y, Ando Y, Kobashi K, et al. Morphology control of diamond films in the region of $\alpha = 1$–1.5 using a 60-kW microwave plasma CVD reactor. Diam Relat Mater 2003, 12: 295–297.

[8] Schelz S, Campillo C, Moisan M. Characterization of diamond films deposited with a 915-MHz scaled-up surface-wave-sustained plasma. Diam Relat Mater 1998, 7: 1675–1683.

[9] Vikharev AL, Gorbachev AM, Kozlov AV, et al. Diamond films grown by millimeter wave plasma-assisted CVD reactor. Diam Relat Mater 2006, 15: 502–507.

[10] Grotjohn T, Liske R, Hassouni K, et al. Scaling behavior of microwave reactors and discharge size for diamond deposition. Diam Relat Mater 2005, 14: 288–291.

[11] Zuo SS, Yaran MK, Grotjohn TA, et al. Investigation of diamond deposition uniformity and quality for freestanding film and substrate applications. Diam Relat Mater 2008, 17: 300–305.

[12] Gicquel A, Hassouni K, Lombardi G, et al. New driving parameters for diamond deposition reactors: Pulsed mode versus continuous mode. Mat Res 2003, 6: 25–37.

[13] Tsai H-Y, Kuo K-L. Nitrogen effect on the diamond deposition processing by 915-MHz microwave plasma enhanced chemical vapor deposition reactor. J Chin Soc Mech Eng 2007, 28: 157–162.

[14] Brandon JR, Coe SE, Sussmann RS, et al. Development of CVD diamond r.f. windows for ECRH. Fusion Eng Des 2001, 53: 553–559.

[15] Windischmann H, Epps GF. Free-standing diamond membranes: Optical, morphological and mechanical properties. Diam Relat Mater 1992, 1: 656–664.

[16] Meykens K, Haenen K, Nesládek M, et al. Measurement and mapping of very low optical absorption of CVD diamond IR windows. Diam Relat Mater 2000, 9: 1021–1025.

[17] Pickles CSJ, Madgwick TD, Sussmann RS, et al. Optical performance of chemically vapour-deposited diamond at infrared wavelengths. Diam Relat Mater 2000, 9: 916–920.

[18] Braz O, Kasugai A, Sakamoto K, et al. High power 170 GHz test of CVD diamond for ECH window. Int J Infrared Milli 1997, 18: 1495–1503.

[19] Heidinger R, Dammertz G, Meier A, et al. CVD diamond windows studied with low- and high-power millimeter waves. IEEE T Plasma Sci 2002, 30: 800–807.

[20] Woerner E, Wild C, Mueller-Sebert W, et al. CVD-diamond optical lenses. Diam Relat Mater 2001, 10: 557–560.

[21] Klein CA. Diamond windows and domes: Flexural strength and thermal shock. Diam Relat Mater 2002, 11: 218–227.

[22] Piazza F, Morell G. Synthesis of diamond at sub 300 ℃ substrate temperature. Diam Relat Mater 2007, 16: 1950–1957.

[23] Stiegler J, Michler J, Blank E. An investigation of structural defects in diamond films grown at low substrate temperatures. Diam Relat Mater 1999, 8: 651–656.

[24] El Hakiki M, Elmazria O, Bénédic F, et al. Diamond film on Langasite substrate for surface acoustic wave devices operating in high frequency and high temperature. Diam Relat Mater 2007, 16: 966–969.

[25] Titus E, Sikder AK, Paltnikar U, et al. Enhancement of (100) texture in diamond films grown using a temperature gradient. Diam Relat Mater 2002, 11: 1403–1408.

[26] Joe R, Badgwell TA, Hauge RH. Atomic carbon insertion as a low-substrate-temperature growth mechanism in diamond CVD. Diam Relat Mater 1998, 7: 1364–1374.

[27] Piazza F, González JA, Velázquez R, et al. Diamond film synthesis at low temperature. Diam Relat Mater 2006, 15: 109–116.

[28] Sun Z, Shi X, Wang X, et al. Morphological features of diamond films depending on substrate temperatures via a low pressure polymer precursor process in a hot filament reactor. Diam Relat Mater 1998, 7: 939–943.

[29] Potocky S, Kromka A, Potmesil J, et al. Investigation of nanocrystalline diamond films grown on silicon and glass at substrate temperature below 400 ℃. Diam Relat Mater 2007, 16: 744–747.

[30] Kromka A, Potocký Š, Čermák J, et al. Early stage of diamond growth at low temperature. Diam Relat Mater 2008, 17: 1252–1255.

[31] Mallik AK, Binu SR, Satapathy LN, et al. Effect of substrate roughness on growth of diamond by hot filament CVD. Bull Mater Sci 2010, 33: 251–255.

[32] Mallik AK, Shivashankar SA, Biswas SK. High

vacuum tribology of polycrystalline diamond coatings. *Sadhana* 2009, **34**: 811–821.

[33] Aleksandrov VD, Sel'skaya IV. Effect of synthesis conditions on the growth rate and structure of diamond films. *Inorg Mater* 2003, **39**: 455–458.

[34] Ralchenko V, Sychov I, Vlasov I, *et al.* Quality of diamond wafers grown by microwave plasma CVD: Effects of gas flow rate. *Diam Relat Mater* 1999, **8**: 189–193.

[35] Zimmer J, Ravi KV. Aspects of scaling CVD diamond reactors. *Diam Relat Mater* 2006, **15**: 229–233.

[36] Mallik AK, Pal KS, Dandapat N, *et al.* Influence of the microwave plasma CVD reactor parameters on substrate thermal management for growing large area diamond coatings inside a 915 MHz and moderately low power unit. *Diam Relat Mater* 2012, **30**: 53–61.

[37] King D, Yaran MK, Schuelke T, *et al.* Scaling the microwave plasma-assisted chemical vapor diamond deposition process to 150–200 mm substrates. *Diam Relat Mater* 2008, **17**: 520–524.

[38] Shenderova O, Hens S, McGuire G. Seeding slurries based on detonation nanodiamond in DMSO. *Diam Relat Mater* 2010, **19**: 260–267.

[39] Young RA. *The Rietveld Method*. Oxford: Oxford University Press, 1993: 1–70.

[40] Rietveld HM. A profile refinement method for nuclear and magnetic structures. *J Appl Cryst* 1969, **2**: 65–71.

[41] Grotjohn TA, Asmussen J. Microwave plasma-assisted diamond film deposition. In *Diamond Films Handbook*. Asmussen J, Reinhard DK, Eds. New York: Marcel Dekker Inc, 2002: 243–260.

[42] Ma J. Exploration of the gas phase chemistry in microwave activated plasmas used for diamond chemical vapour deposition. Ph.D. thesis. Bristol(UK): University of Bristol, 2008.

[43] Raghavan V. *Materials Science and Engineering: A First Course*, 4th edn. New Delhi: Prentice Hall of India Private Limited, 1998: 44.

[44] Wang W-L, Wang S-M, Cho S-Y, *et al.* Fabrication and structural property of diamond nano-platelet arrays on {111} textured diamond film. *Diam Relat Mater* 2012, **25**: 155–158.

[45] Hausmann BJM, Khan M, Zhang Y, *et al.* Fabrication of diamond nanowires for quantum information processing applications. *Diam Relat Mater* 2010, **19**: 621–629.

[46] Stoikou MD, John P, Wilson JIB. Unusual morphology of CVD diamond surfaces after RIE. *Diam Relat Mater* 2008, **17**: 1164–1168.

[47] Li CY, Hatta A. Preparation of diamond whiskers using Ar/O_2 plasma etching. *Diam Relat Mater* 2005,

14: 1780–1783.

[48] Petherbridge JR, May PW, Baines M, *et al.* Observations of nanotube and 'celery' structures following diamond CVD on single crystal diamond substrates. *Diam Relat Mater* 2003, **12**: 1858–1861.

[49] Ando Y, Nishibayashi Y, Sawabe A. 'Nano-rods' of single crystalline diamond. *Diam Relat Mater* 2004, **13**: 633–637.

[50] Shenderova O, McGuire G. Nanocrystalline diamond. In *Nanomaterials Handbook*. Gogotsi Y, Ed. Boca Raton: Taylor & Francis Group, 2006.

[51] Chen H-G, Chang L. Characterization of diamond nanoplatelets. *Diam Relat Mater* 2004, **13**: 590–594.

[52] Erasmus RM, Comins JD, Mofokeng V, *et al.* Application of Raman spectroscopy to determine stress in polycrystalline diamond tools as a function of tool geometry and temperature. *Diam Relat Mater* 2011, **20**: 907–911.

[53] Osipov AS, Nauyoks S, Zerda TW, *et al.* Rapid sintering of nano-diamond compacts. *Diam Relat Mater* 2009, **18**: 1061–1064.

[54] Wang CX, Yang GW. Thermodynamics of metastable phase nucleation at the nanoscale. *Mat Sci Eng R* 2005, **49**: 157–202.

[55] Mochalin VN, Shenderova O, Ho D, *et al.* The properties and applications of nanodiamonds. *Nat Nanotechnol* 2012, **7**: 11–23.

[56] Sails SR, Gardiner DJ, Bowden M, *et al.* Monitoring the quality of diamond films using Raman spectra excited at 514.5 nm and 633 nm. *Diam Relat Mater* 1996, **5**: 589–591.

[57] Donato MG, Faggio G, Marinelli M, *et al.* A joint macro-/micro- Raman investigation of the diamond lineshape in CVD films: The influence of texturing and stress. *Diam Relat Mater* 2001, **10**: 1535–1543.

[58] Morell G, Quiñones O, Díaz Y, *et al.* Measurement and analysis of diamond Raman bandwidths. *Diam Relat Mater* 1998, **7**: 1029–1032.

[59] Windischmann H, Gray KJ. Stress measurement of CVD diamond films. *Diam Relat Mater* 1995, **4**: 837–842.

[60] Chen KH, Lai YL, Lin JC, *et al.* Micro-Raman for diamond film stress analysis. *Diam Relat Mater* 1995, **4**: 460–463.

[61] Pandey M, D'Cunha R, Tyagi AK. Defects in CVD diamond: Raman and XRD studies. *J Alloys Compd* 2002, **333**: 260–265.

[62] Nibennanoune Z, George D, Antoni F, *et al.* Improving diamond coating on Ti6Al4V substrate using a diamond like carbon interlayer: Raman residual stress evaluation and AFM analyses. *Diam Relat Mater* 2012, **22**: 105–112.

[63] Mallik AK. Hot filament CVD growth of polycrystalline diamond films and its characterization. Master thesis. Bangalore: India Institute of Science, 2003: 53–54.

7

Characterization of free carbon in the as-thermolyzed Si–B–C–N ceramic from a polyorganoborosilazane precursor

Adhimoolam Bakthavachalam KOUSAALYA[a], Ravi KUMAR[a,*],
Shanmugam PACKIRISAMY[b]

[a]*Materials Processing Section, Department of Metallurgical and Materials Engineering, Indian Institute of Technology Madras, Chennai 600036, Tamil Nadu, India*
[b]*Analytical Spectroscopy and Ceramics Group, PCM Entity, Vikram Sarabhai Space Centre, Thiruvananthapuram 695022, India*

Abstract: Polyorganoborosilazane (($B[C_2H_4–Si(CH_3)NH]_3$)$_n$) was synthesized via monomer route from a single-source precursor and thermolyzed at 1300 ℃ in argon atmosphere. The as-thermolyzed Si–B–C–N ceramic was characterized using X-ray diffraction (XRD) and Raman spectroscopy. The crystallization behavior of silicon carbide in the as-thermolyzed amorphous Si–B–C–N matrix was understood by XRD studies, and the crystallite size calculated using Scherrer equation was found to increase from 2 nm to 8 nm with increase in dwelling time. Concomitantly, Raman spectroscopy was used to characterize the free carbon present in the as-thermolyzed ceramic. The peak positions, intensities and full width at half maximum (FWHM) of D and G bands in the Raman spectra were used to study and understand the structural disorder of the free carbon. The G peak shift towards 1600 cm^{-1} indicated the decrease in cluster size of the free carbon. The cluster diameter of the free carbon calculated using TK (Tuinstra and Koenl) equation was found to decrease from 6.2 nm to 5.4 nm with increase in dwelling time, indicating increase in structural disorder.

Keywords: synthesis; polyorganoborosilazane; precursor-derived ceramic; Si–B–C–N; Raman spectroscopy; free carbon

1 Introduction

Processing of ceramics through thermolysis of polymeric precursors can be dated back to the late 1950s when carbon composites and carbon fibers were produced from polyacrylonitrile [1,2]. However, the pioneering work of Verbeek and Winter [3,4] and Yajima *et al.* [5–7] on processing ceramics from polymeric precursors in the middle of 1970s made organo-silicon polymers promising candidate materials for producing Si-based binary, ternary and quaternary ceramics. Since these ceramics are obtained from polymeric precursors, it is possible to utilize the advantages of employing polymer processing routes to produce, for instance, fibers [8–11] and coatings [12–15].

* Corresponding author.
E-mail: nvrk@iitm.ac.in

It is now well established that boron incorporated non-oxide silicon boroncarbonitrides (Si–B–C–N) are observed to be stable at temperatures as high as 1800–2200 ℃ in inert atmosphere, rendering them highly attractive [16–19]. Due to their exceptional thermal stability, coupled with high resistance to both crystallization [20] and oxidation [21] at elevated temperatures, excellent electrical properties [22] and high-temperature mechanical properties [23], precursor-derived Si–B–C–N quaternary ceramics (PDCs) have received growing attention in recent decades. With such immense potential, there always exists a need for producing these materials with high yields, for which single-source materials have been preferred due to their ability to control the composition on an atomic scale and also avoid phase separation [9].

However, presence of free carbon is unavoidable in most of these as-thermolyzed ceramics. The nature of bonding in free carbon, and the amount of free carbon together determine the properties of the material. The incorporated free carbon can exist in either of the two hybridized forms: sp^2 and sp^3. The sp^2 hybridized form is carbon bonded to carbon, while the sp^3 hybridized form is carbon bonded to silicon [24]. The presence of free carbon content and the ratio of sp^2 hybridization to sp^3 hybridization play vital role in determining certain properties of the derived ceramic, such as thermal stability and high-temperature mechanical property [25], resistance to crystallization [26], electrical [27] and optical [28] properties. Activity of the free carbon significantly influences thermal stability of the material, with severe mass loss at increased carbon activity [29].

Trassl et al. [30] has shown the segregation of free carbon in SiCN ceramic thermolyzed from polysilazane at 1200 ℃ by the appearance of D and G bands in Raman spectra, which disappear when thermolyzed above 1600 ℃. The increase and decrease in the intensity of D and G bands indicate the ordering of free carbon at first, followed by disordering, resulting in the formation of SiC with consumption of carbon. However, Mera et al. [31] has shown the disordered state of free carbon in SiCN ceramic processed through thermolysis of phenyl-containing poly(silylcarbodiimides) below 1500 ℃. Thermolysis above 1700 ℃ leads to the ordering of free carbon resulting in increase in the intensity of G band. Sarkar et al. [32] has investigated the presence of free carbon via various techniques in SiBCN ceramic from

polysilazane and concluded that there is no ordered network of carbon. Gao et al. [33] has shown the variation of free carbon content in SiBCN and SiCN ceramics obtained from poly(carbodiimides) as precursor. The density of defects decreases with boron content, resulting in the organization of carbon when thermolyzed at 1400 ℃ in contrast to thermolysis at 1100 ℃. The initial precursor (polysilazane and polycarbodiimides) has a determining role in the nature of bonding exhibited by free carbon in the final microstructure of the ceramic, as exemplified by Colombo et al. [26].

While very limited literatures are available on free carbon in Si–B–C–N processed from boron-modified polysilazane [32,33], a systematic study on the exact nature of free carbon, cluster size, their variation and their bonding characteristics at higher temperatures in Si–B–C–N ceramic is lacking to the best of our knowledge. Hence, the focus of this study is to determine the ratio of sp^2 hybridization to sp^3 hybridization in free carbon present in the as-thermolyzed Si–B–C–N ceramic, characterize the cluster size and its influence on the crystallization behavior, using Raman spectroscopy and X-ray diffraction (XRD).

2 Experimental methods

2. 1 Synthesis procedure

Polyorganoborosilazane was synthesized via monomer route from a single-source precursor using standard Schlenk technique in argon (Ar) atmosphere by a two-step synthesis process [16,34]. Initially, monomers containing boron were produced by hydroboration of chlorosilane, followed by ammonolysis of monomers, resulting in condensation polymerization of the reactants and subsequent formation of polyorganoborosilazane.

84.6 g of dichloromethylvinylsilane ((CH_2=CH) CH_3SiCl_2, Alfa Aesar, USA) was dissolved in 90 ml of toluene and hydroborated at 0 ℃ under vigorous stirring for 30 min through drop-wise addition in 100 ml of 2M solution of boranedimethylsulfide (($CH_3)_2SBH_3$) in toluene (Spectrochem, India) in argon atmosphere. The stirring was continued for 16 h at room temperature (25 ℃), and subsequent evaporation of the solvent was observed along with the evaporation of the byproduct (dimethylsulfide (SMe_2)) at 60 ℃

under reduced pressure. The reaction yielded 95 g of the monomer, namely, tris(dichloromethylsilylethyl) borane ($B[C_2H_4Si(CH_3)Cl_2]_3$) in the form of a colorless, oily liquid.

The ammonolysis of the monomer was carried out subsequently for 30 min by drop-wise addition of 15 ml of ammonia to a mixture containing 60 g of monomer dissolved in 90 ml of tetrahydrofuran. The temperature was slowly increased to the boiling point of the monomer (80 ℃) under vigorous stirring. The precipitated ammonium chloride (NH_4Cl) was filtered using Whatman glass micro-fiber filter paper and the filtrate was heated at 80 ℃ under reduced pressure. This yielded 58 g of white solid lumps of polyorganoborosilazane (($B[C_2H_4–Si(CH_3)NH]_3)_n$), which is sensitive to air and soluble in both toluene and tetrahydrafuran. Any increase in temperature beyond 85 ℃ during ammonolysis reaction resulted in degradation of the monomer, as was clearly observed in the form of change in color from colorless to light brown.

After flushing the furnace initially with argon, the as-synthesized polymer was taken in small quantities in an alumina crucible and thermolyzed at 1300 ℃ with a heating rate of 5 ℃/min and dwelling time varying from 0 to 10 h at intervals of 2 h in argon atmosphere, for transforming the polymeric precursor into a ceramic. All the samples were allowed to furnace cool to room temperature at the end of their respective heating cycles. The ceramic yield obtained was ~50% in all cases.

2. 2 Characterization

The obtained ceramics were pulverized using agate mortar and pestle and characterized using XRD and Raman spectroscopy.

The X-ray diffractograms were obtained within the 2θ scan range of 10° to 90° with a step size of 0.05° using X'Pert PRO diffractometer, PANalytical (the Netherlands) with Cu Kα radiation of $\lambda = 0.154\,06$ nm, voltage of 45 kV and current of 30 mA. The background noise was subtracted after the removal of the Kα$_2$ peaks and the peak at $2\theta = 36°$ was chosen for crystal size measurement.

The Raman spectra of the as-thermolyzed ceramics were obtained in the scan range of 100 cm^{-1} to 3000 cm^{-1} from Labram HR800 UV-Raman Spectrometer equipped with charge coupled detector, Horiba Jobin Yvon (Japan) using a He–Ne laser source

with an excitation wave length of $\lambda = 632.8$ nm. An Olympus BX-41 microscope with a 100× magnification was used to image the samples. The spectrometer was calibrated using silicon standard prior to Raman measurement.

3 Results and discussion

The X-ray diffractograms of the ceramic samples, thermolyzed at 1300 ℃ for various dwelling time, are shown in Fig 1. The hump at ~36° in the X-ray diffractogram of the ceramic sample thermolyzed at 1300 ℃ with no dwelling time (0 h) shows silicon carbide (SiC) as the first phase to nucleate from the amorphous Si–B–C–N quaternary ceramic. The nucleated crystals seem to grow with increase in dwelling time from 0 to 10 h, as can be seen in the predominant peaks in XRD of ceramic samples at 35.9°, 60.06° and 72.2°. The peaks are observed to be quite broad, thereby indicating the nano-crystalline nature of SiC. Cai et al. [20] has shown that nucleation of nano-crystallites in as-thermolyzed Si–B–C–N processed from boron-modified polysilazane, starts only at 1350 ℃ and remains completely amorphous below 1350 ℃. However, Gao et al. [33] has shown that the crystallization behavior of powder sample is quite different from that of bulk sample. The powder sample is found to crystallize at lower temperature and faster rate which is due to the sensitivity of the solid state of PDC to the molecular structure of the precursor. The peak at $2\theta = 36°$ was subjected to Pseudo-Voigt profile function fit to determine the full width at half

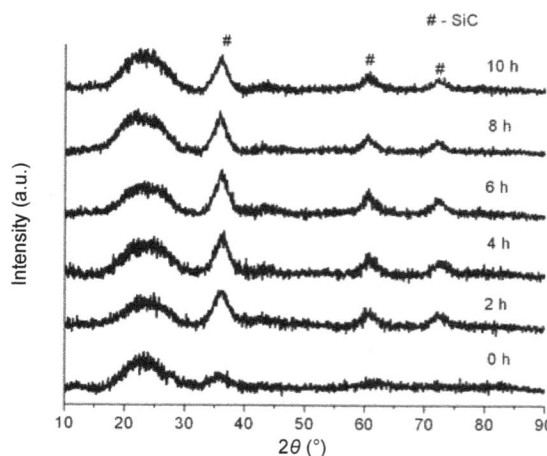

Fig. 1   X-ray diffractograms of the as-thermolyzed Si–B–C–N ceramic at 1300 ℃ for various dwelling time (heating rate of 5 ℃/min).

maximum (FWHM) and the crystallite diameter d was determined using the Scherrer formula (Eq. (1)):

$$d = \frac{K\lambda}{B\cos\theta} \qquad (1)$$

where B is the FWHM; θ is the position of the peak; λ is the wavelength.

The crystallite size is found to be in the range of 2–8 nm and increases linearly with respect to dwelling time as shown in Fig. 2. The amorphous hump found from ~18° to ~28° in all the samples indicates the semi-crystalline nature of the ceramic. Hence, the percentage of crystallinity of the Si–B–C–N ceramic was calculated using Eq. (2):

$$\text{Crystallinty}(\%) = \frac{\sum I_{\text{peaks of SiC}}}{\sum I_{\text{all peaks}}} \qquad (2)$$

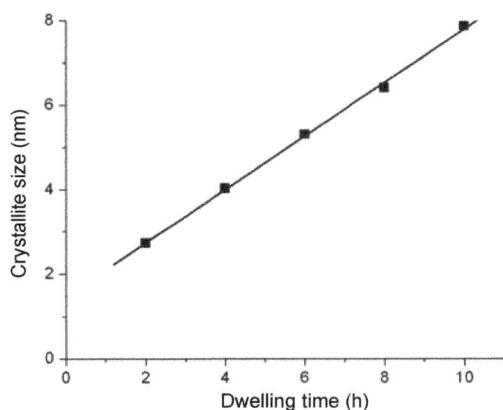

Fig. 2   Crystallite size of SiC of the as-thermolyzed Si–B–C–N ceramic at 1300 ℃ for various dwelling time (heating rate of 5 ℃/min) showing a linear fit.

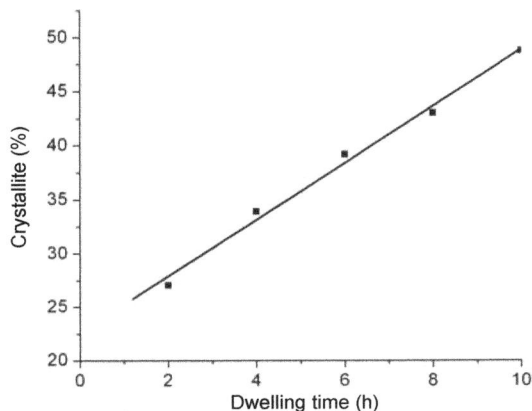

Fig. 3   Percentage of crystallinity in the as-thermolyzed Si–B–C–N ceramic at 1300 ℃ for various dwelling time (heating rate of 5 ℃/min) showing a linear fit.

The percentage of crystallinity is found to increase linearly with increase in dwelling time as shown in Fig. 3. The ceramic sample thermolyzed for 10 h is found to be ~50% crystallized, an increase when compared to ~25% crystallization found for the ceramic sample thermolyzed for 2 h. The increase in both crystal size and crystallinity of SiC indicate the occurrence of reaction between silicon and carbon present in the material, leading to the consumption of carbon. Hence, to determine the nature of the free carbon and its variation with crystallization, Raman spectroscopy was carried out.

Raman spectra of the as-thermolyzed Si–B–C–N ceramic at different dwelling time are shown in Fig. 4. The Raman spectra show two strong peaks at ~1325 cm^{-1} and ~1595 cm^{-1} for all samples, indicating the presence of free carbon. These two peaks are generally denoted as disordered peak (D peak) and graphitic peak (G peak), respectively. The G peak indicates the presence of graphite-like structure of carbon and corresponds to the in-plane bond stretching mode E_{2g} of the sp^2 C–C bond. This is found in pure graphite with no sp^3 hybridized carbon at 1575 cm^{-1} as a single peak, while in diamond with 100% sp^3 hybridized carbon, a single peak at 1332 cm^{-1} is found. Presence of sp^3 bonds (diamond-like arrangement) along with sp^2 bonds in carbon results in a second peak at 1355 cm^{-1} called as D peak along with the G peak at 1575 cm^{-1} [35]. This D peak is due to Raman scattering of the sp^3 hybridized form of carbon bonded to silicon present in the amorphous matrix. This D peak corresponds to the A_{1g} breathing mode of the aromatic ring. When structural disorder is present, the

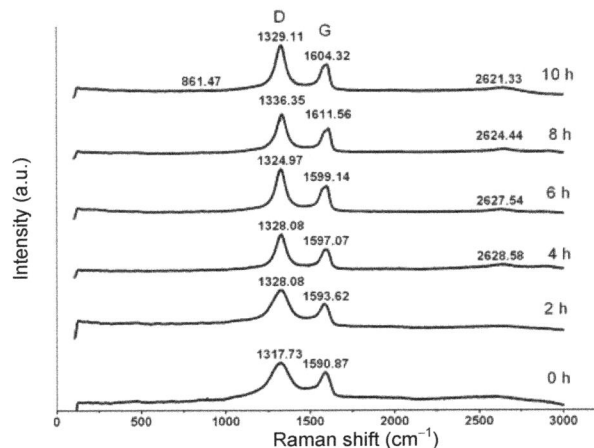

Fig. 4   Raman spectra of the as-thermolyzed Si–B–C–N ceramic at 1300 ℃ for various dwelling time (heating rate of 5 ℃/min).

D peak will shift towards the lower wave number from 1355 cm^{-1} with an increase in intensity and narrowing of the peak. The position of D peak is completely dependent on the excitation energy used which varies from ~1300 cm^{-1} to ~1400 cm^{-1} [36,37].

In general, sp^2 hybridized carbon has a tendency to cluster, and these clusters and their degree of clustering play crucial roles in determining the material properties [38–41]. This cluster diameter (L_a) is inversely proportional to the relative intensity of the D and G peaks and can be determined by the TK (Tuinstra and Koenl) equation [35]:

$$\frac{I_D}{I_G} = \frac{C(\lambda)}{L_a} \qquad (3)$$

where I_D is the peak intensity of D band in Raman spectra; I_G is the peak intensity of G band in Raman spectra; and $C(\lambda)$ is the wavelength-dependent constant [42]:

$$C(\lambda) = C_0 + \lambda_L C_1 \qquad (4)$$

where $C_0 = -12.6$ nm; $C_1 = 0.033$; λ_L is the wavelength used (632.8 nm).

Based on the aforementioned equation, L_a was calculated and observed to decrease from 6.2 nm to 5.4 nm as shown in Fig. 5 with increase in dwelling time. As the cluster diameter decreases, the sp^2 bond in carbon cluster tends to open, indicating the shift from sp^2 hybridization to sp^3 hybridization. This leads to transformation of graphite to nano-crystalline graphite, followed by amorphous carbon and finally forming tetrahedral amorphous carbon, also known as DLC (diamond-like carbon) which is termed *amorphization trajectory* in carbon materials [40,41].

With decrease in cluster size, the position of G peak shifts towards higher wave number, as shown in Fig. 6. The G band is observed to shift from 1590 cm^{-1} to 1611 cm^{-1} with increase in dwelling time from 0 to 8 h indicating the transformation of graphite to nano-crystalline graphite, resulting in disorder of the free carbon. A decrease in shift of G band towards 1600 cm^{-1} for sample thermolyzed at dwelling time of 10 h indicates the transformation of nano-crystalline graphite to amorphous carbon. This shows the increasing disorder in the structure of the free carbon with an increase in the contribution of sp^3 bond in the as-thermolyzed PDC. Unlike the D peak, the position of the G peak does not vary with the excitation energy used but is dependent on the cluster size [35]. The transformation of graphite to nano-crystalline graphite and finally to amorphous carbon indicates the opening up of the sp^2 carbon cluster, resulting in free sites which tend to react with the silicon present in the material, leading to the formation of SiC. This is clearly evident from the XRD data showing the increase in the crystallinity of SiC with increase in dwelling time.

From Fig. 5 it is clear that, the slope is steeper for dwelling time varying from 0 to 4 h, which indicates that the sp^2 hybridized bonds open up at a faster rate. However, for higher dwelling time (beyond 4 h), the relatively shallow slope indicates that the rate of opening up of bonds reduces considerably. Also, Fig. 2 shows that the crystal size of SiC is observed to increase linearly with increase in dwelling time, irrespective of the rate of opening up of sp^2 bonds. This is explained by the fact that at lower dwelling

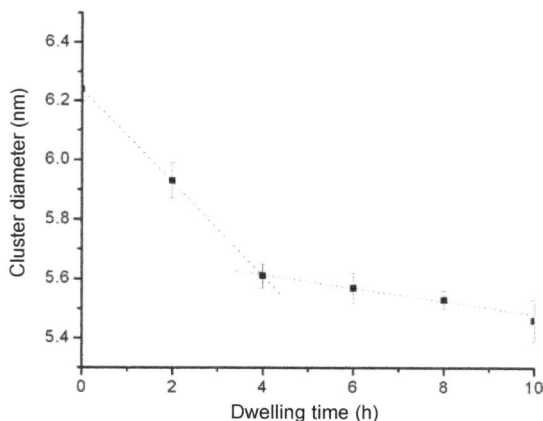

Fig. 5 Cluster size of carbon of the as-thermolyzed Si–B–C–N ceramic at 1300 ℃ for various dwelling time (heating rate of 5 ℃/min) showing change in rate of opening up of bonds in free carbon cluster at 4 h.

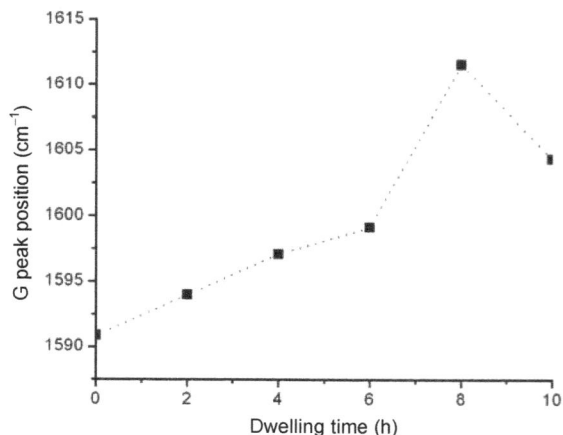

Fig. 6 Shift in the position of G peak for various dwelling time showing the transformation of graphite to nano-crystalline graphite and finally to amorphous carbon.

time (0–4 h), the predominant mechanism of nucleation of SiC crystal is the consumption of the free carbon, while for higher dwelling time varying from 6 h to 10 h, the growth is governed by Ostwald ripening. A low intensity peak can be observed at 861.47 cm^{-1} for the sample thermolyzed at 10 h, corresponding to SiC (Fig. 4). The low intensity is due to the fact that compared to the free carbon, optical absorption of SiC is lower by a factor of 10 [43]. Also, along with the D and G peaks, additional peak is obtained at ~2628 cm^{-1} for all samples and this corresponds to the presence of G′ band, better known as the second-order spectra of crystalline graphite [44].

The FWHM of D and G peaks are calculated by fitting the peaks using Lorentzian fit and the values are listed in Table 1. The FWHM values for both peaks are found to decrease with an increase in disorder of carbon present. In general, FWHM of G peak (FWHM$_G$) is directly related to the structural disorder

present, since it arises due to the distortions in bond length and bond angle. Hence, it tends towards zero if clusters formed are defect-free and unstrained like pure graphite. But, on the contrary, a decrease in FWHM$_G$ is observed with an increase in disorder, whereas relative intensity (I_D/I_G ratio) increases as exemplified in Fig. 7, which is in accordance with the published literature on free carbon in polymer derived ceramics [30,45]. The decrease in FWHM$_G$ with increase in the position of G peak when graphene is doped with a secondary element that can act as electron or hole dopant as shown by Casiraghi et al. [46] further corroborates the aforementioned argument. Hence, the anomalous decrease in FWHM$_G$ with increase in structural disorder observed in the as-thermolyzed ceramic is possibly attributed to the presence of elements like silicon, nitrogen and boron, bonded to each other in various forms [47] within the free carbon phase.

Table 1 Peak positions, FWHM and intensities of D and G bands in the Raman spectra and ratio of their intensities and cluster diameter of carbon

Dwelling time (h)	Position of D peak (cm^{-1})	FWHM of D peak (cm^{-1})	Position of G peak (cm^{-1})	FWHM of G peak (cm^{-1})	I_D (a.u.)	I_G (a.u.)	I_D/I_G	L_a (nm)
0	1321.79	144.94	1590.87	77.08	11 142.39	8364.04	1.33	6.24
2	1324.00	116.90	1593.97	70.05	13 604.00	9684.71	1.40	5.93
4	1327.16	75.11	1597.07	67.22	6 187.14	4158.02	1.48	5.61
6	1325.89	73.09	1599.14	63.61	13 977.10	9368.38	1.49	5.57
8	1332.83	71.63	1611.56	55.29	12 241.79	8138.84	1.50	5.53
10	1327.40	70.81	1604.32	54.57	8 822.94	5776.20	1.52	5.46

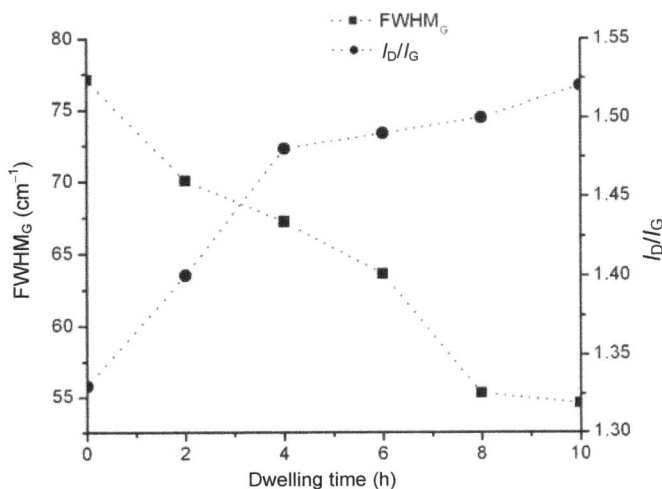

Fig. 7 Anomalous relation between FWHM$_G$ and I_D/I_G as functions of dwelling time.

4 Conclusions

Polyorganoborosilazane ((B[C$_2$H$_4$–Si(CH$_3$)NH]$_3$)$_n$) synthesized from a single-source precursor by monomer route was thermolyzed into Si–B–C–N ceramic and characterized to detect the presence of free carbon and formation of any crystalline phase. The presence of nano-crystalline SiC along with free carbon was observed in the as-thermolyzed ceramic. The cluster size of free carbon was calculated from Raman spectra and found to decrease with increase in dwelling time, thus indicating an increase in the extent of disorder in free carbon phase. XRD studies of the as-thermolyzed Si–B–C–N ceramic revealed nucleation of SiC crystal as the first phase along with increase in crystal size and crystallinity with increase in dwelling time. In this work, the role of free carbon on the crystallization behavior, the bonding character of the free carbon present and the change in the ratio of sp^2 hybridization to sp^3 hybridization of free carbon were studied.

Acknowledgements

We gratefully acknowledge the financial support from the Vikram Sarabhai Space Centre, Thiruvananthapuram through ISRO-IITM cell (Project No. ICSR/ISRO-IITM/MET/08-09/122/RAVK). We also would like to acknowledge the Department of Physics, IIT-M for performing Raman spectroscopy.

References

[1] Houtz RC. "Orlon" acrylic fiber: Chemistry and properties. *Text Res J* 1950, **20**: 786–801.

[2] Ezekiel HM, Spain RG. Preparation of graphite fibers from polymeric fibers. *J Polym Sci Pol Sym* 1967, **19**: 249–265.

[3] Verbeek W. Production of shaped articles of homogeneous mixtures of silicon carbide and nitride. U.S. Patent 3853567 A, Nov. 1973.

[4] Verbeek W, Winter G. Formkoerper aus siliciumcarbid und verfahren zu ihrer herstellung. DE Patent 2236078 A1, July 1974.

[5] Yajima S, Hayashi J, Imori M. Continuous silicon carbide fiber of high tensile strength. *Chem Lett* 1975, **4**: 931–934.

[6] Hayashi J, Omori M, Yajima S. Siliciumcarbidfasern und verfahren zur herstellung derselben. DE Patent 2618150 A1, April 1976.

[7] Yajima S, Hayashi J, Omori M, *et al.* Development of a silicon carbide fiber with high tensile strength. *Nature* 1976, **261**: 683–685.

[8] Bernard S, Weinmann M, Cornu D, *et al.* Preparation of high-temperature stable Si–B–C–N fibers from tailored single source polyborosilazanes. *J Eur Ceram Soc* 2005, **25**: 251–256.

[9] Bernard S, Weinmann M, Gerstel P, *et al.* Boron-modified polysilazane as a novel single-source precursor for SiBCN ceramic fibers: Synthesis, melt-spinning, curing and ceramic conversion. *J Mater Chem* 2005, **15**: 289–299.

[10] Miele P, Bernard S, Cornu D, *et al.* Recent developments in polymer-derived ceramic fibers (PDCfs): Preparation, properties and applications—A review. *Soft Mater* 2007, **4**: 249–286.

[11] Sarkar S, Zhai L. Polymer-derived non-oxide ceramic fibers—Past, present and future. *Mater Express* 2011, **1**: 18–29.

[12] Mucalo MR, Milestone NB, Vickridge IC, *et al.* Preparation of ceramic coatings from pre-ceramic precursors. *J Mater Sci* 1994, **29**: 4487–4499.

[13] Hauser R, Borchard SN, Riedel R, *et al.* Polymer-derived SiBCN ceramic and their potential application for high temperature membranes. *J Ceram Soc Jpn* 2006, **114**: 524–528.

[14] Torrey JD, Bordia RK. Processing of polymer-derived ceramic composite coatings on steel. *J Am Ceram Soc* 2008, **91**: 41–45.

[15] Schütz A, Günthner M, Motz G, *et al.* Characterisation of novel precursor-derived ceramic coatings with glass filler particles on steel substrates. *Surf Coat Technol* 2012, **207**: 319–327.

[16] Riedel R, Kienzle A, Dressler W, *et al.* A silicoboron carbonitride ceramic stable to 2,000 ℃. *Nature* 1996, **382**: 796–798.

[17] Müller A, Gerstel P, Weinmann M, *et al.* Correlation of boron content and high temperature stability in Si–B–C–N ceramics. *J Eur Ceram Soc* 2000, **20**: 2655–2659.

[18] Müller A, Gerstel P, Weinmann M, *et al.* Correlation of boron content and high temperature stability in Si–B–C–N ceramics II. *J Eur Ceram Soc* 2001, **21**: 2171–2177.

[19] Wang Z-C, Aldinger F, Riedel R. Novel silicon–boron–carbon–nitrogen materials thermally stable up to 2200 ℃. *J Am Ceram Soc* 2001, **84**: 2179–2183.

[20] Cai Y, Zimmermann A, Prinz S, *et al.* Nucleation phenomena of nano-crystallites in as-pyrolysed Si–B–C–N ceramics. *Scripta Mater* 2001, **45**: 1301–1306.

[21] Butchereit E, Nickel KG, Müller A. Precursor-derived Si–B–C–N ceramics: Oxidation kinetics. *J Am Ceram Soc* 2001, **84**: 2184–2188.

[22] Ramakrishnan PA, Wang YT, Balzar D, *et al.* Silicoboron–carbonitride ceramics: A class of high-temperature, dopable electronic materials. *Appl Phys Lett* 2001, **78**: 3076.

[23] Ravi Kumar NV, Mager R, Cai Y, *et al.* High temperature deformation behavior of crystallized Si–B–C–N ceramics obtained from a boron modified poly(vinyl)silazane polymeric precursor. *Scripta Mater* 2004, **51**: 65–69.

[24] Saha A, Raj R, Williamson DL. A model for the nanodomains in polymer-derived SiCO. *J Am Ceram*

Soc 2006, **89**: 2188–2195.

[25] Kumar R, Mager R, Phillipp F, *et al*. High-temperature deformation behavior of nanocrystalline precursor-derived Si–B–C–N ceramics in controlled atmosphere. *Int J Mater Res* 2006, **97**: 626–631.

[26] Colombo P, Mera G, Riedel R, *et al*. Polymer-derived ceramics: 40 years of research and innovation in advanced ceramics. *J Am Ceram Soc* 2010, **93**: 1805–1837.

[27] Hermann AM, Wang Y-T, Ramakrishnan PA, *et al*. Structure and electronic transport properties of Si–(B)–C–N ceramics. *J Am Ceram Soc* 2001, **84**: 2260–2264.

[28] Wang Y, Wang K, Zhang L, *et al*. Structure and optical property of polymer-derived amorphous silicon oxycarbides obtained at different temperatures. *J Am Ceram Soc* 2011, **94**: 3359–3363.

[29] Peng J. Thermochemistry and constitution of precursor-derived Si–(B–)C–N ceramics. Ph.D. Thesis. Stuttgart (Germany): Universität Stuttgart, 2002.

[30] Trassl S, Motz G, Rössler E, *et al*. Characterization of the free-carbon phase in precursor-derived SiCN ceramics: I, spectroscopic methods. *J Am Ceram Soc* 2002, **85**: 239–244.

[31] Mera G, Riedel R, Poli F, *et al*. Carbon-rich SiCN ceramics derived from phenyl-containing poly(silylcarbodiimides). *J Eur Ceram Soc* 2009, **29**: 2873–2883.

[32] Sarkar S, Gan Z, An L, *et al*. Structural evolution of polymer-derived amorphous SiBCN ceramics at high temperature. *J Phys Chem C* 2011, **115**: 24993–25000.

[33] Gao Y, Mera G, Nguyen H, *et al*. Processing route dramatically influencing the nanostructure of carbon-rich SiCN And SiBCN polymer derived ceramics. Part I: Low temperature thermal transformation. *J Eur Ceram Soc* 2012, **32**: 1857–1866.

[34] Kumar R, Cai Y, Gerstel P, *et al*. Processing, crystallization and characterization of polymer derived nano-crystalline Si–B–C–N ceramics. *J Mater Sci* 2006, **41**: 7088–7095.

[35] Tunistra F, Koenig JL. Raman spectrum of graphite.

J Chem Phys 1970, **53**: 1126–1130.

[36] Pócsik I, Hundhausen M, Koós M, *et al*. Origin of the D peak in the Raman spectrum of microcrystalline graphite. *J Non-Cryst Solids* 1998, **227–230**: 1083–1086.

[37] Ferrari AC. Raman spectroscopy of graphene and graphite: Disorder, electron–phonon coupling, doping and nonadiabatic effects. *Solid State Commun* 2007, **143**: 47–57.

[38] Jiang T, Wang Y, Wang Y, *et al*. Quantitative Raman analysis of free carbon in polymer-derived ceramics. *J Am Ceram Soc* 2009, **92**: 2455–2458.

[39] Robertson J. Hard amorphous (diamond-like) carbons. *Prog Solid State Ch* 1991, **21**: 199–333.

[40] Ferrari AC, Robertson J. Interpretation of Raman spectra of disordered and amorphous carbon. *Phys Rev B* 2000, **61**: 14095–14107.

[41] Ferrari AC, Robertson J. Raman spectroscopy of amorphous, nanostructured, diamond-like carbon, and nanodiamond. *Phil Trans R Soc Lond A* 2004, **362**: 2477–2512.

[42] Matthews MJ, Pimenta MA, Dresselhaus G, *et al*. Origin of dispersive effects of the Raman D band in carbon materials. *Phys Rev B* 1999, **59**: R6585–R6588.

[43] Ma Y, Wang S, Chen Z. Raman spectroscopy studies of the high-temperature evolution of the free carbon phase in polycarbosilane derived SiC ceramics. *Ceram Int* 2010, **36**: 2455–2459.

[44] Pimenta MA, Dresselhaus G, Dresselhaus MS, *et al*. Studying disorder in graphite-based systems by Raman spectroscopy. *Phys Chem Chem Phys* 2007, **9**: 1276–1290.

[45] Trassl S, Motz G, Rössler E, *et al*. Characterisation of the free-carbon phase in precursor-derived SiCN ceramics. *J Non-Cryst Solids* 2001, **293–295**: 261–267.

[46] Casiraghi C, Pisana S, Novoselov KS, *et al*. Raman fingerprint of charged impurities in grapheme. *Appl Phys Lett* 2007, **91**: 233108.

[47] Ravi Kumar NV, Prinz S, Cai Y, *et al*. Crystallization and creep behavior of Si–B–C–N ceramics. *Acta Mater* 2005, **53**: 4567–4578.

Theoretical investigations on electrocaloric properties of (111)-oriented PbMg$_{1/3}$Nb$_{2/3}$O$_3$ single crystal

Mahmoud Aly HAMAD[*]

Physics Department, Faculty of Science, Tanta University, Tanta, Egypt

Abstract: The electrocaloric (EC) effect accompanied with the ferroelectric to paraelectric phase transition in (111)-oriented PbMg$_{1/3}$Nb$_{2/3}$O$_3$ (PMN) is investigated. It is shown that the largest change ΔT is 0.37 K in 3 kV/cm electric field shift near the Curie temperature of 221 K; that is, the cooling ΔT per unit field (MV/m) is 1.23×10^{-6} m·K/V. This value is significantly larger, and comparable with the value of 0.254×10^{-6} m·K/V for PbZr$_{0.95}$Ti$_{0.05}$O$_3$ thin film under larger electric field shift $\Delta E = 30$ kV/cm. Thus, the EC effect of (111) PMN single crystal provides cooling solutions at low temperatures, and opens more opportunities for practical application in cooling systems.

Keywords: electrocaloric (EC) effect; PbMg$_{1/3}$Nb$_{2/3}$O$_3$ (PMN); model; polarization; entropy change; heat capacity change

1 Introduction

The electrocaloric (EC) effect or magnetocaloric effect provides an efficient approach to realize solid-state cooling devices instead of the existing vapor compression refrigeration [1–16]. Ferroelectric refrigeration is based on the EC effect, an entropy change of ferroelectric material during application or withdrawal of electric field.

PbMg$_{1/3}$Nb$_{2/3}$O$_3$ (PMN) ferroelectric materials with compositions near the morphotropic phase boundary are promising for high-strain actuators/transducers and prototype microelectro-mechanical systems due to their well-known ferroelectric, dielectric and piezoelectric properties [17–23]. In this paper, the dependence of polarization on variation of temperature for (111)-oriented PMN at low electric field is simulated to predict electrocaloric properties under applied electric field shift ΔE (defined as $E_2 - E_1$).

2 Theoretical considerations

According to the phenomenological model [24], the dependence of polarization on variation of temperature and Curie temperature T_C is presented by

$$P = \left(\frac{P_i - P_f}{2}\right)\{\tanh[A(T_C - T)]\} + BT + C \quad (1)$$

where P_i is the initial value of polarization at ferroelectric–paraelectric transition and P_f is the final value of polarization at ferroelectric–paraelectric transition as shown in Fig. 1.

$$A = \frac{2\left(B - \dfrac{\mathrm{d}P}{\mathrm{d}T}\Big|_{T=T_C}\right)}{P_i - P_f}$$

---
* Corresponding author.

E-mail: m_hamad76@yahoo.com

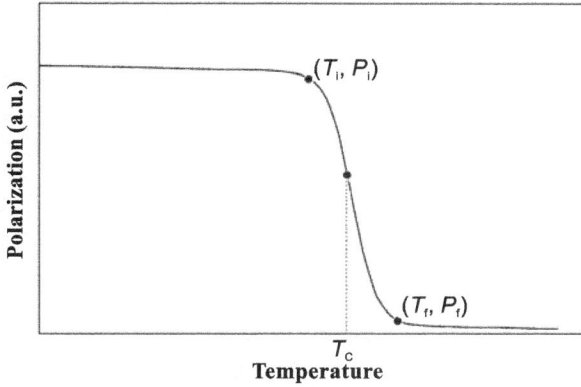

Fig. 1   Dependence of polarization as a function of temperature described by Eq. (1).

B is the polarization sensitivity $\dfrac{dP}{dT}$ at ferroelectric state before transition; $\dfrac{dP}{dT}\bigg|_{T=T_C}$ is the polarization sensitivity $\dfrac{dP}{dT}$ at Curie temperature T_C; and $C = \dfrac{P_f + P_i}{2} - BT_C$.

Equation (1) is determined by the physical mechanism that the dipole-ordered state can be enhanced by decreasing temperature. At the transition temperature, the spontaneous polarization forms surface charges and stray charges accumulate on the surface of ferroelectric material. When there is non-homogeneous distribution of the spontaneous polarization, the surface charges produce an electric field, denoted as depolarization field which is in the opposite direction to the spontaneous polarization.

As a result of this phenomenological model, the electrocaloric entropy change ΔS^E caused by variation of the external electric field from E_1 to E_2 is given by

$$\Delta S^E = \left\{ -A\left(\frac{P_i - P_f}{2}\right) \text{sech}^2[A(T_C - T)] + B \right\} \frac{\Delta E}{\rho} \quad (2)$$

where ρ is the mass density.

A main result of Eq. (2) is that the maximum entropy change ΔS^E_{max} (where $T = T_C$) can be expressed as the following expression:

$$\Delta S^E_{max} = \left[-A\left(\frac{P_i - P_f}{2}\right) + B \right] \Delta E / \rho \quad (3)$$

According to the phenomenological model [24], the full-width at half-maximum (FWHM) δT_{FWHM} can be

calculated as

$$\delta T_{FWHM} = \frac{2}{A} \cosh^{-1}\left[\sqrt{\frac{2A(P_i - P_f)}{A(P_i - P_f) + 2B}} \right] \quad (4)$$

The electrocaloric cooling efficiency is calculated by considering the magnitude of maximum electrocaloric entropy change, $-\Delta S^E_{max}$, and its FWHM (δT_{FWHM}) [25]. A product of $-\Delta S^E_{max}$ and δT_{FWHM} is called the relative cooling power (RCP) based on electrocaloric entropy change:

$$\text{RCP} = -\Delta S^E_{max}(T, \Delta E) \times \delta T_{FWHM}$$
$$= \left(P_i - P_f - 2\frac{B}{A} \right) \frac{\Delta E}{\rho} \times \cosh^{-1}\left[\sqrt{\frac{2A(P_i - P_f)}{A(P_i - P_f) + 2B}} \right] \quad (5)$$

The polarization-related change of heat capacity is given by

$$\Delta C_{P,E} = T \frac{\delta \Delta S^E}{\delta T} \quad (6)$$

According to this phenomenological model, the change of heat capacity is given by

$$\Delta C_{P,E} = -2TA^2 \left(\frac{P_i - P_f}{2}\right) \text{sech}^2[A(T_C - T)]$$
$$\times \tanh[A(T_C - T)] \Delta E / \rho \quad (7)$$

The temperature change of a polar system under adiabatic electric field variation from an initial value E_1 to a final value E_2 can be written in the form:

$$\Delta T = -\frac{T}{C_E \rho} \int_{E_1}^{E_2} \left(\frac{\partial P}{\partial T}\right)_E dE$$
$$= \frac{AT(P_i - P_f)}{2C_E \rho} \{\text{sech}^2[A(T_C - T)] + B\} \Delta E \quad (8)$$

C_E is the heat capacity per mole at constant electric field.

The refrigerant capacity (RC) is the amount of heat that can be transferred in one thermodynamic cycle [26]. Here, RC value can be obtained as [24]

$$\text{RC} = \int_{T_C - \frac{\delta T_{FWHM}}{2}}^{T_C + \frac{\delta T_{FWHM}}{2}} \Delta S^E dT$$
$$= \left[-(P_i - P_f)\tanh\left(A\frac{\delta T_{FWHM}}{2} \right) + B\delta T_{FWHM} \right] \frac{\Delta E}{\rho} \quad (9)$$

From this phenomenological model, it can easily calculate the values of δT_{FWHM}, $|\Delta S^E|_{max}$, $|\Delta T|_{max}$, RCP, RC and ΔC_{min} due to applied electric field shift ΔE.

3 Simulation

In order to apply the phenomenological model, five parameters versus applied electric field were determined as displayed in Table 1. The heat capacity C_P is 320 J/(kg·K) and the mass density ρ is 8.3 g/cm^3 [20]. Figure 2 shows polarization versus temperature for (111) PMN under different electric fields. The symbols represent the experimental data from Ref. [27] and the dashed lines represent the modelled data. It shows a good agreement between the modelled results and experimental data. Figures 3–5 illustrate the predicted entropy changes, heat capacity changes and temperature changes versus temperature due to different electric field shifts calculated from Eqs. (2), (7) and (8), respectively.

Table 2 shows the calculated values of δT_{FWHM}, $|\Delta S^E|_{\max}$, $|\Delta T|_{\max}$, RCP, RC, $\Delta C_{\min}$ and $\Delta C_{\max}$ for (111) PMN due to applied electric field shift ΔE. It is clear that as ΔE increases, the values of $|\Delta T|_{\max}$, $|\Delta S^E|_{\max}$, RCP, RC, $-\Delta C_{\min}$ and $\Delta C_{\max}$ increase.

Under low applied electric field shift $\Delta E = 3$ kV/cm, the peak value of ΔS of (111) PMN is 0.54 J/(kg·K). This indicates that (111) PMN is significant, and comparable with value of the best electrocaloric material (PbZr$_{0.95}$Ti$_{0.05}$O$_3$ thin film) which has peak values of $\Delta S = 0.6$ J/(kg·K) under larger electric field shift $\Delta E = 30$ kV/cm [1].

Table 1   Model parameters for (111) PMN single crystal due to applied electric field

E	T_C	P_i	P_f	B	$\left.\dfrac{dP}{dT}\right|_{T=T_C}$
(kV/cm)	(K)	(C/m^2)	(C/m^2)	(C/(m^2·K))	(C/(m^2·K))
1.5	230	0.128	0.009	−0.000 10	−0.001
2.0	199	0.276	0.088	−0.000 60	−0.015
2.5	209	0.277	0.083	−0.000 55	−0.015
3.0	221	0.280	0.085	−0.000 60	−0.015

Table 2   Calculated values of electrocaloric properties of (111) PMN single crystal at different electric field shifts

ΔE	δT_{FWHM}	$|\Delta S^E|_{\max}$	$|\Delta T|_{\max}$	RCP	RC	$\Delta C_{\min}$	$\Delta C_{\max}$
(kV/cm)	(K)	(J/(kg·K))	(K)	(J/kg)	(J/kg)	(J/(kg·K))	(J/(kg·K))
1.5	127.25	0.02	0.01	2.29	1.83	−0.04	0.05
2.0	11.90	0.36	0.22	4.29	3.44	−7.65	8.29
2.5	12.20	0.45	0.29	5.50	4.41	−10.19	10.23
3.0	12.34	0.54	0.37	6.68	5.35	−12.31	13.32

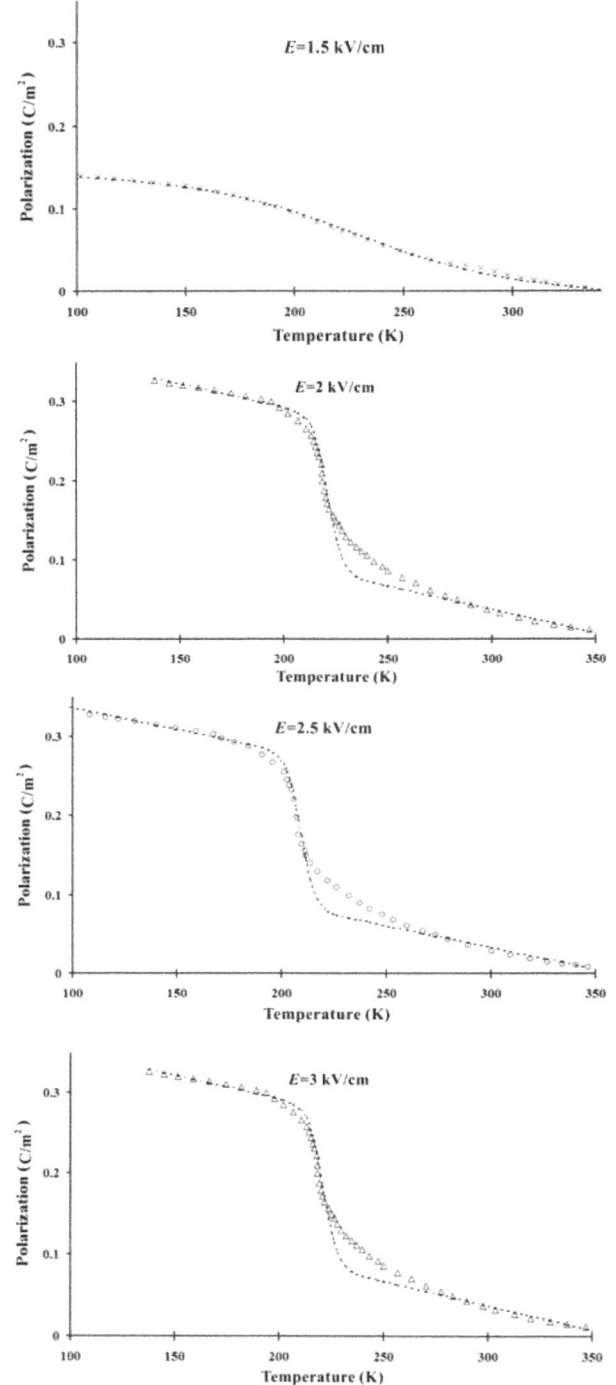

Fig. 2   Polarization variation induced by temperature change under different electric fields for (111) PMN single crystal. The dashed lines are modeled results by Eq. (1) and symbols represent experimental data from Ref. [27].

It is clear that as the temperature is far away from the Curie temperature, the entropy change with increasing temperature is not very noticeable, so the

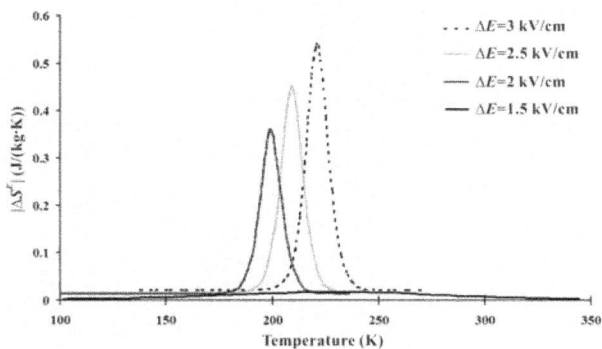

Fig. 3 Absolute values of electrocaloric entropy changes versus temperature due to applied electric field shifts for (111) PMN single crystal. They are obtained by Eq. (2) at different electric field shifts.

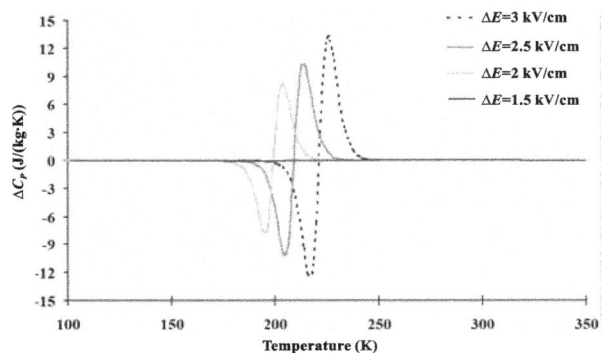

Fig. 4 Heat capacity changes for (111) PMN single crystal due to applied electric field shifts ΔE versus temperature. They are obtained by Eq. (7) at different electric field shifts.

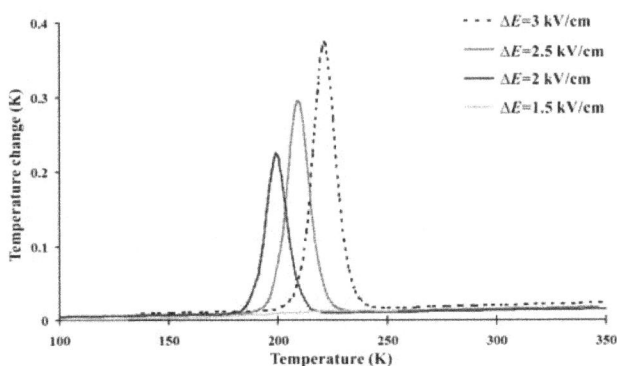

Fig. 5 Electrocaloric temperature changes ΔT due to applied electric field shifts ΔE versus temperature for (111) PMN single crystal. They are obtained by Eq. (8) at different electric field shifts.

electrocaoric temperature change ΔT is small. When the temperature rises close to T_C, the large entropy change is induced by an external strong electric field

during the ferroelectric–paraelectric phase transition, so ΔT becomes large. When the temperature is above and far from T_C, due to the presence of only paraelectric phase, a small entropy change with rising temperature decreases ΔT sharply. This is why a large EC effect and the peak of ΔT occur near T_C, as shown in Fig. 5. Furthermore, the entropy change curves reveal the characteristics of the dipole reorientation by the kinks in the ΔS^E curves. The maxima observed in the ΔS^E curve are associated to a dipole reorientation that occurs continuously. Therefore, the behavior of ΔS^E suggests how to extend the range of temperatures for use in the EC effect.

In Fig. 5, the largest change ΔT is 0.37 K in 3 kV/cm electric field shift; that is, the cooling ΔT per unit field (MV/m) is 1.23×10^{-6} m·K/V. This value is significantly larger, and comparable with the value of 0.254×10^{-6} m·K/V for PbZr$_{0.95}$Ti$_{0.05}$O$_3$ thin film under larger electric field shift $\Delta E = 30$ kV/cm. The results indicate the potential of PMN single crystal to achieve an EC effect because of the large entropy change associated with the electric field-induced dipole ordering–disordering (O–D) processes at temperatures near the O–D transformations. In addition, the electrical properties of ferroelectric single crystals are dependent on many condition parameters, such as composition, preferential orientation, and so on [28].

Finally, in this phenomenological model, it can calculate the electrocaloric properties of PMN single crystal with limited processing time. Moreover, the model does not add any auxiliary computational efforts to the numerical simulation.

4 Conclusions

It is investigated that the ferroelectric–paraelectric transition of (111) PMN single crystal under low applied electric fields, and found that the large EC effect induced by the ferroelectric–paraelectric phase transition at low electric field. The calculations show that the cooling ΔT per unit field of (111) PMN single crystal in 3 kV/cm electric field shift near the Curie temperature of 221 K is larger than that of PbZr$_{0.95}$Ti$_{0.05}$O$_3$ thin film under larger electric field shift $\Delta E = 30$ kV/cm. Thus, the EC effect of (111) PMN single crystal provides cooling solutions at low temperatures. Moreover, the pyroelectric effect could be used for example to recover useful electrical power from waste heat.

References

[1] Hamad MA. Investigations on electrocaloric properties of [111]-oriented $0.955PbZn_{1/3}Nb_{2/3}O_3$–$0.045PbTiO_3$ single crystals. *Phase Transitions* 2013, **86**: 307–314.

[2] Hamad MA. Giant electrocaloric effect of highly (100)-oriented $0.68PbMg_{1/3}Nb_{2/3}O_3$–$0.32PbTiO_3$ thin film. *Phil Mag Lett* 2013, **93**: 346–355.

[3] Hamad MA. Detecting giant electrocaloric properties of ferroelectric SbSI at room temperature. *J Adv Dielect* 2013, DOI: 10.1142/S2010135X13500082.

[4] Hamad MA. Investigations on electrocaloric properties of ferroelectric $Pb(Mg_{0.067}Nb_{0.133}Zr_{0.8})O_3$. *Appl Phys Lett* 2013, **102**: 142908.

[5] Hamad MA. Magnetocaloric effect in $La_{0.65-x}Eu_xSr_{0.35}MnO_3$. *Phase Transitions* 2012, DOI:10.1080/01411594.2013.828056.

[6] Hamad MA. Theoretical work on magnetocaloric effect in $La_{0.75}Ca_{0.25}MnO_3$. *J Adv Ceram* 2012, **1**: 290–295.

[7] Hamad MA. Magnetocaloric effect in $La_{1.25}Sr_{0.75}MnCoO_6$. *J Therm Anal Calorim* 2013, DOI: 10.1007/s10973-013-3362-2.

[8] Hamad MA. Simulation of magnetocaloric effect in $La_{0.7}Ca_{0.3}MnO_3$ ceramics fabricated by fast sintering process. *J Supercond Nov Magn* 2013, DOI: 10.1007/s10948-013-2260-y.

[9] Hamad MA. Magneto-caloric effect in $Ge_{0.95}Mn_{0.05}$ films. *J Supercond Nov Magn* 2013, **26**: 449–453.

[10] Hamad MA. Magnetocaloric effect in $La_{0.7}Sr_{0.3}MnO_3/Ta_2O_5$ composites. *J Adv Ceram* 2013, **2**: 213–217.

[11] Hamad MA. Theoretical work on magnetocaloric effect in ceramic and sol–gel $La_{0.67}Ca_{0.33}MnO_3$. *J Therm Anal Calorim* 2013, **111**: 1251–1254.

[12] Hamad MA. Magnetocaloric effect of perovskite manganites $Ce_{0.67}Sr_{0.33}MnO_3$. *J Supercond Nov Magn* 2013, DOI: 10.1007/s10948-013-2124-5.

[13] Hamad MA. Theoretical investigations on electrocaloric properties of $PbZr_{0.95}Ti_{0.05}O_3$ thin film. *Int J Thermophys* 2013, **34**: 1158–1165.

[14] Hamad MA. Magnetocaloric effect in nanopowders of $Pr_{0.67}Ca_{0.33}Fe_xMn_{1-x}O_3$. *J Supercond Nov Magn* 2013, DOI: 10.1007/s10948-013-2244-y.

[15] Hamad MA. Magnetocaloric effect of perovskite $Eu_{0.5}Sr_{0.5}CoO_3$. *J Supercond Nov Magn* 2013, DOI: 10.1007/s10948-013-2270-9.

[16] Hamad MA. Magnetocaloric effect in (001)-oriented MnAs thin film. *J Supercond Nov Magn* 2013, DOI: 10.1007/s10948-013-2254-9.

[17] Keogh D, Chen Z, Hughes RA, *et al.* (100) $MgAl_2O_4$ as a lattice-matched substrate for the epitaxial thin film deposition of the relaxor ferroelectric PMN–PT. *Appl Phys A* 2010, **98**: 187–194.

[18] Kamzina LS, Snetkova EV, Raevskiĭ IP, *et al.* Evolution of the ferroelectric phase in <001>-oriented $(100-x)PbMg_{1/3}Nb_{2/3}O_3-xPbTiO_3$ single crystals. *Phys Solid State+* 2007, **49**: 762–768.

[19] Yang Y, Liu YL, Ma SY, *et al.* Polarized micro-Raman study of the field-induced phase transition in the relaxor $0.67PbMg_{1/3}Nb_{2/3}O_3$–$0.33PbTiO_3$ single crystal. *Appl Phys Lett* 2009, **95**: 051911.

[20] Correia TM, Young JS, Whatmore RW, *et al.* Investigation of the electrocaloric effect in a $PbMg_{2/3}Nb_{1/3}O_3$–$PbTiO_3$ relaxor thin film. *Appl Phys Lett* 2009, **95**: 182904.

[21] Zeng M, Or SW, Chan HLW. Effect of phase transformation on the converse magnetoelectric properties of a heterostructure of $Ni_{49.2}Mn_{29.6}Ga_{21.2}$ and $0.7PbMg_{1/3}Nb_{2/3}O_3$–$0.3PbTiO_3$ crystals. *Appl Phys Lett* 2010, **96**: 182503.

[22] Rodriguez BJ, Jesse S, Morozovska AN, *et al.* Real space mapping of polarization dynamics and hysteresis loop formation in relaxor-ferroelectric $PbMg_{1/3}Nb_{2/3}O_3$–$PbTiO_3$ solid solutions. *J Appl Phys* 2010, **108**: 042006.

[23] Hamad MA. Room temperature giant electrocaloric properties of relaxor ferroelectric 0.93PMN–0.07PT thin film. *AIP Advances* 2013, **3**: 032115.

[24] Hamad MA. Calculation of electrocaloric properties of ferroelectric $SrBi_2Ta_2O_9$. *Phase Transitions* 2012, **85**: 159–168.

[25] Hamad MA. Magnetocaloric effect in $La_{1-x}Cd_xMnO_3$. *J Supercond Nov Magn* 2013, DOI: 10.1007/s10948-013-2189-1.

[26] Wood ME, Potter WH. General analysis of magnetic refrigeration and its optimization using a new concept: Maximization of refrigerant capacity. *Cryogenics* 1985, **25**: 667–683.

[27] Dkhil B, Kiat JM. Electric-field-induced polarization in the ergodic and nonergodic states of $PbMg_{1/3}Nb_{2/3}O_3$ relaxor. *J Appl Phys* 2001, **90**: 4676.

[28] He Y, Li XM, Gao XD, *et al.* Enhanced electrocaloric properties of PMN–PT thin films with LSCO buffer layers. *Funct Mater Lett* 2011, **4**: 45.

9

Relaxor behavior and Raman spectra of CuO-doped Pb(Mg$_{1/3}$Nb$_{2/3}$)O$_3$–PbTiO$_3$ ferroelectric ceramics

Huiqin LI^a, Jingsong LIU^{a,*}, Hongtao YU^a, Shuren ZHANG^b

^{a}State Key Laboratory Cultivation Base for Nonmetal Composite and Functional Materials, Southwest University of Science and Technology, Mianyang, Sichuan, China
^{b}State Key Laboratory of Electronic Thin Films and Integrated Devices, University of Electronic Science and Technology of China, Chengdu, Sichuan, China

Abstract: In this work, Raman spectra and dielectricity–temperature dependence measurements were used to investigate the B-site order degree in CuO-doped Pb(Mg$_{1/3}$Nb$_{2/3}$)O$_3$–PbTiO$_3$ ferroelectric ceramics. The measurement results indicated a typical relaxor characteristic for all samples. With the increasing of CuO doping content, the B-site order degree increased first and then decreased. However, the frequency dispersion and the relaxation degree decreased first and then increased while the CuO addition content was increasing, which was thought to be strongly correlated with the variations of the B-site order. The opposite variation tendency of the B-site order degree and the relaxation degree revealed that the phase transition dispersity is closely related to the order–disorder behaviors.

Keywords: relaxor ferroelectrics; dielectric properties; Raman spectra

1 Introduction

Lead magnesium niobate Pb(Mg$_{1/3}$Nb$_{2/3}$)O$_3$ (PMN) based relaxor ferroelectrics have attracted extensive attention for their high permittivity, low temperature sensitivity, considerable electrostriction effects and slender hysteresis loops. They have been applied in many areas, such as multilayer ceramic capacitors (MLCC), high-energy pulsed power capacitors, electro-optic devices, micrometric displacement actuators, ferroelectric memories, and so on [1–3]. However, there are some disadvantages existing in such materials, such as high sintering temperature (about 1200 ℃), low phase transition temperature (−12 ℃, 1 kHz), high dielectric loss at negative

temperature, low breakdown strength, and difficulty to realize industrial production. At present, the main modifying method is solid solution doping.

It is easy for PMN and PbTiO$_3$ (PT) to form (1−x)PMN–xPT (PMNT) solid solution. The relaxor behavior does not disappear until near the morphotropic phase boundary [4]. The relaxor characteristic of PMN has been carefully investigated on a consideration of atomic relaxation vibration, desiring to obtain giant dielectric response [5]. The effects of rear-earth doping and mixing with normal ferroelectrics on the relaxor characteristic have been intensively studied, which is aimed to improve the phase transition dispersivity and electrical properties [6,7]. In addition, PbO volatilizes seriously at the sintering temperature (~1200 ℃) and oxygen vacancies are formed. Those defects not only make the ceramic composition deviate from stoichiometric ratio and deteriorate the properties [8], but also pollute the

* Corresponding author.
E-mail: liujingsong@swust.edu.cn

environment and waste energy. Therefore, it is significant to lower the sintering temperature and improve the electrical properties of PMN ceramics [9]. It is implied that doping can improve the properties of PMN ceramics. Wen et al. [10] found that in the case of PMNT ceramics co-doping with La^{3+} and Zn^{2+}, the degree of order–disorder is remarkably increased, and when 2 mol% La^{3+} and 3 mol% Zn^{2+} are co-doped, the maximal relative dielectric constant change decreases from 70% to 23% at 1 kHz (1000 V/mm). Gao et al. [11] prepared Zn^{2+}/Li^+-doped PMNT ceramics, and found that the Curie temperature and dielectric constant decrease, but the ceramic with the Z/L ratio of 1:1 and the amount of 1 wt% has excellent piezoelectric properties. Li et al. [12] reported that temperature dependent diffusive dielectric loss peaks as a doping effect are found for Mn-doped PMNT. In addition, CuO is also one of important dopants and its function is superior to others'. Chou et al. [13] and Zhao et al. [14] reported that CuO doping not only improves the dielectric properties, but it also lowers the sintering temperature of BNT and KNN–LNS ceramics, respectively. However, there are only few works concerned with the properties and Raman spectra of the CuO singly doped PMNT system. Recently, CuO addition was found to be useful for lowering sintering temperature of PMN based ceramics [15,16].

In this work, the CuO-doped PMNT ceramics were prepared by the columbite method [17]. Their microstructures, dielectric properties, phase transition dispersion and B-site order, together with the influence of CuO doping on above features, were all investigated.

2   Experiment

The starting materials were high-purity oxide powders: yellow lead oxide (PbO), niobium oxide (Nb_2O_5), magnesium oxide (MgO), titanium dioxide (TiO_2) and cupric oxide (CuO). The experiments in this paper were based on the composition of 0.94PMN–0.06PT–xCuO ($x = 0$, 0.015, 0.03, 0.06), defined as PMNT, PMNT1, PMNT2 and PMNT3, respectively. Preparing samples contained the following steps. First, PbO and TiO_2 powders were used to synthesize PT. Next, PMNT powders were obtained using the columbite method [17], which involved mixing and batching (800 ℃) MgO, Nb_2O_5, PbO, PT and CuO powders.

Finally, the powders with 5 wt% PVA solution as binder were pressed into pellets with 30 mm in diameter and 5 mm in thickness. The pellets were sintered at different temperatures (1050–1200 ℃) for 2 h in air.

A scanning electron microscope (SEM, Hitachi TM-1000) was used to observe the micro-morphology. The phase identification of the samples was performed using a X-ray diffractometer (XRD, Rigaku D/max-RB). The Raman spectrum measurements were carried out using a Renishaw Ramanor InVia spectrometer. The temperature dependence of dielectricity was measured using an Agilent 4284A LCR meter.

3   Results and discussion

Figure 1 shows a dense section morphology for all samples. With the increasing of CuO content, the sintering temperature is gradually reduced and the particle size becomes uniform and fine. The indexed planes in Fig. 2 indicate that there is only pervoskite phase for all samples. The XRD data illustrates that the CuO doping causes a shifting of diffraction peaks, which indicates a lattice distortion in doped samples. It is possible that the Cu ions incorporate into the lattice and lead to the distortion. As well known, metal cations prefer to take the place of the ions with similar electrovalence and radius. CuO doping makes the cell bigger because Cu^+ ions (radius ~0.77 Å) take the place of Mg^{2+} ions (radius ~0.72 Å). The electron configurations of Cu^+ and Cu^{2+} are $3s^23p^63d^{10}$ and $3s^23p^63d^9$, respectively. According to Hund's rules, a completely filled d orbit is relatively stable, so Cu^+ ion is more stable than Cu^{2+} ion. In fact, CuO will discompose into Cu_2O when the temperature reaches 1273 K [18]. As can be seen from Fig. 3, the binding energy of Cu 2p is significantly decreased, which indicates that the dispersion of CuO into the lattice is increased [19]. In addition, from the peak position and shape, the peak of the sample binding energy is superimposed from the peak of Cu^+ and Cu^{2+} ions, the intensity of the peak spectrum is remarkably lower, and the peak area of Cu^{2+} ion can also be seen lower than that of Cu^+ ion simultaneously. This shows that the content of Cu^{2+} ion is relatively lower compared with that of Cu^+ ion, and the possibility that Cu^+ ion occupies the lattice grid of Mg^{2+} is greater than that of Cu^{2+}. As a result,

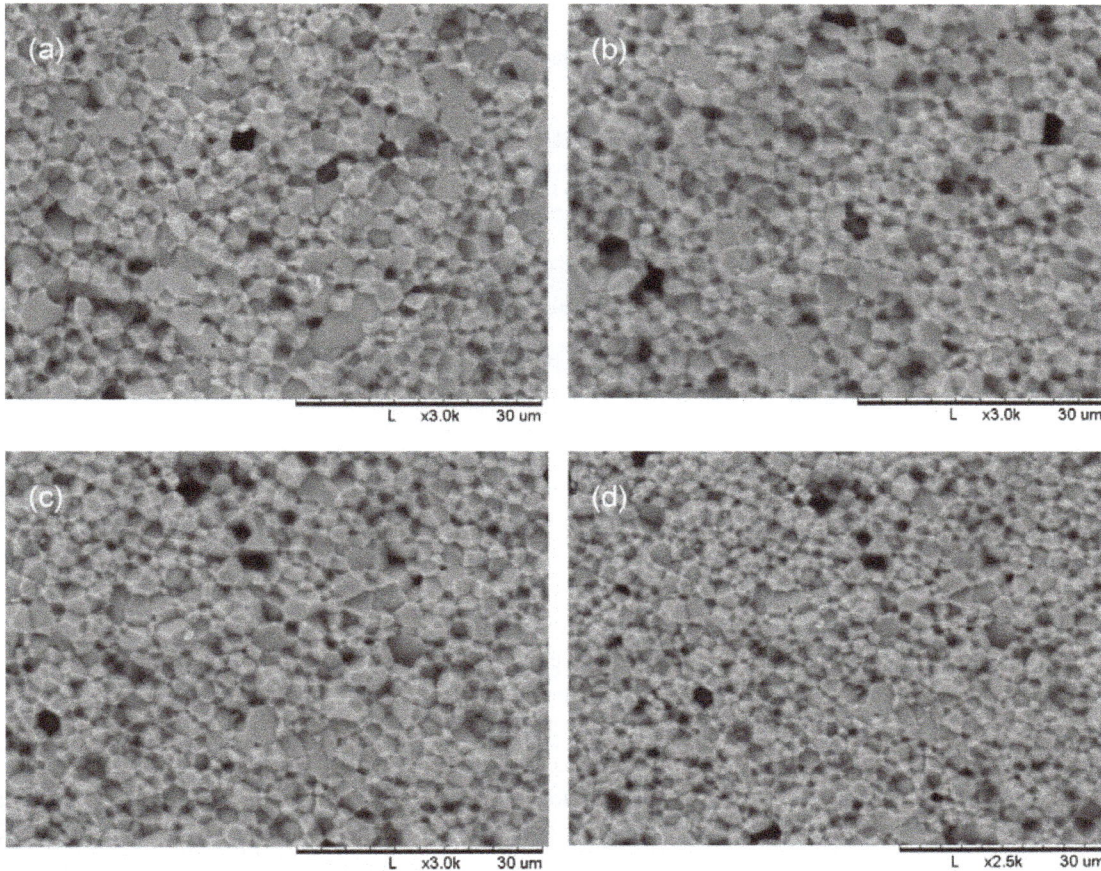

Fig. 1   SEM section images of PMNT–xCuO ceramics: (a) $x = 0$ (1200 ℃); (b) $x = 0.015$ (1080 ℃); (c) $x = 0.03$ (1050 ℃); (d) $x = 0.06$ (1050 ℃).

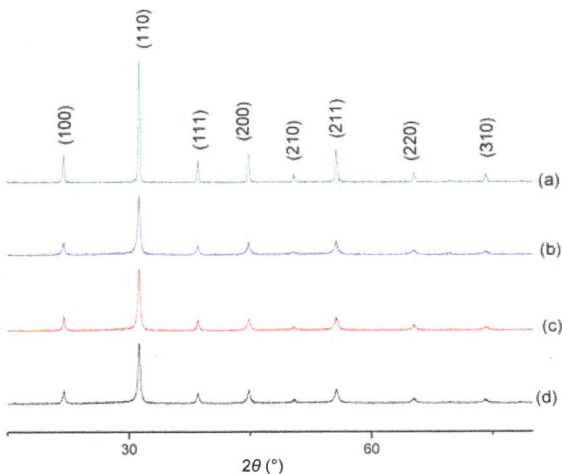

Fig. 2   XRD patterns of PMNT–xCuO ceramic powders: (a) $x = 0$; (b) $x = 0.015$; (c) $x = 0.03$; (d) $x = 0.06$.

Cu^+ exists stably at sintering temperature (1323–1473 K), and an unequivalent replacing affects the structural ordering. The effect of ion replacement on the B-site order is also investigated with Raman analysis.

Figure 4 shows the Raman spectra of all ceramic samples. The characteristic peaks in Fig. 4 and the corresponding peak positions are listed in Table 1. Compared with non-doped sample, frequency shift is found on the doped samples. Especially, the frequency of peak C decreases at first and then increases. For PMN based ferroelectrics, 260 cm^{-1}, 500–600 cm^{-1} and 780 cm^{-1} wave numbers are related to O–B–O bending modes (peak A), Nb–O–Nb stretching modes (peak B) and Nb–O–Mg bending modes (peak C, named A_{1g} mode), respectively [20].

Peak A around 260 cm^{-1} (as shown in Fig. 4 and Table 1) originates from O–B–O bending modes. The frequency shift in Raman spectra reveals that the O–B–O bending vibration is affected when CuO is introduced. Cu^+ ions could increase the composition fluctuation on B-sites, resulting in a change on B-site polarity.

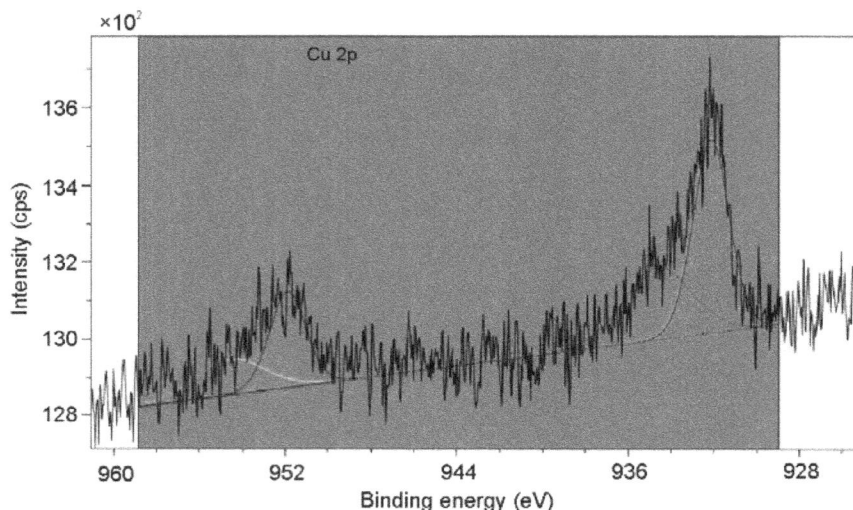

Fig. 3   XPS spectra of Cu 2p of CuO-doped PMNT ceramics.

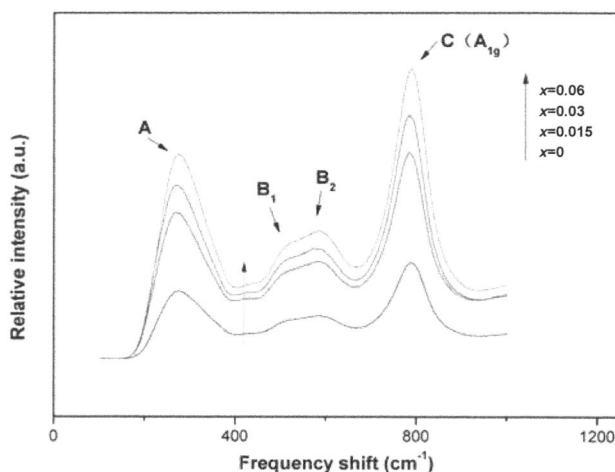

Fig. 4   Raman spectra of PMNT–xCuO ceramics.

Table 1   Raman spectrum peaks of PMNT–xCuO ceramics

Sample	x	Peak A (cm^{-1})	Peak B$_1$ (cm^{-1})	Peak B$_2$ (cm^{-1})	Peak C (cm^{-1})
PMNT	0	270	512	592	788
PMNT1	0.015	272	503	589	785
PMNT2	0.03	273	506	587	783
PMNT3	0.06	275	507	583	787

Nb–O–Nb stretching modes, originating from the polarization displacement of Nb^{5+} ion in oxygen octahedron [21], are corresponding with the peak B in Fig. 4. The peak B does not split when the temperature is above 77 ℃. Below 77 ℃, peak B will split into peaks B$_1$ and B$_2$ (500–600 cm^{-1}) [20]. After CuO doping, the variations of peaks B$_1$ and B$_2$ indicate that CuO doping has an apparent effect on the Nb–O–Nb stretching modes.

Peak C around 780 cm^{-1} is caused by the Nb–O–Mg bending modes (A$_{1g}$ modes). The A$_{1g}$ modes are similar to free oxygen octahedron vibration modes, related to the nano-ordered region in space group $Fm3m$. As an effective indicator of B-site order, the A$_{1g}$ modes reflect the change of oxygen octahedron in pervoskite structure.

There are two different models: the space charge model and the random layer model, used to study PMN Raman behaviors [22]. In the space charge model, B^{I} and B^{II} sublattices are occupied by Mg^{2+} and Nb^{5+}, respectively. Nanoscale ordered regions emerge with a local Mg^{2+}/Nb^{5+} ratio of 1:1, instead of the 1:2 stoichiometry at B-site. Then a super structure of $[\text{Pb}^{2+}(\text{Mg}_{1/2}^{2+}\text{Nb}_{1/2}^{5+})\text{O}_3^{2-}]^{-0.5}$ with negative charge forms, which is twice bigger than the normal lattice period. This will prevent the growth of ordered structure because of unbalanced charges. In the random layer model, the B-site 1:1 ordered structure is corresponding to Pb[(Mg$_{2/3}$Nb$_{1/3}$)$_{1/2}$Nb$_{1/2}$]O$_3$ stoichiometry, and the B^{I} and B^{II} sublattices are occupied by Mg^{2+}/Nb^{5+} and Nb^{5+}, respectively. Thus the ordered regions are consistent with the 1:2 stoichiometry. Jiang et $al.$ [23] employed a simple harmonic oscillator model to discuss the effect of B-site order on A$_{1g}$ mode frequency ω:

$$\omega = \sqrt{k/m^*} \qquad (1)$$

where k represents the force constant related to the O–O and O–B band strengths; m^* is the reduced mass of this mode. In the random layer model, the space group of PMN is explained by the stoichiometry of $Pb[(Mg_{2/3}Nb_{1/3})_{1/2}Nb_{1/2}]O_3$. B^I sublattice represents $2/3Mg^{2+} + 1/3Nb^{5+}$, and its average atomic mass is 47, while B^{II} sublattice represents Nb^{5+} with atomic mass of 93. In this work, the distribution of Mg^{2+}/Nb^{5+} on B-site is changed by Cu^+ doping. The introduction of Cu^+ ions leads to Nb^{5+} entering B^I sublattice when a little amount of CuO is doped, which would increase the m^* value and B-site order. The decrease of A_{1g} mode frequency in PMNT1 and PMNT2 is related to the increase of m^* value (Table 1). However, the A_{1g} mode frequency increases when excessive CuO is doped, because the replacement of Mg^{2+} by Cu^+ makes the composition deviate from stoichiometry of $Mg^{2+}:Nb^{5+} = 1:2$. This serious unbalanced stoichiometry also prevents Nb^{5+} entering B^I sublattice, and decreases B-site order. By applying Eq. (1) and the random layer model, the peak shifting of Raman spectra is explained. In a word, the A_{1g} mode frequency variations indicate that B-site order increases at first and then decreases with the introduction of Cu^+ ions. Subsequently, the dielectric properties are discussed in terms of B-site order.

As shown in Fig. 4, we can also find that the peak intensities become stronger with increasing CuO doping concentration. Especially, the peak intensities of B_1 and B_2 are become stronger and sharper compared with those of the non-doped sample that are broad; it may be closely related to the structure order. Siny et al. [24] proposed that samples with high order degrees exhibit strong Raman peaks and Raman active modes, which are also closely related to the structure order. So the change of the peak intensity is attributed to the B-site order degree. In addition, the intensity of the Raman spectrum peak is also correlated with the polarizability and it has effect on the intensity of the peak simultaneously. When doping content increases, it will raise the atomic distance and lead to the change of the polarizability. While the polarizability increases with increasing of atomic distance, the vibration intensity of the modes also will increase. Therefore, the intensity of the Raman spectrum peak will become stronger.

Figure 5 shows the temperature dependences of dielectric constant and loss at 1 kHz. The dielectric properties of all samples are listed in Table 2. The

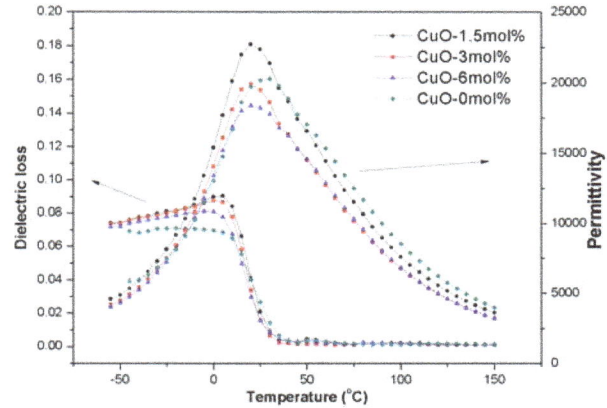

Fig. 5 Temperature dependences of dielectric constant and loss at 1 kHz.

Table 2 Dielectric properties and diffusive factors of PMNT–xCuO at 1 kHz

Sample	x	ε_m	T_m (℃)	δ (℃)	ε at 25 ℃
PMNT	0	20296	30	41.44	20203
PMNT1	0.015	22708	20	40.99	22362
PMNT2	0.03	19942	20	42.88	19506
PMNT3	0.06	18411	20	44.04	18237

results indicate that the peak temperatures (T_m) of doped samples are about 10 ℃ lower than that of non-doping. This shift is probably related to the compositional differences and microstructure characteristic, particularly the degree of crystallinity and grain size [25]. Therefore, the changes in grain size, shape and composition by CuO doping result in the change of T_m. With the increasing of doping amount, the maximum dielectric constant of samples shows a tendency to increase at first and then decrease. However, the dielectric loss keeps nearly unchanged above the room temperature. The maximum dielectric constant reaches the highest (22 708) when the CuO content is 1.5 mol%, but it decreases to 19 942 and 18 411 when the CuO content is 3 mol% and 6 mol%, respectively (Table 2). The reason of this phenomenon is that a minor amount of CuO doping causes the B-site order to increase. Meanwhile, a lot of unbalanced space charges form, which contribute to the polarization and dielectric constant. Unfortunately, excessive doping leads to incomplete ion replacement and decreasing of B-site order. Consequently, the dielectric constant is reduced. On the other hand, the dramatic decline of dielectric loss around phase transition temperature is due to the polarization disappearance. It can be seen that the dielectric loss is reduced to less than 1% for all doped samples while the temperature is above 30 ℃ (Fig. 5). Attributed to the suppression of PbO volatilization and lowering of sintering temperature, the dielectric loss retains relatively low when the temperature is above the room temperature. However, the dielectric loss of the doped ceramic samples is higher than that of non-doped samples at negative temperature. It may originate from the increase in the oxygen vacancies. As mentioned above, Cu^+ ions dissolved into lattice will substitute for Mg^{2+} ions on the B-site and oxygen vacancies are simultaneously generated with the increasing of CuO concentration, and the appearance of oxygen vacancy can lead to lattice distortion and decreasing of B-site order [26]. In addition, the vacancies in the system can be polarized under an alternating electric field. Therefore, the dielectric loss increases [27].

One characteristic of relaxor ferroelectrics is diffusive phase transition. The relationship between maximum dielectric constant and dispersion degree could be used to analyze the order–disorder behavior of ferroelectric ceramics. The diffusive factor δ can be calculated from the quadratic formula [28]:

$$\frac{1}{\varepsilon} = \frac{1}{\varepsilon_m} + \frac{1}{2\varepsilon_m \delta^2}(T - T_m)^2 \qquad (2)$$

where ε and ε_m are the relative and maximum dielectric constants, respectively; T and T_m are the temperatures corresponding to ε and ε_m, respectively. The curve of $(1/\varepsilon - 1/\varepsilon_m)$ vs. $(T - T_m)^2$ is linear fitted, as shown in Fig. 6. The diffusive factor δ could be calculated with the slope of the curve. It is obvious that the dispersities of samples decrease at first and then increase with the CuO content increasing (Table 2 and Fig. 6). According to the order–disorder transition model [29], the dispersion and dielectric

relaxation degree decrease with the B-site order increasing. In this work, the related B-site order increases at first and then decreases, which is confirmed by the Raman analysis. Additionally, it could be seen that the variation tendency of the maximum dielectric constant is opposite to that of phase dispersion degree, which is another evidence that the dispersion degree is closely related to the order–disorder behaviors.

4   Conclusions

CuO-doped PMN–PT ferroelectric ceramics with a single pervoskite-type phase were fabricated by solid state reaction method. The Raman measurement results showed that the A_{1g} peak firstly shifted to low frequency and then to high frequency with increasing CuO content, which meant the B-site order degree increased and then decreased. The variations in the B-site order degree resulted in the changes in the relaxor behavior. The relaxation degree and the dispersion degree decreased first and then increased with the increasing of CuO addition content. This paper further provided an evidence on the relationship between the ferroelectric relaxation degree and the B-site order degree in the PMN based ceramics.

Acknowledgements

This research was supported by the Open Foundation of State Key Laboratory of Electronic Thin Films and Integrated Devices (KFJJ201207).

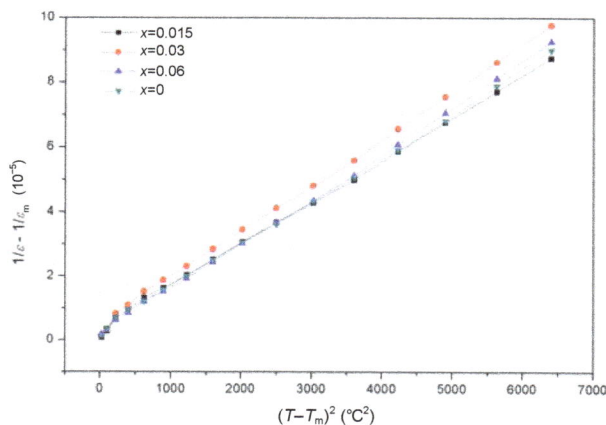

Fig. 6   Linear fitting graph of $(1/\varepsilon - 1/\varepsilon_m)$ vs. $(T - T_m)^2$.

References

[1] Haertling GH. Ferroelectric ceramics: History and technology. *J Am Ceram Soc* 1999, **82**: 797–818.

[2] Swartz SL, Shrout TR, Schulze WA, *et al*. Dielectric properties of lead–magnesium niobate ceramics. *J Am Ceram Soc* 1984, **67**: 311–314.

[3] Uchino K. Electrostrictive actuators: Materials and application. *Am Ceram Soc Bull* 1986, **65**: 647–652.

[4] Lin D, Li Z, Li F, *et al*. Characterization and

piezoelectric thermal stability of PIN–PMN–PT ternary ceramics near the morphotropic phase boundary. *J Alloys Compd* 2010, **489**: 115–118.

[5] Prosandeev S, Raevski IP, Malitskaya MA, *et al.* Condensation of the atomic relaxation vibrations in lead–magnesium–niobate at $T = T^*$. *J Appl Phys* 2013, **114**: 124103.

[6] Sun E, Zhang R, Wu F, *et al.* Complete matrix properties of $[001]_c$ and $[011]_c$ poled 0.33Pb$(In_{1/2}Nb_{1/2})O_3$–0.38Pb$(Mg_{1/3}Nb_{2/3})O_3$–0.29PbTiO$_3$ single crystals. *J Alloys Compd* 2013, **553**: 267–269.

[7] Chang W-Y, Huang W, Bagal A, *et al.* Study on dielectric and piezoelectric properties of 0.7Pb$(Mg_{1/3}Nb_{2/3})O_3$–0.3PbTiO$_3$ single crystal with nano-patterned composite electrode. *J Appl Phys* 2013, **114**: 114103.

[8] Wang C, Zhang M, Xia W. High-temperature dielectric relaxation in Pb$(Mg_{1/3}Nb_{2/3})O_3$–PbTiO$_3$ single crystals. *J Am Ceram Soc* 2013, **96**: 1521–1525.

[9] Chen J, Qiu S, Chen X, *et al.* Preparations and characterizations of perovskite 0.80PMN–0.20PT ceramic by using a one-step calcination method. *J Alloys Compd* 2010, **497**: 155–158.

[10] Wen X, Feng C, Chen L, *et al.* Effect of order–disordered nano-domains on the dielectric and electrical properties of PMNT ceramics. *J Alloys Compd* 2006, **422**: 244–248.

[11] Gao F, Hong R, Liu J, *et al.* Effects of ZnO/Li$_2$O codoping on microstructure and piezoelectric properties of low-temperature sintered PMN–PNN–PZT ceramics. *Ceram Int* 2009, **35**: 1863–1869.

[12] Li X, Zhao X, Ren B, *et al.* Microstructure and dielectric relaxation of dipolar defects in Mn-doped $(1–x)$Pb$(Mg_{1/3}Nb_{2/3})O_3$–xPbTiO$_3$ single crystals. *Scripta Mater* 2013, **69**: 377–380.

[13] Chou C-S, Liu C-L, Hsiung C-M, *et al.* Preparation and characterization of the lead-free piezoelectric ceramic of Bi$_{0.5}$Na$_{0.5}$TiO$_3$ doped with CuO. *Powder Technol* 2011, **210**: 212–219.

[14] Zhao Y, Zhao Y, Huang R, *et al.* Microstructure and piezoelectric properties of CuO-doped 0.95(K$_{0.5}$Na$_{0.5}$)NbO$_3$–0.05Li(Nb$_{0.5}$Sb$_{0.5}$)O$_3$ lead-free ceramics. *J Eur Ceram Soc* 2011, **31**: 1939–1944.

[15] Chao X, Ma D, Gu R, *et al.* Effects of CuO addition on the electrical responses of the low-temperature sintered Pb(Zr$_{0.52}$Ti$_{0.48}$)O$_3$–Pb(Mg$_{1/3}$Nb$_{2/3}$)O$_3$–Pb(Zn$_{1/3}$Nb$_{2/3}$)O$_3$ ceramics. *J Alloys Compd* 2010, **491**: 698–702.

[16] Wang L, Mao C, Wang G, *et al.* Effect of CuO addition on the microstructure and electric properties of low-temperature sintered 0.25PMN–0.40PT–0.35PZ ceramics. *J Am Ceram Soc* 2013, **96**: 24–27.

[17] Swartz SL, Shrout TR. Fabrication of perovskite lead magnesium niobate. *Mater Res Bull* 1982, **17**: 1245–1250.

[18] Hoang NN, Huynh DC, Nguyen TT, *et al.* Synthesis and structural characterization of uranium-doped Ca$_2$CuO$_3$, a one-dimensional quantum antiferromagnet. *Appl Phys A* 2008, **92**: 715–725.

[19] Wang Z, Liu Q, Yu J, *et al,* Surface structure and catalytic behavior of silica-supported copper catalysts prepared by impregnation and sol–gel methods. *Appl Catal A: Gen* 2003, **239**: 87–94.

[20] Husson E, Abello L, Morell A. Short-range order in PbMg$_{1/3}$Nb$_{2/3}$O$_3$ ceramics by Raman spectroscopy. *Mater Res Bull* 1990, **25**: 539–545.

[21] Siny IG, Lushnikov SG, Katiyar RS, *et al.* PbMg$_{1/3}$Nb$_{2/3}$O$_3$ as a model object for light scattering experiments. *Ferroelectrics* 1999, **226**: 191–215.

[22] Chen J, Chan HM, Harmer MP. Ordering structure and dielectric properties of undoped and La/Na-doped Pb(Mg$_{1/3}$Nb$_{2/3}$)O$_3$. *J Am Ceram Soc* 1989, **72**: 593–598.

[23] Jiang F, Kojima S, Zhao C, *et al.* Chemical ordering in lanthanum-doped lead magnesium niobate relaxor ferroelectrics probed by A$_{1g}$ Raman mode. *Appl Phys Lett* 2001, **79**: 3938–3940.

[24] Siny IG, Tao R, Katiyar RS, *et al.* Raman spectroscopy of Mg–Ta order–disorder in BaMg$_{1/3}$Ta$_{2/3}$O$_3$. *J Phys Chem Solids* 1998, **59**: 181–195.

[25] Li T, Liu J, Li H, *et al.* Dielectric behavior and Raman spectra of lanthanum-doped lead magnesium niobate ceramics. *J Mater Sci: Mater El* 2011, **22**: 1188–1194.

[26] Cai W, Fu CL, Gao JC, *et al.* Dielectric properties and microstructure of Mg doped barium titanate ceramics. *Adv Appl Ceram* 2011, **110**: 181–185.

[27] Liang X, Wu W, Meng Z, *et al.* Dielectric and tunable characteristics of barium strontium titanate modified with Al$_2$O$_3$ addition. *Mat Sci Eng B* 2003, **99**: 366–369.

[28] Cao L, Yao X, Xu Z. Effect of Ta substitution on microstructure and electrical properties of 0.80Pb(Mg$_{1/3}$Nb$_{2/3}$)O$_3$–0.20PbTiO$_3$ ceramics. *Ceram Int* 2004, **30**: 1369–1372.

[29] Zhong WL. *Physics of the Ferroelectric*. Beijing: Science Press, 2000.

10

Electrical properties of bulk and nano Li$_2$TiO$_3$ ceramics: A comparative study

Umasankar DASH^{a,*}, Subhanarayan SAHOO^{b}, Paritosh CHAUDHURI^{c},
S. K. S. PARASHAR^{a}, Kajal PARASHAR^{a}

^{a}School of Applied Sciences, KIIT University, Bhubaneswar-751 024, India
^{b}Department of Electrical and Electronics Engineering, Trident Academy of Technology, Bhubaneswar-751 024, India
^{c}Institute for Plasma Research, Bhat, Gandhinagar-382 428, India

Abstract: Nanocrystalline and bulk Li$_2$TiO$_3$ having monoclinic structure were prepared by mechanical alloying as well as conventional ceramic route. Complex impedance analysis in the frequency range of 100 Hz–1 MHz over a wide range of temperature (50–500 ℃) indicates the presence of grain boundary effect along with the bulk contribution. The frequency-dependent conductivity plots exhibit power law dependence, suggesting three types of conduction in the material: low-frequency (100 Hz–1 kHz) conductivity showing long-range translational motion of electrons (frequency independent), mid-frequency (1–10 kHz) conductivity showing short-range hopping of charge carriers and high-frequency (10 kHz–1 MHz) conductivity showing conduction due to localized orientation of hopping mechanism. The electrical conductivity measurement of nanocrystalline and bulk Li$_2$TiO$_3$ with temperature shows the negative temperature coefficient of resistance (NTCR) behavior. The activation energy (0.77 eV for nano sample and 0.88 eV for bulk sample) study shows the conduction mechanism in both samples. The low activation energies of the samples suggest the presence of singly ionized oxygen vacancies in the conduction process.

Keywords: nanocrystalline; complex impedance spectroscopy (CIS); AC conductivity; X-ray diffraction (XRD)

1 Introduction

Lithium-based ceramics such as (Li$_2$TiO$_3$, Li$_2$SiO$_4$, Li$_2$ZrO$_3$ and LiAlO$_3$) have been considered as candidates for tritium breeding material in D–T fusion reactors. Li$_2$TiO$_3$ has attracted many researchers due to

its strong tritium releasing capacity, high chemical stability, low activation energy and good lithium density. This material shows acceptable mechanical strength when used in a fusion blanket. Much effort has been given to study the properties of Li$_2$TiO$_3$ in order to establish it as a blanket material [1–5]. As we know, ^{6}Li is used as a fuel in the fusion reaction for the generation of tritium. The reaction is as follows:

$$^{6}\text{Li} + \text{n} \rightarrow (\text{He} + 2.1\,\text{MeV}) + (\text{T} + 2.7\,\text{MeV}) \quad (1)$$

---

* Corresponding author.
E-mail: dashumasankar@gmail.com

Lithium density in the blanket is an important parameter for the economy of fusion reactor. To date, much work has already been done on the fabrication of both pellets and pebbles of Li_2TiO_3, characterizing tritium releasing behavior and developing the database. Different techniques are routinely utilized to produce Li_2TiO_3 nanoceramic materials, such as sol–gel [6–10], extrusion spherodization [11,12], wet chemical method [13], solution combustion [14], polymer solution [15] and many more. All these chemical and mettalo–organic methods have drawbacks with respect to large-scale production, reproducibility, maintenance of stoichiometry and homogeneity in composition, and furthermore require extra high-temperature calcinations. In contrast to the above mentioned techniques, high-energy ball milling (HEBM) provides an environmentally friendly way to prepare Li_2TiO_3 nanoparticles with exceptionally high conductivity, sinterabillity and good lithium density.

A few works have been reported for the synthesis of Li_2TiO_3. The feasibility of preparing Li_2TiO_3 that can be compacted and sintered to high density with relatively low sintering temperature and short sintering time in order to have an ideal microstructure and reduce the overall production costs, has been studied by HEBM technique. In this contribution, we discuss the electrical conductivity and complex modulus spectroscopy of both nanocrystalline and bulk Li_2TiO_3.

2 Experimental procedure

2.1 Conventional solid state technique

Lithium carbonate and titanium dioxide powder was taken in a stoichiometric ratio for the preparation of lithium titanate. The mixture was grinded in a motor pestle with the addition of acetone. The mixture was dried and calcined at different temperatures. Phase analysis of the powder was done by using Cu Kα X-ray diffractometry (XRD). Single-phase monoclinic structure was observed at 1000 ℃.

2.2 Mechanical alloying route

Lithium carbonate and titanium dioxide powder was taken in BPR (ball to powder ratio) of 40:1. The mixture was milled in a high-energy planetary ball mill for 10 h. The mixture was allowed to cool every hour. The milled powder was subjected to calcination at different temperatures. Calcined powder was

characterized with respect to phase identification and lattice parameter measurement using Cu Kα XRD. The calcined powder was pressed uniaxially at 41 MPa with 3 wt% PVA solution as binder. The green pellets were sintered at 1000 ℃ for 2 h. The microstructure of the sintered pellets was studied by scanning electron microscope (SEM, JEOL). The complex impedance spectroscopy (CIS) measurement was done by using LCR meter (Hioki 3532, Japan) in the frequency range of 100 Hz–1 MHz.

3 Results and discussion

Figure 1 shows the X-ray analysis of bulk and nanocrystalline of lithium titanate calcined at 1000 ℃ and 700 ℃ respectively. It is observed that the single-phase monoclinic structure is observed at 1000 ℃ in the case of bulk sample and 700 ℃ for nano sample. In the case of nano sample, it is observed from the XRD pattern that the formation of lithium titanate starts at 500 ℃ and the single phase of lithium titanate is formed at 700 ℃. The powder calcined at 700 ℃ demonstrates the monoclinic phase with space group $C2/c$ and is found to be well matched with JCPDS No. 33-0831 with $a = 5.069$ Å, $b = 8.799$ Å and $c = 9.759$ Å. The crystallite size is calculated by using Scherer's equation, which results in an average crystallite size of 88 nm in the case of ball-milled sample after removing crystal strain and instrumental broadening.

Figure 2 shows the differential scanning calorimetry (DSC) curve for the mixture of Li_2CO_3 and TiO_2 with the stoichiometric ratio. The broad endothermic range

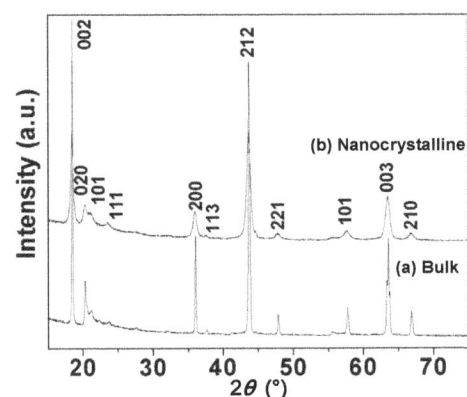

Fig. 1   XRD patterns of Li_2TiO_3: (a) bulk and (b) nanocrystalline.

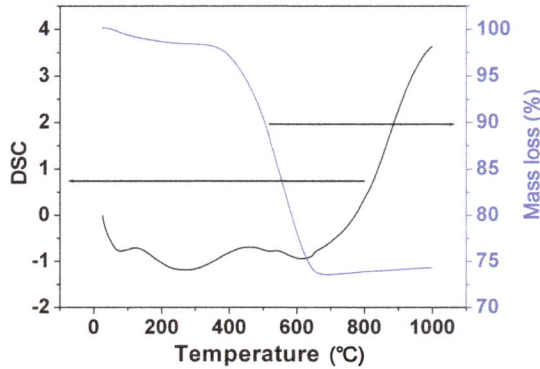

Fig. 2 DSC and TGA curves of the 10-h milled sample.

of 300–700 ℃ is present which is associated with the reaction of Li_2CO_3 and TiO_2 as follows:

$$Li_2CO_3 + TiO_2 \rightarrow Li_2TiO_3 + CO_2 \qquad (2)$$

This reaction is completed at 700 ℃. The three broad endothermic peaks at 125 ℃, 270 ℃ and 450 ℃ are indicated as the removal of bonded water and the decomposition of lithium carbonate and titanium dioxide, respectively. However, it exhibits a highly exothermic event with a major weight loss (about 85%) starting at 292 ℃, and there is no further thermal event up to 400 ℃. The endotherm from 80 ℃ to 292 ℃ is less than 5%, and mass loss occurs up to 660 ℃ as shown by thermogravimetric analysis (TGA). The most important evidence from the thermal studies is the absence of enthalpy changes at high temperatures, which implies that the reaction is complete and no organic matter or un-reacted phase is present in the sample, and there is no evidence of a phase transition taking place in the sample up to the temperature of 700 ℃.

Figure 3 shows the microstructure of bulk and nanocrystalline Li_2TiO_3. The micrographs reveal well defined grains with limited porosity. It is also clear from the micrographs that there is a variation in grain morphology from the bulk sample to the nano sample. In the bulk sample, neck formation occurs and interparticle distance decreases. Grain boundary distance decreases and bulk transport occurs. Bulk transport mechanism includes volume diffusion, grain boundary diffusion, plastic flow and viscous flow. Plastic flow, which is the most important, occurs during the heating period where the initial dislocation of density is large. It is usually best verified by compacted powder samples.

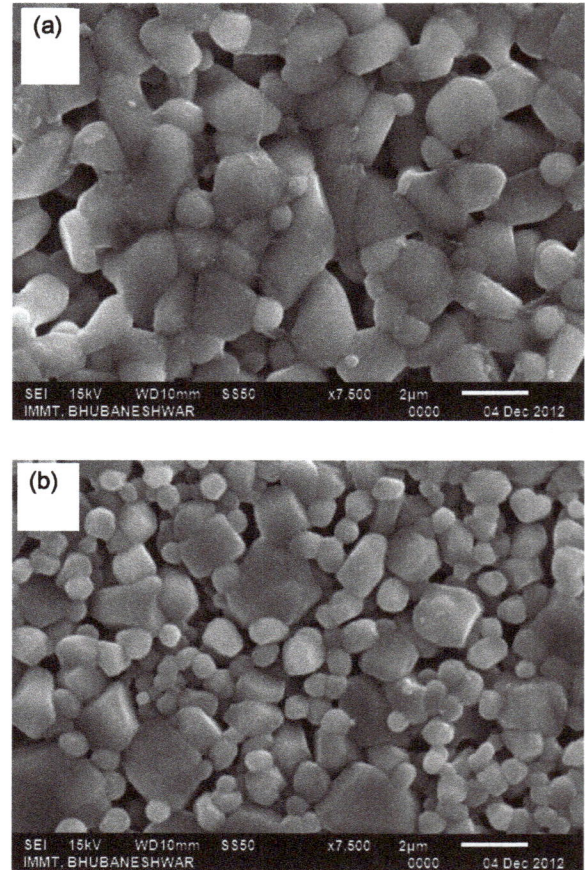

Fig. 3 SEM micrographs of Li_2TiO_3: (a) bulk and (b) nanocrystalline.

Complex impedance spectroscopy (CIS) is a useful technique for measurement of electrical response in material. This analysis enables us to resolve the contribution of various processes, such as bulk, grain boundary and electrode effects in the frequency domain [16–18]. Electrical AC data can be represented in any one of the four interrelated formulas:

Relative permittivity: $\varepsilon^* = \varepsilon' - j\varepsilon''$ \qquad (3)

Impedance: $Z^* = Z' - jZ'' = 1/(j\omega C_0 \varepsilon^*)$ \qquad (4)

Electric modulus: $M^* = M' + jM'' = 1/\varepsilon^*$ \qquad (5)

Admittance: $Y^* = Y' + jY'' = j\omega C_0 \varepsilon^*$ \qquad (6)

$\tan\delta = \varepsilon''/\varepsilon' = M'/M'' = Z'/Z'' = Y'/Y''$ \qquad (7)

where $\omega = 2\pi f$ is the angular frequency; $C_0 = \varepsilon_0 A d^{-1}$ is the geometrical capacitance; ε_0 is the permittivity of free space; and d and A are the thickness and area of the pellet, respectively.

Figure 4 shows the complex plots of the lithium metatitante sample. The electrical behavior of the system has been studied over a wide range of temperature starting from 50 ℃ to 500 ℃ and frequency from 100 Hz to 1 MHz. The CIS technique helps us to find various microscopic elements such as intragrain, intergrain, electrode effect and relaxation process. The figures show the Cole–Cole plots of nanocrystalline and bulk Li_2TiO_3 at different temperatures. At low temperature, we have noticed that there is only one semicircle present in the plot which shows only bulk property of the material. After increasing the temperature, the curve bends which forms a depression in the real axis, confirming relaxation in the material. This non-ideal behavior could be attributed to several factors such as grain orientation, grain boundary stress strain phenomena and atomic defect distribution. The semicircular arc of the CIS can be expressed as an equivalent circuit consisting of RC circuits. The presence of two semicircular arcs in the impedance spectrum indicates the presence of both bulk and grain boundary effects in the material. The high-frequency curves are attributed to the bulk property, whereas the low-frequency curves are attributed to grain boundary effect. It has been seen that the sample resistance decreases with increase in

measuring temperature which is an effect analogous to the negative temperature coefficient of resistance (NTCR) of the material. The two equivalent circuits containing RC elements are frequently encountered to describe the grain and grain boundary phenomena in the material. In the circuit model, both grain and grain boundary behaviors are assumed to follow a Debye-like behavior. By considering well-known Brick layer model where the conduction through the grain and grain boundary dominates, the equivalent electrical equation can be written as

$$Z^* = Z' - jZ'' = \frac{1}{R_g^{-1} + j\omega C_g} + \frac{1}{R_{gb}^{-1} + j\omega C_{gb}} \quad (8)$$

$$Z' = \frac{R_g}{1 + (\omega R_g C_g)^2} + \frac{R_{gb}}{1 + (\omega R_{gb} C_{gb})^2} \quad (9)$$

$$Z'' = R_g \left[\frac{\omega R_g C_g}{1 + (\omega R_g C_g)^2} \right] + R_{gb} \left[\frac{\omega R_{gb} C_{gb}}{1 + (\omega R_{gb} C_{gb})^2} \right] \quad (10)$$

At high temperature, the grain boundary plays an important role. The value of Z'' can be written as

$$Z'' = R_{gb} \left[\frac{\omega R_{gb} C_{gb}}{1 + (\omega R_{gb} C_{gb})^2} \right] \quad (11)$$

Figure 5 shows the variations of real part of the impedance with frequency at different temperatures. It is observed from the figures that magnitude of Z' decreases with both frequency and temperature, indicating an increase in AC conductivity with rise in temperature and frequency. The Z' values for all temperatures merge at high frequency. This is due to the release of space charges as a result of reduction in barrier properties of the material with rise in temperature [19,20]. Further at low frequency, the values of Z' decrease with rise in temperature showing NTCR-type behavior similar to that of semiconductors. It is noticed that before merging into a single line, real part of impedance gives a dip which is associated with the paraelectric phase of the material. This dip may be due to charge carrier hopping associated with the material.

Figure 6 shows the variations of imaginary part of the impedance with frequency at different temperatures. The spectrum is characterized by a few important features, such as (1) appearance of peaks in the spectrum, (2) peak broadening with increase in temperature and (3) asymmetric peak broadening. The curves show that the value of Z'' reaches a maximum peak (Z''_{max}) at certain frequencies and falls down with

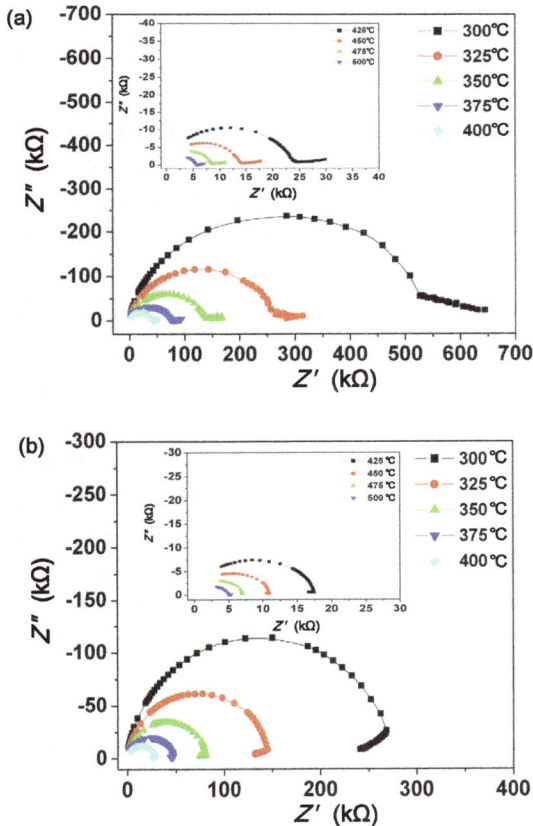

Fig. 4 Variations of real (Z') and imaginary (Z'') parts of complex impedance of Li_2TiO_3 at different temperatures: (a) bulk and (b) nanocrystalline .

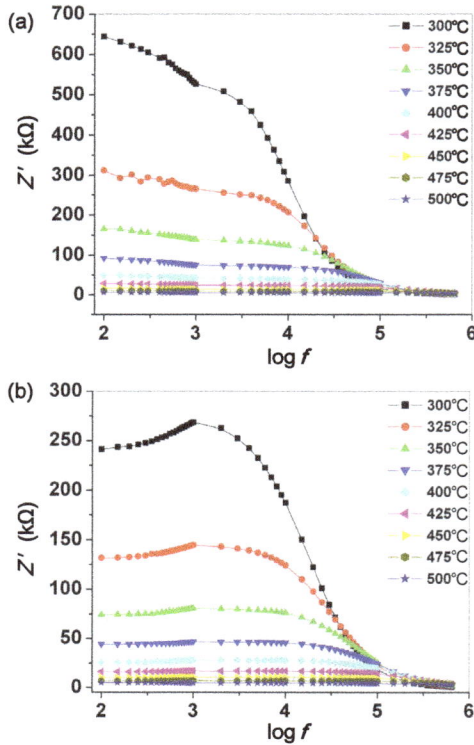

Fig. 5 Variations of real part (Z') of complex impedance of Li_2TiO_3 with log$f$ at different temperatures: (a) bulk and (b) nanocrystalline.

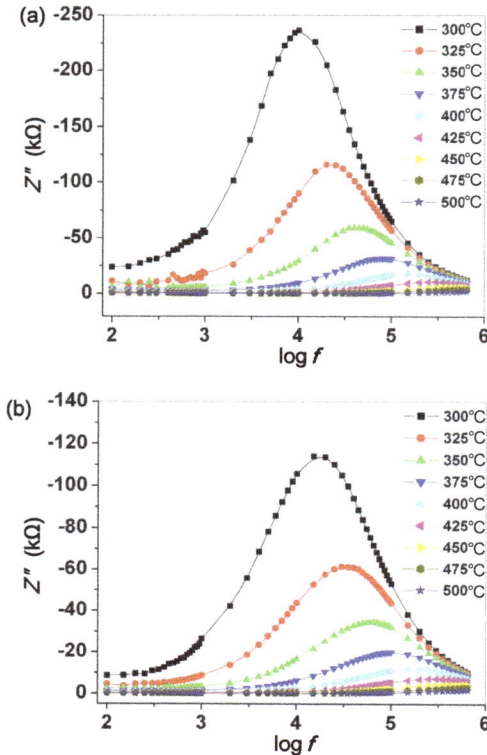

Fig. 6 Variations of imaginary part (Z'') of complex impedance of Li_2TiO_3 with log$f$ at different temperatures: (a) bulk and (b) nanocrystalline.

the value of the peak shifts towards higher frequency, which indicates active conduction through the grain boundary. The magnitude of grain boundary also decreases with increase in temperature, which shows loss in resistivity property in the material. With increase in temperature, the value of Z'' merges in the high-frequency region. This may be the temperature-dependent relaxation phenomenon in which the hopping of charge carriers and polarons is dominating in the polycrystalline material [21,22]. From the graphs, it shows that with the increase in temperature, peak broadening occurs which shows the temperature-dependent relaxation phenomenon in the material. The asymmetric broadening of peaks suggests the spread of relaxation time in the material.

Figure 7 shows variations of AC conductivity with frequency at different temperatures. The conductivity has increased monotonically with frequency in the measured frequency range from 100 Hz to 1 MHz. The slope of the conductivity gives the amount of charge carriers present in the material. The value of AC conductivity can be easily understood from the power law given by Ref. [23]:

$$\sigma(\omega) = A(T)\omega^n + \sigma(T) \qquad (12)$$

where $A(T)$ is a pre-exponential function dependent on temperature; $\sigma(T)$ corresponds to the DC conductivity; and n is a universal factor. It is a function of both temperature and frequency and corresponds to short-range hopping of charge carriers through trap sites separated by energy barriers of different heights. Figure 8 shows variations of AC conductivity with temperature at different frequencies. The bulk electrical conductivity can be written as

$$\sigma_b = \frac{1}{R_b} \times \frac{l}{A} \qquad (13)$$

where l is the thickness; A is the area of the electrode deposited in the sample. The values of bulk resistance R_b and grain boundary resistance R_{gb} are obtained from the high-frequency and low-frequency intercepts of the semicircle on the real axis in the CIS plot, respectively. Figure 9 shows the variations of DC conductivity of both bulk and nanocrystalline Li_2TiO_3.

When the starting reactants are used to prepare Li_2TiO_3, there is a distinct possibility that the resulting compound will be oxygen rich and cations will act as acceptor sources causing conduction in the material. In ceramic samples, some conduction mechanisms associated with material properties are as follows:

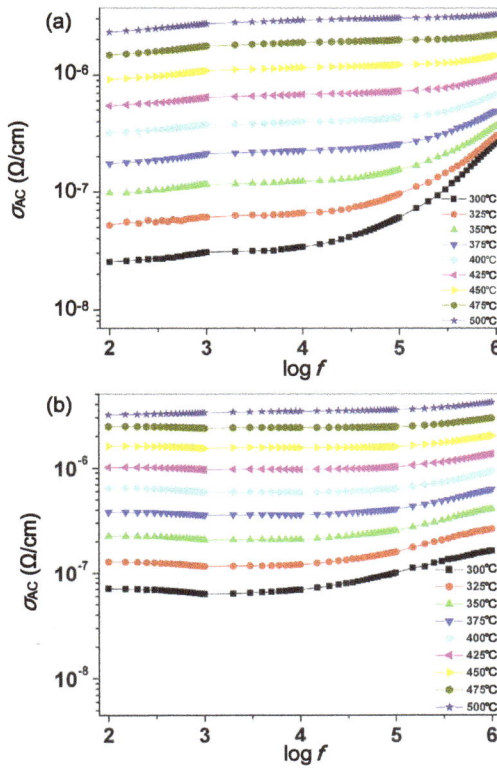

Fig. 7 Variations of AC conductivity of Li_2TiO_3 with $\log f$ at different temperatures: (a) bulk and (b) nanocrystalline.

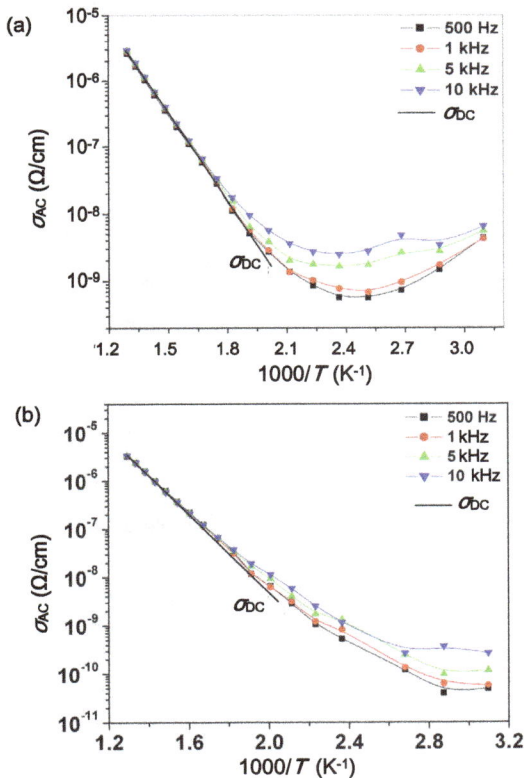

Fig. 8 Arrhenius plots of AC conductivity of Li_2TiO_3: (a) bulk and (b) nanocrystalline.

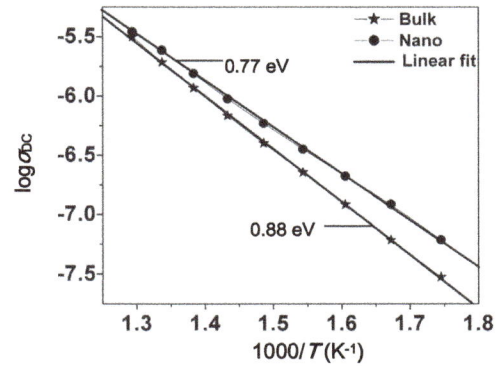

Fig. 9 Variations of DC conductivity of bulk and nanocrystalline Li_2TiO_3 with inverse of temperature.

(1) The conduction mechanism due to the small polarons associated with lattice strain accompanied by free charges.

(2) Ionic conductivity, which is an intrinsic conduction mechanism in dielectric materials at high temperatures.

Each one of the above mentioned mechanisms contributes to the total electrical conductivity of the compound and is thermally activated. In order to explain the conductivity as a function of temperature and frequency, we have to investigate lithium incorporation and microstructure of the compound [23]. In ceramic samples, oxygen vacancies are usually considered as one of the mobile charge carriers. The ionization of oxygen vacancies creates conducting electrons, which are easily thermally activated. From the results of conduction and the value of the activation energy for conduction, it clearly suggests a possibility that the conduction of charge carriers in the high-temperature range may be oxygen vacancies [24]. The conductivity variation indicates an increase of conductivity with rise in temperature with a typical Arrhenius-type behavior having linear dependence on logarithm of frequency. The type of temperature dependence of DC conductivity indicates that the electrical conduction in the material is a thermally activated process [25]. The activation energy is calculated from the linear portion of the plot of DC conductivity versus $10^3/T$. The activation energy calculated from bulk and grain boundary effects is higher as compared to bulk sample. Bulk resistance (R_b), bulk capacitance (C_b) and relaxation frequency (f_0) of Li_2TiO_3 sample are shown in Table 1.

Table 1   Bulk resistance (R_b), bulk capacitance (C_b) and relaxation frequency (f_0) of Li$_2$TiO$_3$ nanoceramic sample and $\omega R_b C_b$ as product

Parameter	System	Temperature (℃)								
		300	325	350	375	400	425	450	475	500
R_b (10^{4}Ω)	Bulk	55	27	13.5	7.25	4.1	2.4	1.4	0.85	0.52
	Nano	28	14	8.1	4.8	2.9	1.8	1.1	0.7	0.52
C_b (10^{-11}F)	Bulk	72.3568	1.17914	1.68449	87.8261	86.2791	1.20594	1.51604	2.08084	3.06124
	Nano	1.13703	1.17914	1.96524	73.6967	1.09782	1.36055	1.80892	2.39375	3.06124
f_0 (10^{4}Hz)	Bulk	4	5	7	25	45	55	75	90	100
	Nano	5	6.5	10	45	50	65	80	95	100
$\omega R_b C_b$	Bulk	0.9999	0.9999	0.9999	0.9999	0.9999	0.9999	0.9999	0.9999	0.9999
	Nano	0.9999	0.9999	0.9999	0.9999	0.9999	0.9999	0.9999	0.9999	0.9999

Electric modulus formula is proposed to understand the electrical conduction phenomena occurred in the material. The complex electric modulus M^* is defined in terms of complex dielectric constant ε^* and represented as

$$M^* = M' + iM'' = 1/\varepsilon^* \tag{14}$$

$$M^* = M' + iM'' = \frac{\varepsilon_r'}{(\varepsilon_r')^2 + (\varepsilon_r'')^2} + i\frac{\varepsilon_r''}{(\varepsilon_r')^2 + (\varepsilon_r'')^2} \tag{15}$$

where $M' = \omega C_0 Z''$ and $M'' = \omega C_0 Z'$; C_0 is the geometrical capacitance.

Figure 10 shows the variation of real part of complex modulus with frequency at different temperatures. M' approaches to zero at all temperatures suggesting the suppression of electrode polarization. M' reaches a maximum value corresponding to $M_\infty = (\varepsilon_\infty)^{-1}$ due to the relaxation process. It is also observed that the value of M_∞ decreases with increase in temperature. Figure 11 shows the variation of imaginary part of complex modulus with frequency at different temperatures. The modulus peak shifts towards higher frequencies with increase in temperature. This evidently suggests the involvement of temperature-dependent relaxation phenomenon occurs in the material. The frequency region below the M'' peak indicates the long range drift of Li$^+$ ions, whereas above the peak the ions are spatially confined to potential wells and are free to move within the wells. The frequency range where the imaginary peak occurs shows the long range to short range mobility of charge carriers. The electric modulus (M^*) could be expressed in terms of Fourier transform of a relaxation function $\varphi(t)$:

$$M^* = M_\infty \left[1 - \int_0^\infty \exp(-\omega t)\left(-\frac{d\varphi}{dt}\right)dt \right] \tag{16}$$

where the function $\varphi(t)$ is the time evolution of the

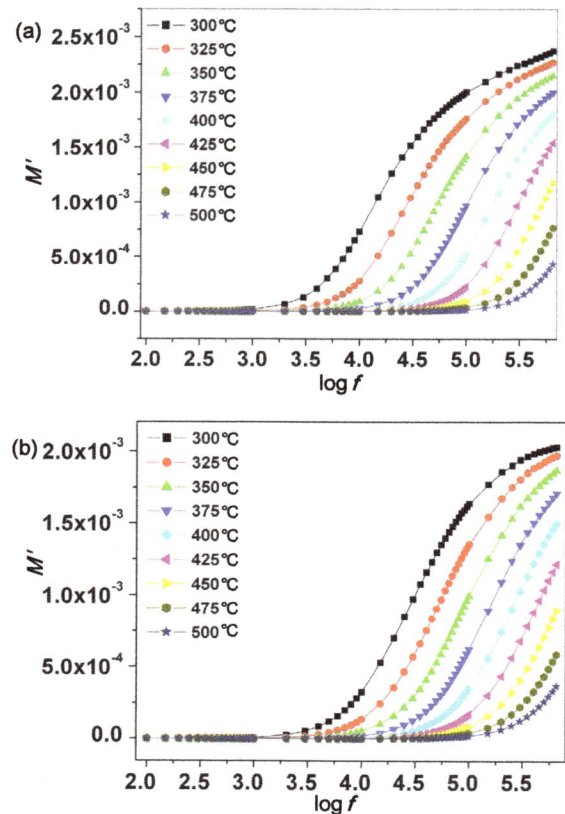

Fig. 10   Variation of real part of modulus of Li$_2$TiO$_3$ with log$f$ at different temperatures: (a) bulk and (b) nanocrystalline.

electric field within the material and is usually taken as the Kohlrausch–Williams–Watts (KWW) function:

$$\varphi(t) = \exp\left[-\left(\frac{t}{\tau_m}\right)^\beta \right] \tag{17}$$

where τ_m is the conductivity relaxation time; the exponent β $(0 < \beta < 1)$ indicates the deviation from Debye-type relaxation. The imaginary part of electric modulus (M'') is defined as

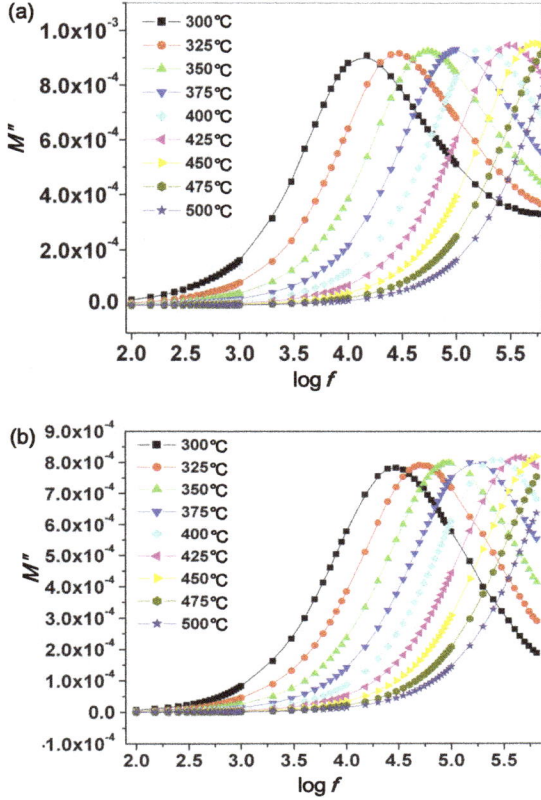

Fig. 11 Variation of imaginary part of modulus of Li$_2$TiO$_3$ with log$f$ at different temperatures: (a) bulk and (b) nanocrystalline.

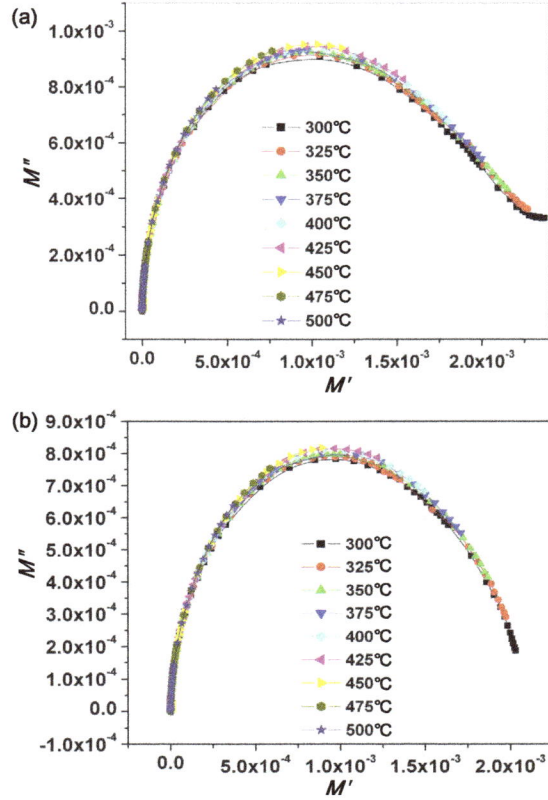

Fig. 12 Complex modulus plot of Li$_2$TiO$_3$ at different temperatures: (a) bulk and (b) nanocrystalline.

$$M'' = \frac{M''_{max}}{(1-\beta) + \frac{\beta}{1+\beta}\left[\beta\left(\frac{\omega_{max}}{\omega}\right) + \left(\frac{\omega}{\omega_{max}}\right)^{\beta}\right]} \quad (18)$$

where M''_{max} is the peak value of M''; ω_{max} is the corresponding frequency.

Figure 12 shows the variation of M'' with M' at different temperatures from 300 ℃ to 500 ℃. It shows a complete semicircle overlapping each other. The overlapping in the case of M'/M'' shows a strong evidence of long range of conductivity. It also shows that all dynamic processes occurring at different frequencies exhibit the same thermal activation energy.

4   Conclusions

In summary, we have performed a comparative investigation of bulk and nanocrystalline Li$_2$TO$_3$ ceramics. We find that (1) single-phase monoclinic structure is exhibited at 700 ℃ (nano sample) as compared to bulk sample (1000 ℃); (2) charge carriers are more prominent in the nano sample for which it gives more conduction; (3) activation energy, AC conductivity, DC conductivity are greatly improved in the case of nano sample which in turn improves conductivity; (4) bulk resistance and grain boundary resistance of nano sample are found to be lower than those of bulk sample. From the electrical properties, it shows that charge carriers are confined in their respective potential well. It also shows that due to oxide ions conduction process occurs in the material. Two semicircles reflect both bulk and grain boundary conduction in the material.

Acknowledgements

We thank the Board of Research in Fusion Science and Technology (BRFST), Institute for Plasma Research Gandhinagar India for support of the research (Grant No. NFP/MAT/F10-01).

References

[1] Ohno H, Konishi S, Nagasaki T, et al. Correlation behavior of lithium and tritium in some solid breeder materials. *J Nucl Mater* 1985, **133–134**: 181–185.

[2] Roux N, Tanaka S, Johnson C, et al. Ceramic breeder material development. *J Nucl Mater* 1998, **41**: 31–38.

[3] Roux N, Avon J, Floreancing A, et al. Low-temperature tritium releasing ceramics as potential materials for the ITER breeding blanket. *J Nucl Mater* 1996, **233–237**: 1431–1435.

[4] Hofmann P, Dienst W. Compatibility studies of metallic materials with lithium-based oxides. *J Nucl Mater* 1988, **155–157**: 485–490.

[5] Rasneur B, Charpin J. Chemical properties of lithium ceramics: Reactivity with water and water vapour. *J Nucl Mater* 1988, **155–157**: 461–465.

[6] Vittal Rao TV, Bamankar YR, Mukerjee SK, et al. Preparation and characterization of Li_2TiO_3 pebbles by internal gelation sol–gel process. *J Nucl Mater* 2012, **426**: 102–108.

[7] Wu X, Wen Z, Han J, et al. Fabrication of Li_2TiO_3 pebbles by water based sol–gel method. *Fusion Eng Des* 2008, **83**: 112–116.

[8] Wu X, Wen Z, Lin B, et al. Sol–gel synthesis and sintering of nano-size Li_2TiO_3 powder. *Mater Lett* 2008, **62**: 837–839.

[9] Deptuła A, Łada W, Olczak T, et al. Preparation of lithium titanate by sol–gel method. *Nukleonika* 2001, **46**: 95–100.

[10] Deptuła A, Brykała M, Łada W, et al. Preparation of spherical particles of Li_2TiO_3 (with diameters below 100 μm) by sol–gel process. *Fusion Eng Des* 2009, **84**: 681–684.

[11] Lulewicz D, Roux N. Fabrication of Li_2TiO_3 pebbles by the extrusion–spheronisation–sintering process. *J Nucl Mater* 2002, **307–311**: 803–806.

[12] Mandal D, Sathiyamoorthy D, Govardhana Rao V. Preparation and characterization of lithium–titanate pebbles by solid-state reaction extrusion spheronization techniques for fusion reactor. *Fusion Eng Des* 2012, **87**: 7–12.

[13] Tsuchiya K, Kawamura H, Takayama T, et al. Control of particle size and density of Li_2TiO_3 pebbles fabricated by indirect wet process. *J Nucl Mater* 2005, **345**: 239–244.

[14] Jung C-H. Sintering characterization of Li_2TiO_3 ceramic breeder powders prepared by the solution combustion synthesis process. *J Nucl Mater* 2005, **341**: 148–152.

[15] Jung C-H, Lee SJ, Waltraud M, et al. A polymer solution technique for the synthesis of nano-sized Li_2TiO_3 ceramic breeder powders. *J Nucl Mater* 2008, **373**: 194–198.

[16] Sinclair DC, West AR. Impedance and modulus spectroscopy of semiconducting $BaTiO_3$ showing positive temperature coefficient of resistance. *J Appl Phys* 1989, **66**: 3850.

[17] Lanfredi S, Rodrigues ACM. Impedance spectroscopy study of the electrical conductivity and dielectric constant of polycrystalline $LiNbO_3$. *J Appl Phys* 1999, **86**: 2215.

[18] Barsoukov E, Macdonald JR. *Impedance Spectroscopy*, 2nd edn. Hoboken, NJ: John Wiley & Sons, 2005.

[19] Barranco AP, Piñar FC, Martínez OP, et al. AC behaviour and conductive mechanisms of 2.5 mol% La_2O_3 doped $PbZr_{0.53}Ti_{0.47}O_3$ ferroelectric ceramics. *J Eur Ceram Soc* 1999, **19**: 2677–2683.

[20] Bharadwaj SSN, Victor P, Venkateswarulu P, et al. AC transport studies of La-modified antiferroelectric lead zirconate thin films. *Phys Rev B* 2002, **65**: 174106.

[21] Cao W, Gerhardt R. Calculation of various relaxation times and conductivity for a single dielectric relaxation process. *Solid State Ionics* 1990, **42**: 213–221.

[22] Gerhardt R. Impedance and dielectric spectroscopy revisited: Distinguishing localized relaxation from long-range conductivity. *J Phys Chem Solids* 1994, **55**: 1491–1506.

[23] Argall F, Jonscher AK. Dielectric properties of thin films of aluminium oxide and silicon oxide. *Thin Solid Films* 1968, **2**: 185–210.

[24] Fehr Th, Schmidbauer E. Electrical conductivity of Li_2TiO_3 ceramics. *Solid State Ionics* 2007, **178**: 35–41.

[25] León C, Rivera A, Várez A, et al. Origin of constant loss in ionic conductors. *Phys Rev Lett* 2001, **86**: 1279–1282.

Structural, spectroscopic, and dielectric characterizations of Mn-doped 0.67BiFeO₃–0.33BaTiO₃ multiferroic ceramics

Qiming HANG^a, Wenke ZHOU^a, Xinhua ZHU^{a,*}, Jianmin ZHU^a,
Zhiguo LIU^b, Talaat AL-KASSAB^c

^{a}*National Laboratory of Solid State Microstructures, School of Physics, Nanjing University, Nanjing 210093, China*
^{b}*National Laboratory of Solid State Microstructures, Department of Materials Science and Engineering, Nanjing University, Nanjing 210093, China*
^{c}*King Abdullah University of Science & Technology (KAUST), Physical Sci. and Eng., Thuwal 23955-6900, Kingdom of Saudi Arabia*

Abstract: 0.67BiFeO₃–0.33BaTiO₃ multiferroic ceramics doped with x mol% MnO₂ (x = 2–10) were synthesized by solid-state reaction. The formation of a perovskite phase with rhombohedral symmetry was confirmed by X-ray diffraction (XRD). The average grain sizes were reduced from 0.80 μm to 0.50 μm as increasing the Mn-doped levels. Single crystalline nature of the grains was revealed by high-resolution transmission electron microscopy (HRTEM) images and electron diffraction patterns. Polar nano-sized ferroelectric domains with an average size of 9 nm randomly distributed in the ceramic samples were revealed by TEM images. Ferroelectric domain lamellae (71° ferroelectric domains) with an average width of 5 nm were also observed. Vibrational modes were examined by Raman spectra, where only four Raman peaks at 272 cm^{-1} (E-4 mode), 496 cm^{-1} (A_1-4 mode), 639 cm^{-1}, and 1338 cm^{-1} were observed. The blue shifts in the E-4 and A_1-4 Raman mode frequencies were interpreted by a spring oscillator model. The dieletric constants of the present ceramics as a function of the Mn-doped levels exhibited a V-typed curve. They were in the range of 350–700 measured at 10³ Hz, and the corresponding dielectric losses were in range of 0.43–0.96, approaching to 0.09 at 10⁶ Hz.

Keywords: multiferroic ceramics; dielectric properties; Raman spectra; microstructure

1 Introduction

Recently, multiferroic materials have received much attention due to their promising applications in novel electronic devices (e.g., multiple-state memories and new data-storage media), since multiferroics with multiple (charge, spin) order parameters can offer an exciting way of coupling between the electronic and magnetic orderings [1,2]. Besides the potential applications, the fundamental physics of multiferroic materials is also interesting and fascinating. However, it is generally difficult to find materials that are magnetic as well as ferroelectric because ferroelectricity occurs as the metal ions have empty d-orbitals, whereas a partly filled d shell is necessary

---

* Corresponding author.
E-mail: xhzhu@nju.edu.cn

for magnetism to occur in transition metal ions [3]. As one except, $BiFeO_3$ is perhaps the only multiferroic material that is both magnetic and strong ferroelectric at room temperature, which has been widely investigated in the past few years [2]. However, the inherent problems in $BiFeO_3$ such as high leakage current, structural instability, and the formation of multiphase system during the synthesis, have limited its applications [4]. To solve these problems, some attempts have been made including small dopants (e.g., lanthanides such as La, Ce, Eu) at the Bi/Fe sites to improve the dielectric properties of $BiFeO_3$ [5,6]. There are also several reports on the synthesis of solid solution of $BiFeO_3$ with other ABO_3 perovskite materials such as $BaTiO_3$ [4,7–9] and $PbTiO_3$ [10] to stabilize the perovskite structure and enhance the electrical insulating properties of $BiFeO_3$. For example, in the $BiFeO_3$–$BaTiO_3$ solid solution system, anomalous dielectric and magnetic behaviors were observed at concentration of 33 mol% of $BaTiO_3$, which was closely related to the structural transformation between the rhombohedral and cubic phases. The composition of $0.67BiFeO_3$–$0.33BaTiO_3$ solid solution exhibited the spontaneous polarizations of 35 $\mu C/cm^2$ at room temperature [8]. Although the perovskite structure of these ceramics has been found to be stable and the spontaneous magnetic moment is increased at room temperature, the insulation resistance of $BiFeO_3$–ABO_3 ceramics has not been improved sufficiently and their ferromagnetism is still weak. Previous investigations on the ferroelectric and ferromagnetic properties of $BiFeO_3$–$BaTiO_3$ solid solution have demonstrated that their dielectric properties were complicated by high dielectric losses and high conductivity, which were ascribed to the valence fluctuation of Fe and the formation of oxygen vacancies [4,8]. Recently, it has been reported that Mn-doping can significantly reduce the leakage current and dielectric losses of the $BiFeO_3$–$BaTiO_3$ solid solution [11,12]. However, up to now, the effects of the Mn-doped levels on the microstructure and dielectric properties of the $BiFeO_3$–$BaTiO_3$ ceramic system are not systematically investigated, especially in the ferroelectric nanodomain structures.

Here we report on the structural, spectroscopic, and dielectric characterizations of the Mn-doped $0.67BiFeO_3$–$0.33BaTiO_3$ multiferroic ceramics with x mol% MnO_2 ($x = 2, 3, 4, 5, 6, 8, 10$), which were synthesized by the solid-state reaction route.

2 Experimental details

Multiferroic ceramics of $0.67BiFeO_3$–$0.33BaTiO_3$ doped with x mol% MnO_2 ($x = 2, 3, 4, 5, 6, 8, 10$) were prepared by a conventional solid-state reaction method. The stoichiometric amounts of the analytical-grade starting materials, Bi_2O_3 (with 5 mol% excess), Fe_2O_3, MnO_2, $BaCO_3$, and TiO_2 were weighted and mixed in a ball mill for 24 h in agate container with agate balls. After drying at 120 ℃, the mixed powders were ground, then pressed into disc shape and pre-sintered at 850 ℃ for 2 h in air. Subsequently, the pre-sintered disc samples were ball-milled again for 24 h to get crushed powders. After drying, the crushed powders mixed with 5 wt% polyvinyl alcohol binders were pressed into cylindrical pellets with a diameter of 10 mm and thickness of about 2 mm. Disk pellets were sintered at 950 ℃ for 1 h. To measure the dielectric properties, the specimens were electroded with postfire silver paste on both sides. The dielectric properties of the sintered ceramics were measured as a function of frequency using an HP4192A impedance analyzer controlled by a computer.

The crystal structures of the sintered pellets were examined by X-ray diffraction (XRD, Philips X'Pert MRD four-circle diffractometer, Almelo, the Netherlands). To obtain the structural information of the Mn-doped $0.67BiFeO_3$–$0.33BaTiO_3$ multiferroic ceramics from vibrational spectra, Raman scattering was carried out at room temperature with a Jobin Yvon HR800 spectrometer (JY Ltd., France), and the visible laser light (wavelength 514.5 nm) was used as the excitation source. The grain morphology, ferroelectric nanodomain structures, crystal lattice structures, and grain boundary structures were examined by a field-emission transmission electron microscopy (FEI, Titan S/TEM, operated at 300 kV).

3 Results and discussion

3.1 Phase structure

Figure 1 shows the XRD patterns of the $0.67BiFeO_3$–$0.33BaTiO_3$ ceramics doped with x mol% MnO_2 ($x = 2, 3, 4, 5, 6, 8, 10$), which indicate the formation of high-purity perovskite structure. All the XRD patterns can be indexed based on the data of $BiFeO_3$ with rhombohedrally distorted perovskite structure. To demonstrate the characteristic splitting of the diffraction peaks such as (006) and (202) peaks in the XRD patterns, the local XRD pattern around $2\theta = 40°$ is shown as inset in Fig. 1. From the inset, the characteristic splitting of the diffraction peaks associated with deviation from cubic symmetry can be clearly observed, although the present Mn-doped $0.67BiFeO_3–0.33BaTiO_3$ ceramics have small rhombohedral distortions. With increasing the Mn-doped levels from 2 mol% to 10 mol%, the changes of the lattice parameters a and c of the Mn-doped $0.67BiFeO_3–0.33BaTiO_3$ ceramics are shown in Fig. 2(a), and Fig. 2(b) demonstrates the change of the unit cell volume. Basically, the lattice parameter a is decreased with increasing the Mn-doped levels, whereas the lattice parameter c is first increased at Mn-doping level of 2 mol%, then decreases, and then remains almost constant value of 1.383 nm as the Mn-doped levels are in the range of 3–10 mol%. The unit cell volume is first decreased as increasing the Mn-doped levels, and reaches a minimum value at the Mn-doped level of 8 mol%, and then increases as further increasing the Mn-doped levels. The above phenomena can be understood by considering the following reaction:

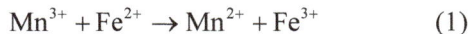

$$Mn^{3+} + Fe^{2+} \rightarrow Mn^{2+} + Fe^{3+} \tag{1}$$

which depresses the electron hopping from Fe^{3+} to Fe^{2+} and reduces the levels of oxygen vacancies. Since the ionic radius of Mn^{3+} ions (72.0 pm, coordination number 6 (low spin)) is smaller than that of Fe^{2+} ions

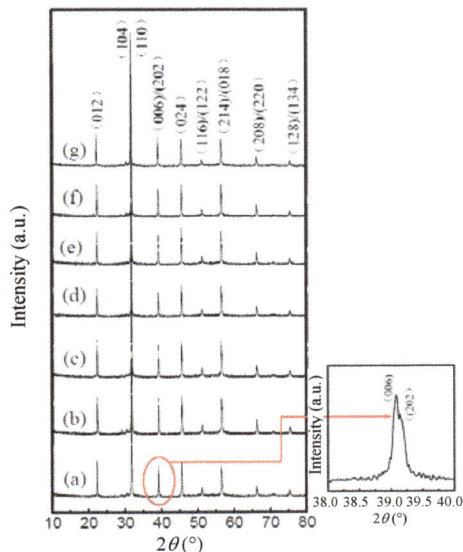

Fig. 2 Composition dependence of (a) the lattice parameters a and c of the Mn-doped $0.67BiFeO_3–0.33BaTiO_3$ ceramics, and (b) the volume of unit cell.

(75.0 pm, coordination number 6 (low spin)) [13], therefore as increasing the Mn-doped concentration to 8 mol%, the doped Mn^{3+} ions possessing smaller ionic radius will evidently reduce the unit cell volume; whereas as further increasing the Mn-doped levels, more Mn^{3+} (72.0 pm, coordination number 6 (low spin)) and Mn^{2+} (81.0 pm, coordination number 6 (low spin)) may substitute the Fe^{3+} (69.0 pm, coordination number 6 (low spin)) and Ti^{4+} (74.5 pm, coordination number 6) ions [13], respectively. Thus, the volume of the unit cell is increased, as demonstrated in Fig. 2(b).

3.2 Raman spectra

Figure 3 illustrates the Raman spectra of the Mn-doped $0.67BiFeO_3–0.33BaTiO_3$ multiferroic ceramics, where only four characteristic Raman peaks near 272 cm^{-1}, 496 cm^{-1}, 639 cm^{-1} and 1338 cm^{-1}, are clearly observed. A very small Raman peak at around 870 cm^{-1} in curve (d) is due to the artifact. It is known that there are 13 Raman active modes ($4A_1 + 9E$) for perovskite $BiFeO_3$ with distorted rhombohedral ($R3c$) structure [14]. Since the Mn-doped $0.67BiFeO_3–0.33BaTiO_3$ multiferroic ceramics have the same symmetry, the same set of the Raman active modes should be expected in the present case. However, actually only four Raman peaks are observed. It is known that the Raman spectra of crystals with relatively small unit cells and a limited number of

Fig. 1 XRD patterns of the $0.67BiFeO_3–0.33BaTiO_3$ ceramics doped with x mol% MnO_2: (a) $x=2$, (b) $x=3$, (c) $x=4$, (d) $x=5$, (e) $x=6$, (f) $x=8$, and (g) $x=10$.

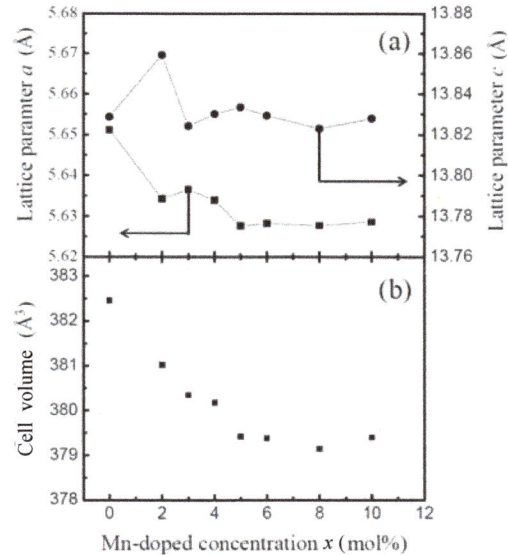

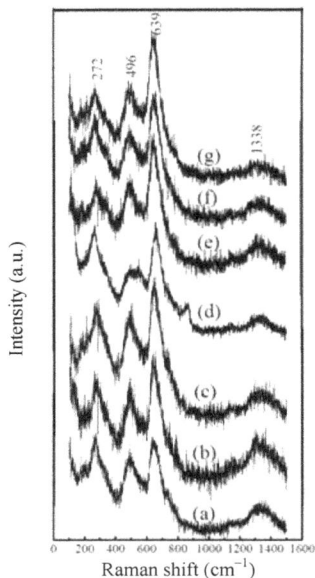

Fig. 3 Raman spectra of the $0.67BiFeO_3$–$0.33BaTiO_3$ multiferroic ceramics doped with x mol% MnO_2: (a) $x=2$, (b) $x=3$, (c) $x=4$, (d) $x=5$, (e) $x=6$, (f) $x=8$, and (g) $x=10$.

atoms, usually exhibit all of the Raman modes (like the case of $BiFeO_3$), as expected from symmetry arguments. However, in a complex structure with a large number of atoms in a unit cell, experimental results account for fewer of the calculated Raman modes to appear [15]. Therefore, some Raman active modes predicted for the rhombohedral symmetry are not observed in the Mn-doped $0.67BiFeO_3$–$0.33BaTiO_3$ multiferroic ceramics, which is due to the too small rhombohedral distortions to activate these Raman modes, or namely the symmetry-forbidden scattering. The Raman peaks near 272 cm^{-1} and 496 cm^{-1} can be assigned to the E-4 and A_1-4 modes of $BiFeO_3$ structure, respectively. Similar Raman modes were reported previously in $BiFeO_3$ crystal by Fukumura et al. [16] (e.g., E-4 mode 279 cm^{-1}, and A_1-4 mode 490 cm^{-1}), and in the $0.70BiFeO_3$–$0.30BaTiO_3$ solid solution by Ianculescu et al. [17]. As shown in Fig. 3, the Raman frequencies of the E-4 and A_1-4 modes are shifted slightly to high frequency as increasing the Mn-doped levels up to 10 mol%. As viewed from a spring oscillator model for interatomic forces, the frequency (f) of the E-4 or A_1-4 Raman mode is dependent on several local factors, such as the magnitude of the force constant (k) of the spring oscillator, ionic mass (m) of the doped-element, and the specific bond length betweenthe oxygen atom

and the doped element. So one can describe the mode frequency f by the following equation [18]:

$$f \propto \sqrt{\frac{k}{m}} \propto \frac{1}{\text{bond length between atoms}} \quad (2)$$

Due to the smaller molar weight of the Mn atoms replacing the Fe atoms with larger molar weight, the effective mass of the spring oscillator will become smaller, which results in higher vibrating frequency for the mode of E-4 or A_1-4, as expected from Eq. (2). The bond length between the oxygen and the doped atom is also expected to be reduced, leading to a smaller volume of the unit cell of the Mn-doped $0.67BiFeO_3$–$0.33BaTiO_3$ multiferroic ceramics. This is consistent with the above XRD results. The observed Raman band centered near 639 cm^{-1}, is attributed from the symmetrical stretching vibrations of the basal oxygen ions of the MnO_6 (and perhaps FeO_6) octahedra in the Mn-doped $0.67BiFeO_3$–$0.33BaTiO_3$ ceramics. With increasing the Mn-doped level from 2 mol% to 10 mol%, the intensity of this Raman band is increased, as observed in Fig. 3. That can be attributed to the structural distortion of $(Mn,Fe)O_6$ octahedra (i.e., Jahn–Teller distortion) due to the substitution of Fe by Mn. The very wide Raman peak near 1338 cm^{-1}, may be attributed to the random occupancy of the cations at A site by Bi/Ba, and at B site by Fe, Mn and Ti ions, and also the microscopic strains in the Mn-doped $0.67BiFeO_3$–$0.33BaTiO_3$ ceramics. Considering a unit cell containing either FeO_6, TiO_6, or MnO_6 octahedron, due to the different magnitudes of the Fe–O, Ti–O, and Mn–O bond force constants, the bond distances can fluctuate from one unit cell to the other, which lead to different characteristic vibrational frequencies for these three octahedrals, resulting in a wider Raman band [19]. Similar phenomenon was reported in the La-doped $BiFeO_3$–$PbTiO_3$ mixed crystal system [20].

3.3 Microstructures revealed by TEM

TEM images of typical ceramic grains in the $0.67BiFeO_3$–$0.33BaTiO_3$ ceramics doped with x mol% MnO_2 ($x=3$, 4, 6, and 10) are shown in Fig. 4, which demonstrate that the grains basically have rounded grain morphology. Some grains exhibit grey TEM contrast whereas others have a bright fluctuating mottled contrast, especially clear in Figs. 4(b) and 4(d). The grains contact tightly with each other and the grain boundaries are very clean; as a consequence, triple-grain boundary regions are rarely observed. The

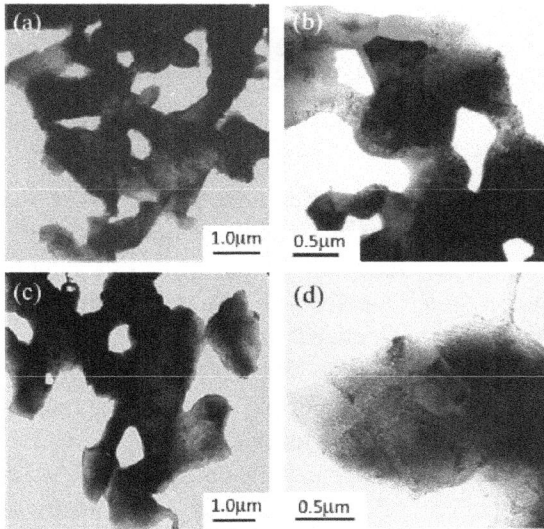

Fig. 4   Bright-field TEM images of the typical grains in the $0.67BiFeO_3$–$0.33BaTiO_3$ ceramics doped with x mol% MnO_2: (a) $x=3$, (b) $x=4$, (c) $x=6$, and (d) $x=10$.

average grain size is measured to be 0.80 μm in the $0.67BiFeO_3$–$0.33BaTiO_3$ solid solution doped with 3 mol% MnO_2 (Fig. 4(a)), and 0.60 μm for the samples doped with 4 mol% MnO_2 and 6 mol% MnO_2, and 0.50 μm for the sample doped with 10 mol% MnO_2. Therefore, the average grain size shows a tendency to be decreased as increasing the Mn-doped levels. That can be ascribed to the fact that the grain growth during the sintering process is suppressed by a diffusion-controlled process through the formed liquid phases by the aid of the doped MnO_2. The HRTEM image and the selected area electron diffraction (SAED) pattern taken from a single grain in the $0.67BiFeO_3$–$0.33BaTiO_3$ ceramics doped with x mol% MnO_2 ($x=3$, 4, 6, and 10) are shown in Fig. 5. All the HRTEM images clearly reveal two-dimensional lattice fringes (as indicated in Fig. 5), indicating well developed atomic arrangements in the Mn-doped ceramics. The SAED patterns shown as insets in Fig. 5 also demonstrate that the grains have a single crystalline nature, and their lattice constants determined from the SAED patterns match well with the XRD analysis.

Figure 6 shows the bright- and dark-field TEM images of the domain structures in the $0.67BiFeO_3$–$0.33BaTiO_3$ solid solution doped with 3 mol% MnO_2. Nano-sized ferroelectric domains are clearly observed in the dark-field TEM image (Fig. 6(b)), which exhibit bright TEM contrast and are

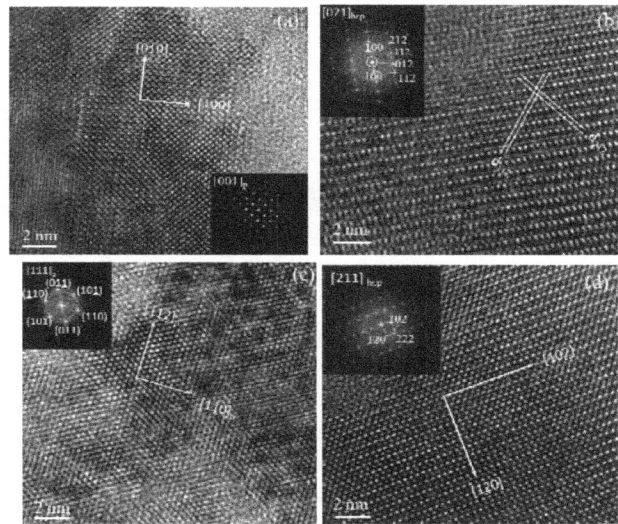

Fig. 5   HRTEM images taken from single grain in the $0.67BiFeO_3$–$0.33BaTiO_3$ ceramics doped with x mol% MnO_2: (a) $x=3$, (b) $x=4$, (c) $x=6$, and (d) $x=10$. The insets are the SAED patterns taken from a single grain or the fast Fourier patterns of the corresponding HRTEM images of different grains. The two-dimensional lattice fringes are indicated.

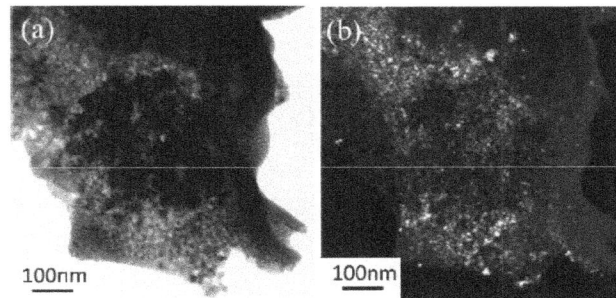

Fig. 6   (a) Bright- and (b) dark-field TEM images of the ferroelectric domain structures observed in the $0.67BiFeO_3$–$0.33BaTiO_3$ solid solution doped with 3 mol% MnO_2.

randomly distributed in the ceramic sample. Their sizes are in the range of 4–16 nm with an average size of 9 nm. Such polar nano-sized ferroelectric domains in the randomly substituted mixed-crystal systems can lead to the relaxor dielectric behavior of the mixed-crystal system. Microscopic compositional fluctuations or chemical ordering over a few nanometer length scale can result in the formation of nano-sized ferroelectric domains although the XRD yields an average cubic structure consistent with the average stoichiometry. Within the nano-sized ferroelectric domains, the local structure could be different due to the chemical ordering. Furthermore, nano-sized ferroelectric domain lamellae are also observed in the rhombohedral $0.67BiFeO_3–0.33BaTiO_3$ solid solution doped with 10 mol% MnO_2. Figure 7(a) shows the TEM image of the nano-sized ferroelectric domains lamellae, as marked by a box. The enlarged TEM image of the nano-sized ferroelectric domain lamellae is shown in Fig. 7(b), and their average width is measured to be 5 nm. The HRTEM images obtained from the local areas marked by boxes A and B in Fig. 7(b), are shown in Figs. 7(c) and 7(d), respectively. In the ferroelectric materials with rhombohedral structure like $BiFeO_3$, the ferroelectric polarizations point along the $[111]_p$ (where the subscript p represents in pseudocubic setting) directions of the perovskite unit cell, with two antiparallel polarities for each direction. Therefore, eight different polar domains can be observed in $BiFeO_3$ with rhombohedral structure. Separating adjacent domains, there are three possible types of ferroelectric domain walls, namely 71°, 109°, and 180° domain walls [2]. Since the 109° domain energy is nearly three times that of 71° domain, the latter is observed to occur more frequently than the former [21]. As shown in Figs. 7(c) and 7(d), the angle between the

ferroelectric polarizations in the adjacent nano-sized ferroelectric domain lamellae is 70.5°, so they are named as 71° ferroelectric domains.

3.4 Dielectric properties

The dielectric constants and dielectric losses of the Mn-doped $0.67BiFeO_3–0.33BaTiO_3$ ceramics were measured as a function of frequency at room temperature, which are shown in Figs. 8(a) and 8(b), respectively. It is noticed that all the dielectric constants and dielectric losses are decreased as increasing the frequency up to 10^5 Hz, and then remain almost constant in the frequency range of $10^5–10^6$ Hz. The fast decrease of the dielectric constants in the low frequency region (below 10^5 Hz), can be attributed to the space charge polarization induced by oxygen vacancies in the perovskite Mn-doped $0.67BiFeO_3–0.33BaTiO_3$ ceramics [4,8,9]. The dielectric constants of Mn-doped $0.67BiFeO_3–0.33BaTiO_3$ ceramics measured at 10^4 Hz, 10^5 Hz and 10^6 Hz as a function of the Mn-doped level, are shown as inset of Fig. 8(a). It is demonstrated that the Mn-doped $0.67BiFeO_3–0.33BaTiO_3$ ceramics at the level of 6 mol% have the lowest value of the dielectric constant, which indicates the sample with high level of oxygen vacancies. It is assumed that the doped Mn atoms in the $0.67BiFeO_3–0.33BaTiO_3$ solid solution could lead to a reaction of Eq. (1), thus suppressing the electron hopping from Fe^{3+} to Fe^{2+} ions and reducing the concentrations of the oxygen vacancies due to the donor doping with Mn. However, since the Mn atoms enter into the B-site of perovskite structure in the $0.67BiFeO_3–0.33BaTiO_3$ solid solution, there is also the possibility of substitution for Ti atoms at B-site, which can be represented as

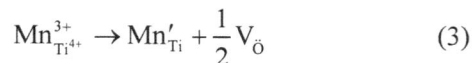

$$Mn_{Ti^{4+}}^{3+} \rightarrow Mn_{Ti}' + \frac{1}{2}V_{\ddot{O}} \qquad (3)$$

The above reaction can lead to the formation of oxygen vacancies, resulting in high dielectric loss and fast decrease of the dielectric constant at low frequency region through the space charge polarization mechanism. Therefore, in the Mn-doped $0.67BiFeO_3–0.33BaTiO_3$ ceramics, the concentrations of oxygen vacancies are controlled by the balance between the reactions of Eqs. (1) and (2). At low Mn-doped levels, Mn atoms prefer to depressing the electron hopping $(Fe^{3+} \rightarrow Fe^{2+})$ and reducing the concentrations of oxygen vacancies via reaction of Eq. (1), whereas at high Mn-doped levels, some Mn atoms can substitute

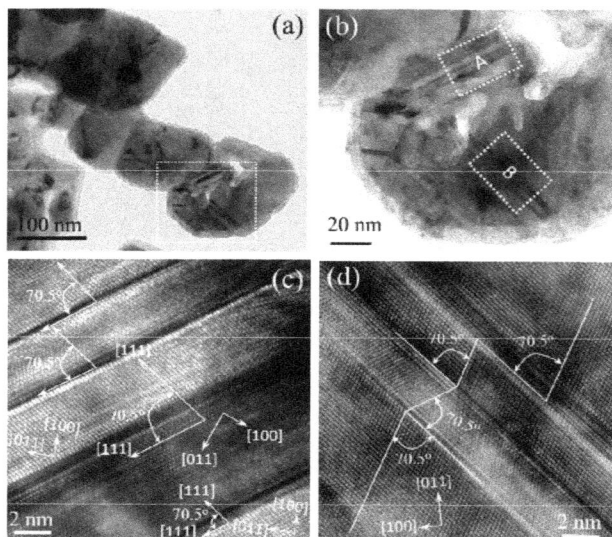

Fig. 7 (a) TEM image of nano-sized ferroelectric domains lamellae (with average width of 5 nm) observed in the $0.67BiFeO_3–0.33BaTiO_3$ solid solution doped with 10 mol% MnO_2, and (b) enlarged TEM image of the nano-sized ferroelectric domain lamellae. (c) and (d) HRTEM images of the nano-sized ferroelectric domain lamellae obtained from the local areas marked by boxes A and B in (b), respectively.

Ti atoms at B-site, creating oxygen vacancies via reaction of Eq. (3). That is the reason why the dieletric constants of Mn-doped $0.67BiFeO_3$–$0.33BaTiO_3$ ceramics as a function of the Mn-doped levels exhibit a V-typed curve. The dielectric constants (measured at 10^3 Hz) for the $0.67BiFeO_3$–$0.33BaTiO_3$ ceramics doped with 4 mol% and 10 mol% MnO_2 are in the range of 600–700, which are much higher than that for the $0.67BiFeO_3$–$0.33BaTiO_3$ ceramics doped with x mol% MnO_2 (x = 3, 5, and 6) (dielectric constants in the range of 350–475). However, the dielectric losses of the $0.67BiFeO_3$–$0.33BaTiO_3$ ceramics doped with 4 mol% and 10 mol% MnO_2 are also very high, which are in the range of 0.8–0.9 (measured at 10^3 Hz) (Fig. 8(b)). Considering the trade-off between the dielectric constant and dielectric loss of the Mn-doped $0.67BiFeO_3$–$0.33BaTiO_3$ ceramics, the optimized Mn-doped level is determined to be 3 mol%, which has both relatively higher dielectric constant and smaller dielectric loss, as demonstrated in Fig. 8.

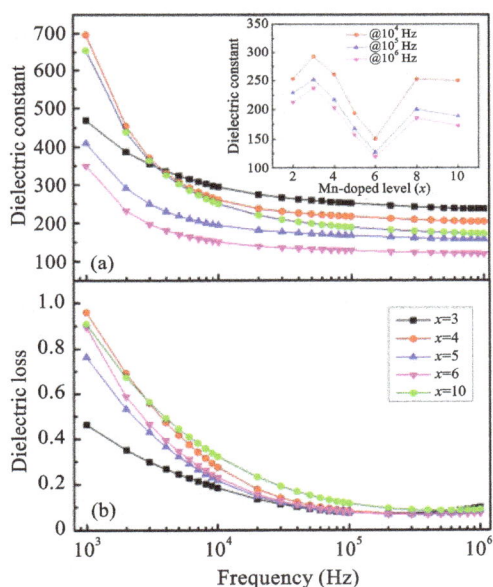

Fig. 8 (a) Room-temperature dielectric constants and (b) dielectric losses of the $0.67BiFeO_3$–$0.33BaTiO_3$ ceramics doped with different levels of MnO_2 as a function of the measured frequency. Inset in (a) is the dielectric constants of the Mn-doped $0.67BiFeO_3$–$0.33BaTiO_3$ ceramics as a function of the Mn-doped level.

4 Conclusions

Perovskite $0.67BiFeO_3$–$0.33BaTiO_3$ multiferroic

ceramics doped with x mol% MnO_2 (x = 2, 3, 4, 5, 6, 8, 10), have been synthesized by the solid-state reaction. Their structural, spectroscopic, and dielectric properties were well characterized. XRD patterns revealed that the Mn-doped $0.67BiFeO_3$–$0.33BaTiO_3$ ceramics were rhombohedral phase, and their lattice parameters a and c, and the volume of the unit cell, were closely dependent upon the Mn-doping contents. The Raman spectra revealed that only four characteristic Raman peaks near 272 cm^{-1} (A_1-4 mode), 496 cm^{-1} (E-4 mode), 639 cm^{-1}, and 1338 cm^{-1}, were observed; whereas some other Raman active modes predicted for the rhombohedral symmetry were not observed due to the too small rhombohedral distortions in the present multiferroic ceramics. A slight blue shifts in the frequencies of the E-4 and A_1-4 Raman modes were observed as increasing the Mn-doped levels, which could be interpreted by using a spring oscillator mode. TEM images revealed that the grains basically had rounded grain boundaries, and their average grain sizes were decreased from 0.80 μm to 0.50 μm with increasing the Mn-doped levels from 3 mol% to 10 mol%. The grains had a single crystalline nature as confirmed by HRTEM images and SAED patterns. Dark-field TEM images showed that the polar nano-sized ferroelectric domains with average size of 9 nm were randomly distributed in the ceramic sample. In addition, nano-sized ferroelectric domain lamellae with an average width of 5 nm were also revealed by HR-TEM images, which were identified as 71° ferroelectric domains. The dielectric constants and dielectric losses of the Mn-doped $0.67BiFeO_3$–$0.33BaTiO_3$ multiferroic ceramics were decreased fast as increasing the frequency in the low frequency region (below 10^5 Hz), and then they remained stable in the frequency range of 10^5–10^6 Hz. These phenomena were ascribed to the space charge polarization induced by the oxygen vacancies in the ceramic sysytem. The dieletric constants of Mn-doped $0.67BiFeO_3$–$0.33BaTiO_3$ ceramics as a function of the Mn-doped levels exhibited a V-typed curve. The dielectric constants (measured at 10^3 Hz) of the Mn-doped $0.67BiFeO_3$–$0.33BaTiO_3$ ceramics were in the range of 350–700, and the dielectric losses in the range of 0.43–0.96 (at 10^3 Hz), which approached to 0.09 at 10^6 Hz.

Acknowledgements

This work is financially supported by the National

Natural Science Foundation of China (Grant Nos. 10874065, 11174122 and 11134004), the National Basic Research Program of China (Grant Nos. 2009CB929503 and 2012CB619400), key project from Ministry of Science and Technology of the People's Republic of China (Grant No. 2009ZX02101-4), and Analysis & Test Fund of Nanjing University. T. Al-Kassab acknowledges the generous support of the KAUST baseline funds.

References

[1]   Cheong S-W, Mostovoy M. Multiferroics: A magnetic twist for ferroelectricity. *Nat Mater* 2007, **6**: 13–20.

[2]   Catalan G, Scott JF. Physics and applications of bismuth ferrite. *Adv Mater* 2009, **21**: 2463–2485.

[3]   Hill NA. Why are there so few magnetic ferroelectrics? *J Phys Chem B* 2000, **104**: 6694–6709.

[4]   Mahesh Kumar M, Srinivas A, Suryanarayana SV. Structure property relations in $BiFeO_3/BaTiO_3$ solid solutions. *J Appl Phys* 2000, **87**: 855.

[5]   Yan F, Lai MO, Lu L. Domain structure and piezoelectric response in lanthanide rare earth-substituted multiferroic $BiFeO_3$ thin films. *J Phys D: Appl Phys* 2012, **45**: 325001.

[6]   Kan D, Pálová L, Anbusathaiah V, *et al.* Universal behavior and electric-field-induced structural transition in rare-earth-substituted $BiFeO_3$. *Adv Funct Mater* 2010, **20**: 1108–1115.

[7]   Ueda K, Tabata H, Kawai T. Coexistence of ferroelectricity and ferromagnetism in $BiFeO_3$–$BaTiO_3$ thin films at room temperature. *Appl Phys Lett* 1999, **75**: 555.

[8]   Itoh N, Shimura T, Sakamoto W, *et al.* Fabrication and characterization of $BiFeO_3$–$BaTiO_3$ ceramics by solid state reaction. *Ferroelectrics* 2007, **356**: 19–23.

[9]   Hang Q, Xing Z, Zhu X, *et al.* Dielectric properties and related ferroelectric domain configurations in multiferroic $BiFeO_3$–$BaTiO_3$ solid solutions. *Ceram Int* 2012, **38**: S411–S414.

[10] Cheng J-R, Li N, Cross LE. Structural and dielectric properties of Ga-modified $BiFeO_3$–$PbTiO_3$ crystalline solutions. *J Appl Phys* 2003, **94**: 5153.

[11] Liu X-H, Xu Z, Qu S-B, *et al.* Ferroelectric and ferromagnetic properties of Mn-doped $0.7BiFeO_3$–$0.3BaTiO_3$ solid solution. *Ceram Int* 2008, **34**: 797–801.

[12] Leontsev SO, Eitel RE. Dielectric and piezoelectric properties in Mn-modified $(1-x)BiFeO_3$–$xBaTiO_3$ ceramics. *J Am Ceram Soc* 2009, **92**: 2957–2961.

[13] Shannon RD. Revised effective ionic radii and systematic studies of interatomic distances in halides and chalcogenides. *Acta Cryst* 1976, **A32**: 751–767.

[14] Haumont R, Kreisel J, Bouvier P. Raman scattering of the model multiferroic oxide $BiFeO_3$: Effect of temperature, pressure and stress. *Phase Transitions* 2006, **79**: 1043–1064.

[15] White WB. The structure of particles and the structure of crystals: Information from vibrational spectroscopy. *J Ceram Process Res* 2005, **6**: 1–9.

[16] Fukumura H, Harima H, Kisoda K, *et al.* Raman scattering study of multiferroic $BiFeO_3$ single crystal. *J Magn Magn Mater* 2007, **310**: e367–e369.

[17] Ianculescu A, Mitoseriu L, Chiriac H, *et al.* Preparation and magnetic properties of the $(1-x)BiFeO_3$–$xBaTiO_3$ solid solutions. *J Optoelectron Adv M* 2008, **10**: 1805–1809.

[18] Suen WP. B-site cation mixed multiferroic perovskite materials. Ph.D. Thesis. Hong Kong, China: The Hong Kong Polytechnic University, 2010.

[19] Pagès O, Postnikov AV, Kassem M, *et al.* Unification of the phonon mode behavior in semiconductor alloys: Theory and *ab initio* calculations. *Phys Rev B* 2008, **77**: 125208.

[20] Mishra KK, Sivasubramanian V, Sarguna RM, *et al.* Raman scattering from La-substituted $BiFeO_3$–$PbTiO_3$. *J Solid State Chem* 2011, **184**: 2381–2386.

[21] Randall CA, Barber DJ, Whatmore RW. Ferroelectric domain configurations in a modified-PZT ceramic. *J Mater Sci* 1987, **22**: 925–931.

12

Morphology and dielectric properties of single sample $Ni_{0.5}Zn_{0.5}Fe_2O_4$ nanoparticles prepared via mechanical alloying

Rafidah HASSAN[a,*], Jumiah HASSAN[a,b], Mansor HASHIM[a,b],
Suriati PAIMAN[a], Raba'ah Syahidah AZIS[a,b]

[a]Physics Department, Faculty of Science, Universiti Putra Malaysia, 43400 UPM Serdang, Selangor, Malaysia
[b]Institute of Advanced Technology, University Putra Malaysia, 43400 UPM Serdang, Selangor, Malaysia

Abstract: Nickel–zinc ferrite nanoparticles are important soft magnetic materials for high and low frequency device application and good dielectric materials. Nickel–zinc ferrite nanoparticles with composition $Ni_{0.5}Zn_{0.5}Fe_2O_4$ were prepared using mechanical alloying to analyze the effect of sintering temperature on microstructure evolution of a single sample with dielectric properties. The single sample with nanosized pellet was sintered from 600 ℃ to 1200 ℃ and analyzed by X-ray diffraction (XRD) to investigate the phases of the powders and by field emission scanning electron microscopy (FESEM) for the morphology and microstructure analyses. Dielectric properties such as dielectric constant (ε') and dielectric loss (ε'') were studied as functions of frequency and temperature for $Ni_{0.5}Zn_{0.5}Fe_2O_4$. The dielectric properties of the sample were measured using HP 4192A LF impedance analyzer in the low frequency range from 40 Hz to 1 MHz and at temperature ranging from 30 ℃ to 250 ℃. The results showed that single phase $Ni_{0.5}Zn_{0.5}Fe_2O_4$ cannot be formed by milling alone and therefore requires sintering. The crystallization of the ferrite sample increased with increasing sintering temperature, while the porosity decreased and the density and average grain size increased. Evolution of the microstructure resulted in three activation energies of grain growth, where above 850 ℃ there was a rapid grain growth in the microstructure. Dielectric constant and loss factor decreased with the increase in frequency. The optimum sintering temperature of $Ni_{0.5}Zn_{0.5}Fe_2O_4$ was found to be 900 ℃ which had high dielectric constant and low dielectric loss.

Keywords: mechanical alloying; sintering temperature; dielectric properties

1 Introduction

During the past two decades, a much deeper appreciation of solid phenomena has been gained by intensive research into the dielectric properties of materials. One outcome is the present wide range of ferrites which have given rise to new techniques in the overall improvement of communication and computer systems. Dielectric material measurement can provide critical design parameter information for many electronic applications [1]. Nanocrystalline ferrites are important materials because of their electrical, dielectric, magnetic and optical properties which make them suitable for electronic and storage devices. The properties of ferrite materials can be affected by chemical composition and microstructure of the materials which are related to the manufacturing process. The microstructure of ferrite is mainly determined by several factors such as the quality of the

* Corresponding author.
E-mail: rafidahhassan89@yahoo.com

raw materials, sintering temperature, milling parameter and sintering condition. Nickel–zinc ferrites are the most popular composition of soft ferrites and the most versatile of all ferrites because of their many technological applications. Ni–Zn ferrites have unique high dielectric constant which enables them to design useful electronic devices and also be used in high frequency applications as core materials for power transformers and circuit inductors in the megahertz frequency region [2].

Ni–Zn ferrites exhibit higher resistivity than Mn–Zn, and are therefore more suitable for frequencies above 1 MHz. They are more stable than the other types of ferrites, easily manufactured and low cost, and have excellent desirable magnetic properties [1]. Dielectric permittivity can be written as $\varepsilon^* = \varepsilon' - j\varepsilon''$, where ε' is the dielectric constant which is the ability of a material to polarize and store a charge within it, whereas the imaginary part ε'' is the loss factor which is a measure of the loss of power usually in the form of heat [3]. Nanoparticles can be synthesized by various physical, chemical and mechanical methods. One of the various techniques for synthesizing the material is mechanical alloying which is used to reduce grain size, mix powder uniformly and produce non-equilibrium structure materials. Mechanical alloying via high energy ball milling has now become one of the conventional methods for producing nano/non-crystalline materials. It is a modified solid state technique used with the advantage of cutting down mixing time. This technique enables economical and rapid preparation of metastable and amorphous alloys, nanocomposites, ceramics and other valuable powders. It is simple and quickly obtained after relatively short laboratory experimentation [4]. During mechanical alloying, materials in powder form will undergo severe collisions between the balls and vial wall of the grinding media by the process of high energy collision that results in a reduction of particle size. The particles themselves which normally possess a distribution of size can be nanoparticles if their average characteristic dimension (diameter for spherical particles) is less than 100 nm. Spinel ferrite nanoparticles have great importance in nanoscience and nanotechnology for technological applications because of their outstanding properties such as nanometer size and large surface area to volume ratio. It has been studied that in Ni–Zn ferrites, the electrical

and magnetic properties of ferrites depend on the stoichiometric composition [5]. The nickel–zinc ferrite with the well known composition of $Ni_{0.5}Zn_{0.5}Fe_2O_4$ is chosen in this study. This composition has high resistivity, good soft magnetic property, low dielectric loss, good mechanical hardness and chemical stability. The study of effect of temperature on structural and dielectric properties is carried out on the sintered Ni–Zn ferrite pellets.

2   Materials and method

The starting raw materials NiO (99.7%), ZnO (99.9%) and Fe_2O_3 (99.7%) with high purity were weighed according to the stoichiometric equation below:

$$0.5NiO + 0.5ZnO + Fe_2O_3 \rightarrow Ni_{0.5}Zn_{0.5}Fe_2O_4$$

Ferrite nanoparticles $Ni_{0.5}Zn_{0.5}Fe_2O_4$ were prepared by mechanical alloying consisting of a mixture of metallic oxides. The chemicals were mixed with the molar ratio of 0.5:0.5:1 and the ball-to-powder mass-charge ratio (BPR) of 10:1. These compositions were milled for 24 h using a SPEX 8000D high energy ball milling. The initial particle size of the powders was confirmed using transmission electron microscopy (TEM). The powders were mixed with 1–2 wt% polyvinyl alcohol (PVA) as binder and lubricated with 0.3 wt% zinc stearate. The alloyed powders were then uniaxially pressed into a single pellet at a pressure of 4 tonnes to yield a 2.0 g pellet of 18 mm in diameter and 2.0 mm in thickness. A single pellet sample was subjected to repeated sintering process from 600 ℃ to 1000 ℃ at 50 ℃ interval. The sample sintered at different sintering temperature was the same sample. It was sintered for 10 h at the rate of 4 ℃/min. The resulting pellet was used for the characteristic measurement. The sample was examined with X-ray diffraction (XRD, Philips X'pert diffractometer model 7602 EA Almelo) using Cu Kα radiation source with $\lambda = 1.5418$ Å to identify the formed phases. The X-ray powder diffraction pattern was recorded at room temperature in a 2θ scanning range from 20° to 80°.

The field emission scanning electron microscopy (FESEM, FEI NOVA NanoSEM 230) revealed the surface structure of the obtained ferrite and the average grain size was measured by the mean linear intercept method. TEM was carried out on the as-milled powders to confirm the particle size. The density of the sintered pellet for every sintering temperature was obtained using the Archimedes principle with water as the fluid medium. The theoretical density of $Ni_{0.5}Zn_{0.5}Fe_2O_4$ was calculated by taking its molecular weight to be 237.73 g. The weight of eight molecules in one unit cell is $8 \times 237.73/N_A$ g, where N_A is the Avogadro's number. The volume of a cube with side length a is a^3. The unit cell edge a_0 of $Ni_{0.5}Zn_{0.5}Fe_2O_4$ is 8.3827 Å; therefore $a^3 = 589.0495$ Å^3. As 1 Å^3 = 10^{-24} cm^3, the theoretical density (mass/volume) is equal to 5.3573 g/cm^3. For the dielectric measurements, the Ni–Zn ferrite pellet was sandwiched between two brass electrodes. The dielectric properties of the pellet were determined using Agilent 4294A precision impedance analyzer and carried out in an LT furnace in the frequency range of 40 Hz–1 MHz over the measuring temperature range of 30–250 ℃.

3   Results and discussion

The TEM image of the as-milled raw powders in Fig. 1 shows that the particles are in the nanometer range with particle sizes ranging from 8 nm to 21 nm, and the particles are nearly spherical and agglomerated. The milling time 24 h is chosen because the average grain size can be reduced to less than 100 nm during the milling process due to the high impact of the milling. Mechanical alloying process reaches a steady state when the particles have homogenous shape and size. A steady state is achieved after 30 h milling when there is no change in crystallite size and the size remains constant [6]. There is also a relationship between the temperature of the outer wall and the

milling time, and the temperature of the vial increases slowly as the milling time is extended. Kinetic energy of the ball and the properties of the material can be controlled by the temperature during the milling process. The diffusivity and the phase transformation caused by milling are also affected by the temperature of the powders. It is assumed that higher temperature results in phases which need higher atomic mobility, while at lower temperature the formation of amorphous phases is expected if the energy is sufficient [7].

The XRD patterns of the $Ni_{0.5}Zn_{0.5}Fe_2O_4$ ferrite nanoparticles are shown in Fig. 2. For the as-milled sample in Fig. 2(a), only raw starting material peaks are observed. The results show that $Ni_{0.5}Zn_{0.5}Fe_2O_4$ could not be formed during milling alone and therefore requires sintering, suggesting that the thermal energy should be supplied to the materials during the sintering process to complete the reaction. In ceramics, sintering is always an important processing parameter that

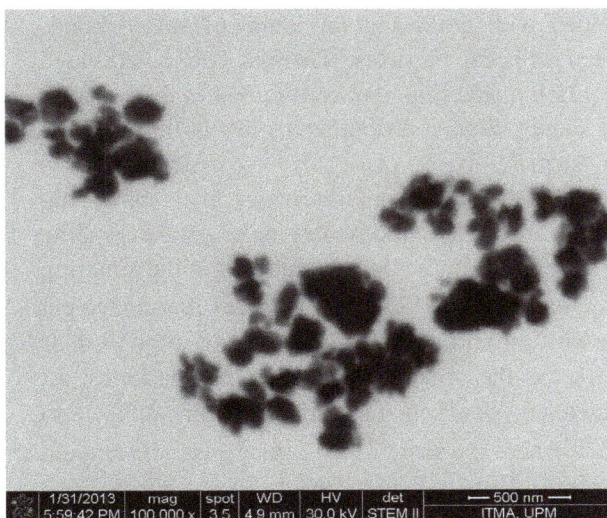

Fig. 1   TEM image of the as-milled raw powders.

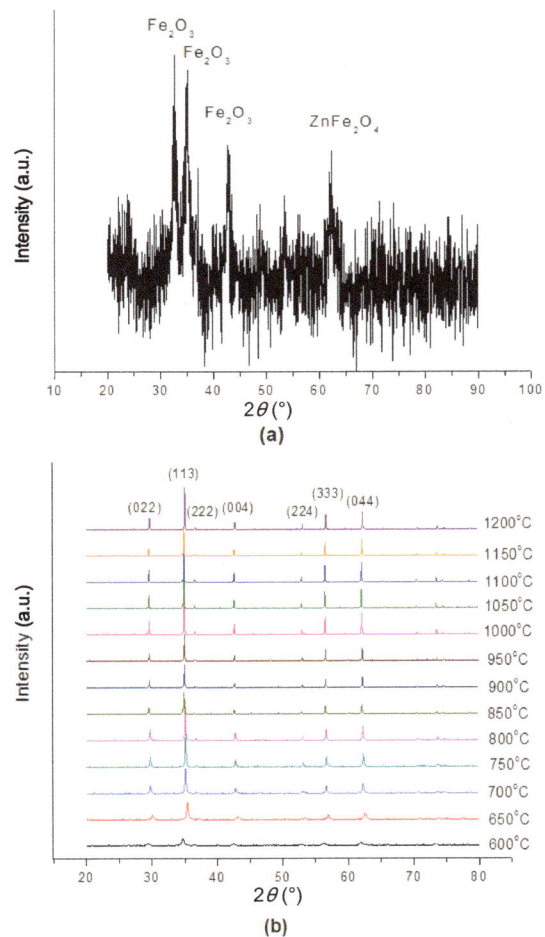

Fig. 2   (a) XRD pattern of the as-milled powders; (b) XRD spectra of $Ni_{0.5}Zn_{0.5}Fe_2O_4$ after sintering from 600 ℃ to 1200 ℃.

influences ceramics' microstructure evolution, grain growth and densification [8]. As shown in Fig. 2(b), when the sample is sintered at 600 ℃, the single phase crystallization of $Ni_{0.5}Zn_{0.5}Fe_2O_4$ is detected indicating the formation of nickel–zinc ferrite. All peaks match the standard pattern of nickel–zinc ferrite 00-008-0234 of the ICDD database that shows $Ni_{0.5}Zn_{0.5}Fe_2O_4$ and can be clearly indexed to the seven major peaks of the spinel ferrites, which are (022), (113), (222), (004), (224), (333) and (044) planes of a cubic unit cell, corresponding to spinel structure. It confirms that the single sample sintered Ni–Zn ferrite has single phase cubic spinel structure and the entire peaks observed match well with those of Ni–Zn ferrites. The full crystallization is achieved at as low as 600 ℃ exhibiting the advantage of mechanical alloying. Moreover, the major change observed is the increment in the intensity of the ferrite sample as the sintering temperature increases. The increase in the intensity means that either the contents of the phases are increased or the concentration of the atoms in the alloy is higher. All the XRD peaks become sharper and narrower with increase in temperature. This indicates the increase in particle size and the enhancement of crystallinity.

It also shows the relationship of porosity and density with sintering temperature in Table 1. The results obtained reveal a linear increase in density with sintering temperature. It is found that there is more densification or less porosity at the higher sintering temperature. The true porosity of the sample is calculated using X-ray density, and the density increases while porosity decreases with sintering temperature. This is because both porosity and density are inversely proportional to each other and are fractions of theoretical density of the material. The sintered density of the sample is between 4.575 g/cm^3 and 5.179 g/cm^3 or about 86%–97% of the theoretical density of 5.3573 g/cm^3. The sintering of the nanocrystalline material influences the particle size, shape and crystallization. The total pores are directly related to the density. Therefore higher heat treatment is required to remove a fraction of the pores. Sintering is the control of both densification and grain growth as densification is the act of reducing porosity in the sample thereby making it denser and thus changing the pore structure which affects ferrite properties.

The FESEM micrographs and the process of grain growth are taken on the surface microstructure of

Table 1   Average grain size, porosity and measured density of $Ni_{0.5}Zn_{0.5}Fe_2O_4$

Sintering temperature (℃)	Measured density (g/cm^3)	Theoretical density (%)	Porosity (%)	Average grain size (nm)
600	4.575	85.878	14.121	88.479
650	4.659	87.455	12.545	89.697
700	4.666	87.587	12.413	100.578
750	4.718	88.562	11.437	101.316
800	4.740	88.976	11.024	103.755
850	4.828	90.628	9.372	114.202
900	4.868	91.378	8.622	120.282
950	4.879	91.585	8.415	162.777
1000	4.896	91.904	8.096	188.845
1050	4.925	92.448	7.552	215.061
1100	4.966	93.218	6.782	239.727
1150	5.066	95.095	4.905	453.838
1200	5.179	97.216	2.784	645.744

sintered $Ni_{0.5}Zn_{0.5}Fe_2O_4$ with sintering temperature from 600 ℃ to 1200 ℃ by using a scanning electron microscope as shown in Fig. 3. The average grain size of a sintered body is measured over 200 grains by the linear intercept method. From Table 1, it is found that the average grain size increases with sintering temperature and it shows the microstructure evolution of the sample. The densification rate decreases as the distance of the defects to the grain boundaries increases with increasing grain size. Grain growth also gives pore coalescence where smaller pores are merged together into larger ones and this also reduces the densification rate and explains the density results obtained. The increasing average grain size may be due to mass transport mechanism in the sample during the sintering process. Sintering could be defined as removal of the pores between starting particles combined with grain and formation of strong bonds between adjacent particles. Thermal energy provided by increasing sintering temperature makes the particles move closer, thus contributing to the initial stage of grain growth. It is obviously seen that there is inhomogeneity in the microstructure that shows big size pores meaning that probably only a few crystallite sizes are present in the bulk sample. At initial sintering temperature of 600–750 ℃, it involves rearrangement of the powder particles and formation of strong bonds or necks at the contact points between particles. At intermediate sintering temperature from 800 ℃ to 1075 ℃ where the size of the necks grows, the amount of porosity decreases and particles move closer. The final stage of sintering at 1100 ℃, 1150 ℃ and 1200 ℃

exhibits that the pores are spherical and closed, removed slowly by diffusion process of pore vacancies and significant grain growth is clearly seen at this stage. As the sintering temperature increases, the driving force promoting neck growth increases [9]. Neck growth among the grains contributes to the increase in the average grain size.

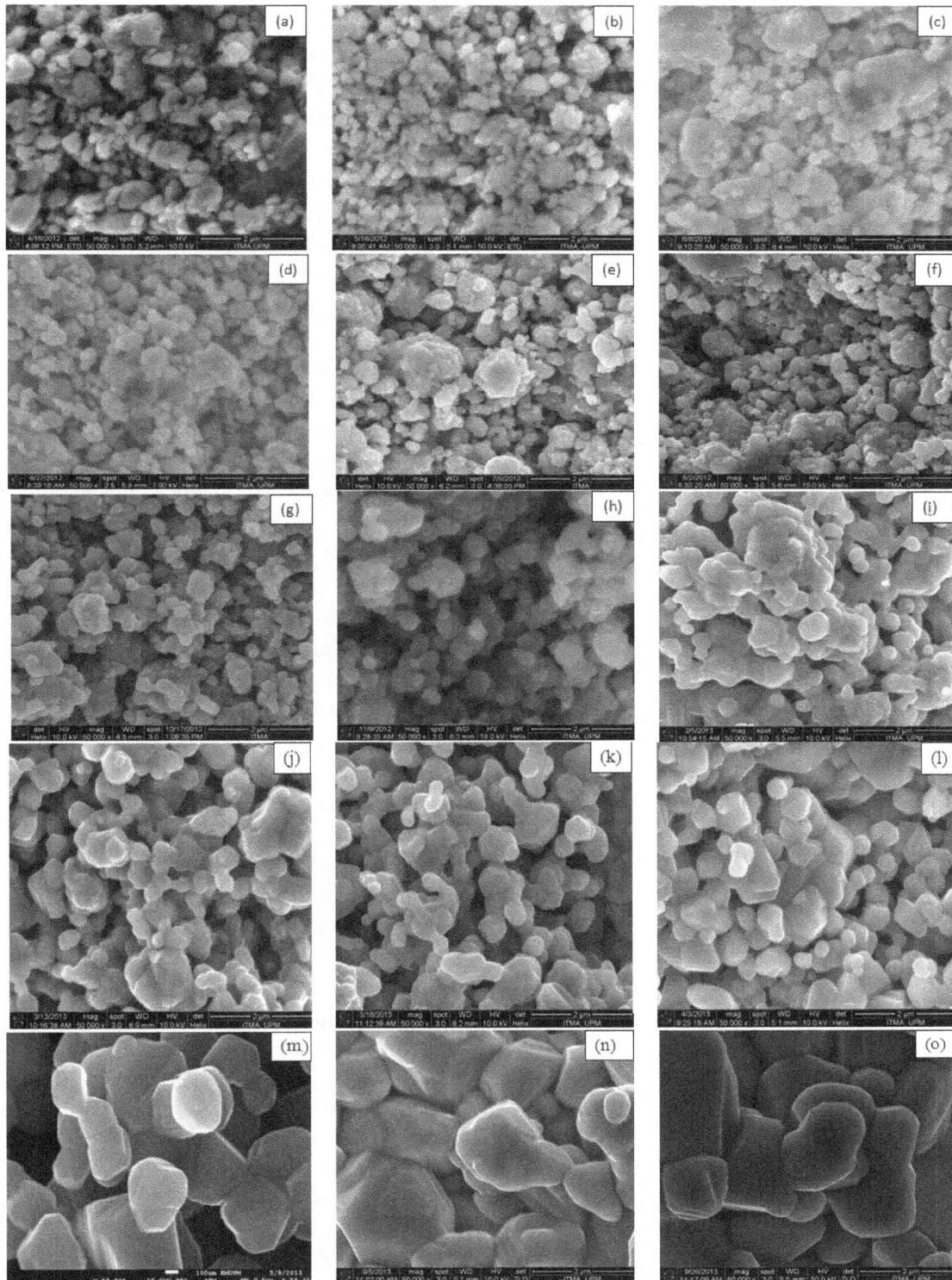

Fig. 3   SEM images of $Ni_{0.5}Zn_{0.5}Fe_2O_4$ at different sintering temperatures: (a) 600 ℃, (b) 650 ℃, (c) 700 ℃, (d) 750 ℃, (e) 800 ℃, (f) 850 ℃, (g) 900 ℃, (h) 950 ℃, (i) 1000 ℃, (j) 1025 ℃, (k) 1050 ℃, (l) 1075 ℃, (m) 1100 ℃, (n) 1150 ℃ and (o) 1200 ℃.

Table 1 shows the average grain size, porosity and measured density of the sample. The X-ray density and porosity listed in Table 1 are obtained by using Eqs. (1) and (2), respectively:

$$p_{xrd} = 8M / (N_A a^3) \qquad (1)$$

where p_{xrd} is the X-ray density; M is the molecular weight of the sample; N_A is the Avogadro's number; and a is the lattice constant. The porosity (P) of the sample is calculated by using the equation:

$$P = (1 - p_{exp} / p_{xrd}) \times 100\% \qquad (2)$$

where p_{exp} is the experimental density determined from the Archimedes principle. The theoretical density for $Ni_{0.5}Zn_{0.5}Fe_2O_4$ is 5.3573 g/cm^3.

The activation energy for grain growth can be predicted from the behavior of particle growth by using the Arrhenius equation below [10]:

$$d \ln k / dT = Q / (RT^2) \qquad (3)$$

where k is the specific reaction rate constant; Q is the activation energy; T is the absolute temperature; and R is the ideal gas constant. The value of k can be directly related to grain size [9], which results in the equation:

$$\log D = (-Q / 2.303R) \times 1 / T \qquad (4)$$

where T is the absolute temperature and D is the grain size. From Eq. (4), there are three best fitted straight-line plots of grain size where $\log D$ versus $1/T$ is shown in Fig. 4. Three slopes of the line are also obtained which are y_1, y_2 and y_3 of $-Q / 2.303R$, and the value of the activation energies of grain growth Q can be calculated from the Arrhenius plots of y_1, y_2 and y_3 which are 4.56 kJ/mol, 27.94 kJ/mol and 109.139 kJ/mol, respectively. The activation energies are increasing with rise in sintering temperature as the average grain size is increased from nano size to micron size where the grain growth is depending on pore removal that supports the view that a metastable network of pores and boundaries occurs especially in the intermediate and final stages of sintering. All of these values are much lower than that reported by previous researcher [11] where the average grain size of Ni–Zn ferrites at different sintering temperatures is larger and also lower than that obtained by Rao et al. for Ni–Zn ferrites prepared using conventional ceramic method [12]. The low activation energies are due to the smaller size of starting powders milled by the mechanical alloying that reduces their dimensions into the nanosized range. This causes the surface area of the starting powders to be increased and lowers the activation energy. Higher activation energy is required in order to continue the growth of grains which is the result of atom diffusion in grain boundary and an energy barrier that must be overcome for a reaction to occur. Furthermore, the sample has been sintered repeatedly, causing it to have a thermal history and higher activation energy required for the further stage of the grain growth. The rates controlling transport mechanism for densification, grain growth and coarsening have different activation energies and hence dominate in different temperature regimes. In general, the coarsening mechanism dominates at lower temperature due to lower activation energy. The heating rate of sintering is controlled to maintain a constant densification rate. In constant heating rate, contribution from surface diffusion is minimized since the time of surface diffusion regime is comparatively small.

The dielectric properties ε' and ε'' in the frequency range 40 Hz–1 MHz with measuring temperature from 30 ℃ to 250 ℃ at each sintering temperature are illustrated in Figs. 5 and 6, respectively. Both ε' and ε'' curves for all the samples sintered at 600–1000 ℃ show similar trends, decreasing with increasing frequency. This decrease is rapid at lower frequency and becomes slower at higher frequency which is a normal dielectric behavior in ferrites having mobile charge carriers. In Fig. 5, the dielectric dispersion curve and the decrease observed at lower frequency region can be explained on the basis of Koop's phenomenological theory [13] due to Maxwell–Wagner interfacial type of polarization [14]. The dispersion of ε' at low frequency is due to the interfacial polarization and existence of depletion layers near the sample–electrode contacts. It is assumed that there should be two layers which are well

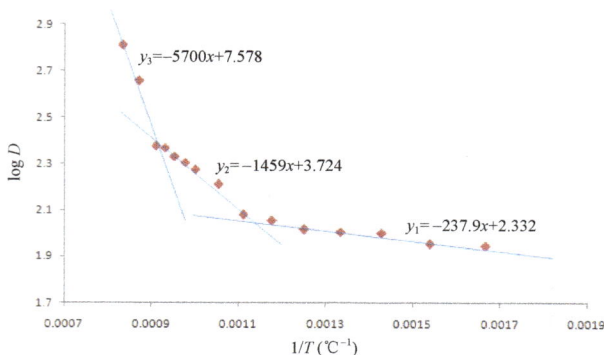

Fig. 4   Arrhenius plot of activation energies for grain growth of $Ni_{0.5}Zn_{0.5}Fe_2O_4$.

conducting materials (ferrite grains) and separated by the layers of lower conductivity (grain boundaries) that are effective at higher and lower frequencies respectively in the inhomogeneous dielectric structure. It can be also explained on the basis of space charge polarization resulted from difference among the conductivities of the various phases present. At low frequency electron hopping occurs between Fe^{3+} and Fe^{2+} on the localized sites. The electrons attain the grain boundary through hopping of charge carriers which results in the interfacial polarization. However, as the frequency is increased, the probability of electrons reaching the grain boundary decreases, which results in a decrease in the interfacial polarization. Therefore, the dielectric constant decreases with increasing frequency. Besides this, the sintered ferrite material consists of cracks, pores and other defects, and due to inhomogeneity, there are regions of different permittivity. Thus the conducting grains and less conducting boundaries are separated by voids or pores which were explained by Maxwell and Wagner.

The sample shows high values of ε' for frequencies lower than 10^3 Hz, but at high frequencies (10^3–10^6 Hz) as the temperature increases ε' decreases. At low frequencies the charges have time to accumulate at the borders of the conducting regions causing ε' to increase, while at higher frequencies the charges do not have time to accumulate and polarization does not occur since the charge displacement is small compared to the dimensions of the conducting region [15]. When orientational polarization is dominated in the system, the temperature randomizes it and tends to decrease the dielectric constant. Therefore, the overall dielectric constant is going to decrease in the system. However, if polarization is predominantly governed by the space polarization then dielectric constant of the system is going to increase with increasing temperature because the rate of interfacial charge accumulation increases. As a result, the present sample exhibits the orientational polarization predominantly. At low frequencies, the dielectric constant is mainly due to the movement of free charges which increases with the rise in temperature and further increases the value of dielectric constant. But at high frequencies and high temperature, dispersion losses increase due to the thermal vibrations and less time is available for the material to respond to the applied electric field. The rapid increase in the dielectric constant with increase

in temperature at low frequency suggests that the effect of temperature is more pronounced on the interfacial polarization rather than on the dipolar polarization. The electron exchanges between the ferrous and ferric ions, which produce local displacements in the direction of the applied external fields, determine the polarization in ferrites. Above certain frequencies of the electric field, this electronic exchange cannot follow the alternating field and this causes a decrease in the dielectric constant.

From the graphs, the dielectric constant initially increases gradually with temperature at lower frequency of 100 Hz and then it begins to decrease at 1 kHz to 1 MHz, but up to particular measuring temperature 250 ℃, it becomes constant and slowly decreases. This is attributed to the transition of the sample due to a magnetic transition from ferromagnetic to paramagnetic that changes the behavior of the dielectric constant with temperature. A further increase in temperature adds to the random vibrational motion of the molecules, which becomes less susceptible to the orientation in the field direction and hence the dielectric constant decreases. The space charge polarization resulting from electron displacement on application of electric field and the following charge build up at the insulating grain boundary is a major contributor to the dielectric constant in ferrites. Space charge polarization is expected when there are large numbers of Fe^{2+} ions in the ferrite. It is due to the ease of electron transfer between Fe^{3+} and Fe^{2+} ions and consequently higher dielectric constant. Now, with increasing sintering temperature, partial reduction of Fe^{3+} to Fe^{2+} takes place. Thus, the value of dielectric constant increases with increasing sintering temperature.

The dielectric loss factor decreases with increasing frequency and increases with temperature from 30 ℃ to 250 ℃ as shown in Fig. 6. The frequency and temperature effects on the ε'' illustrate the interfacial polarization of the grain boundaries within the sample. These frequency and temperature dependences are due to the conversion of the movement of phonons into the vibrations of the lattice. The inconsistent lattice vibrations cause instability in the interfacial polarization. Therefore, ε'' increases. For the lower frequency, the polarization is increased by the electric field and also by the increase in the number of charge carriers with increasing temperature. Dielectric loss in ferrites is also a result of the lag in polarization with

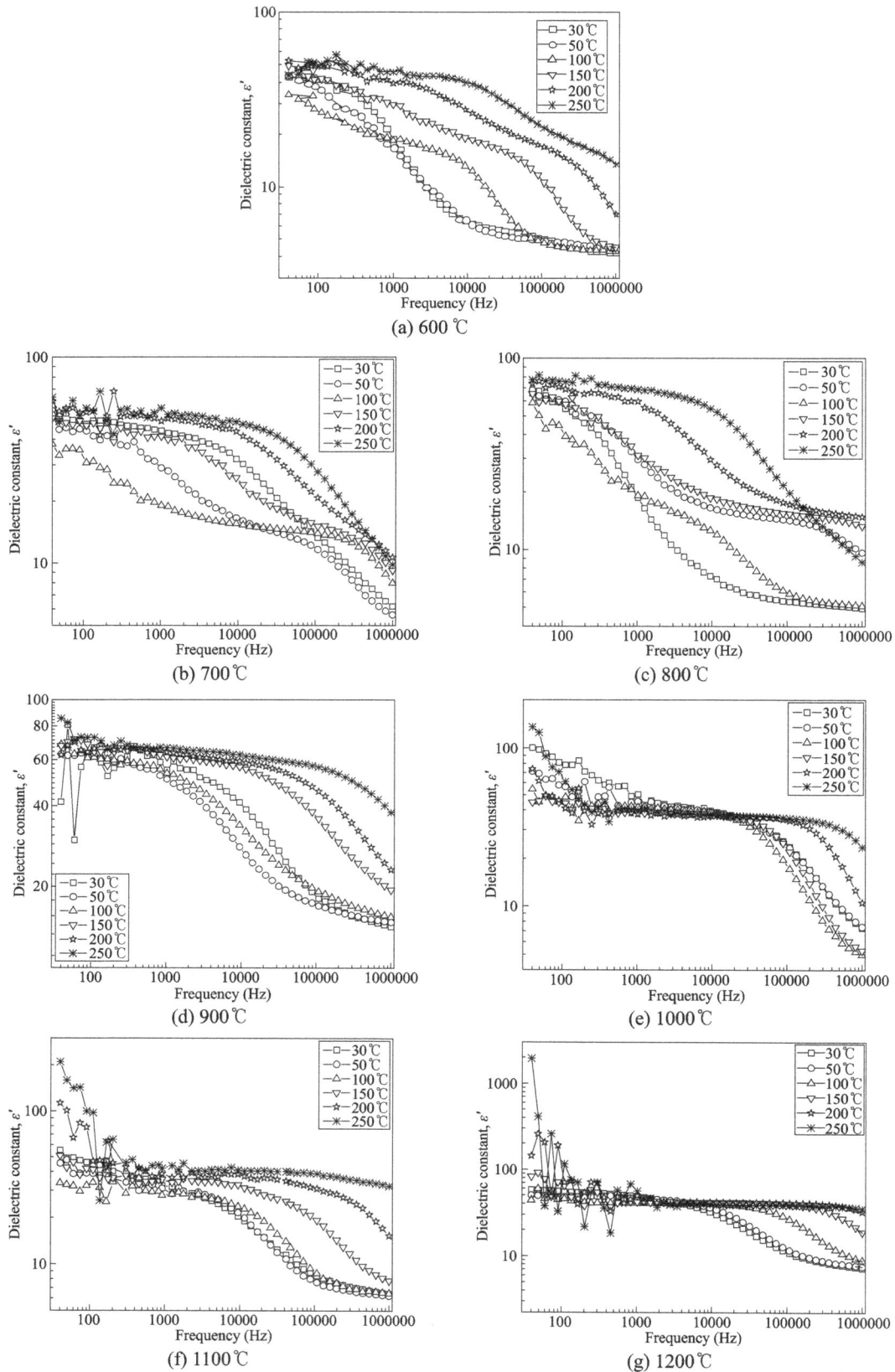

Fig. 5   Dielectric constant of $Ni_{0.5}Zn_{0.5}Fe_2O_4$ sintered at different temperatures: (a) 600 ℃, (b) 700 ℃, (c) 800 ℃, (d) 900 ℃, (e) 1000 ℃, (f) 1100 ℃ and (g) 1200 ℃.

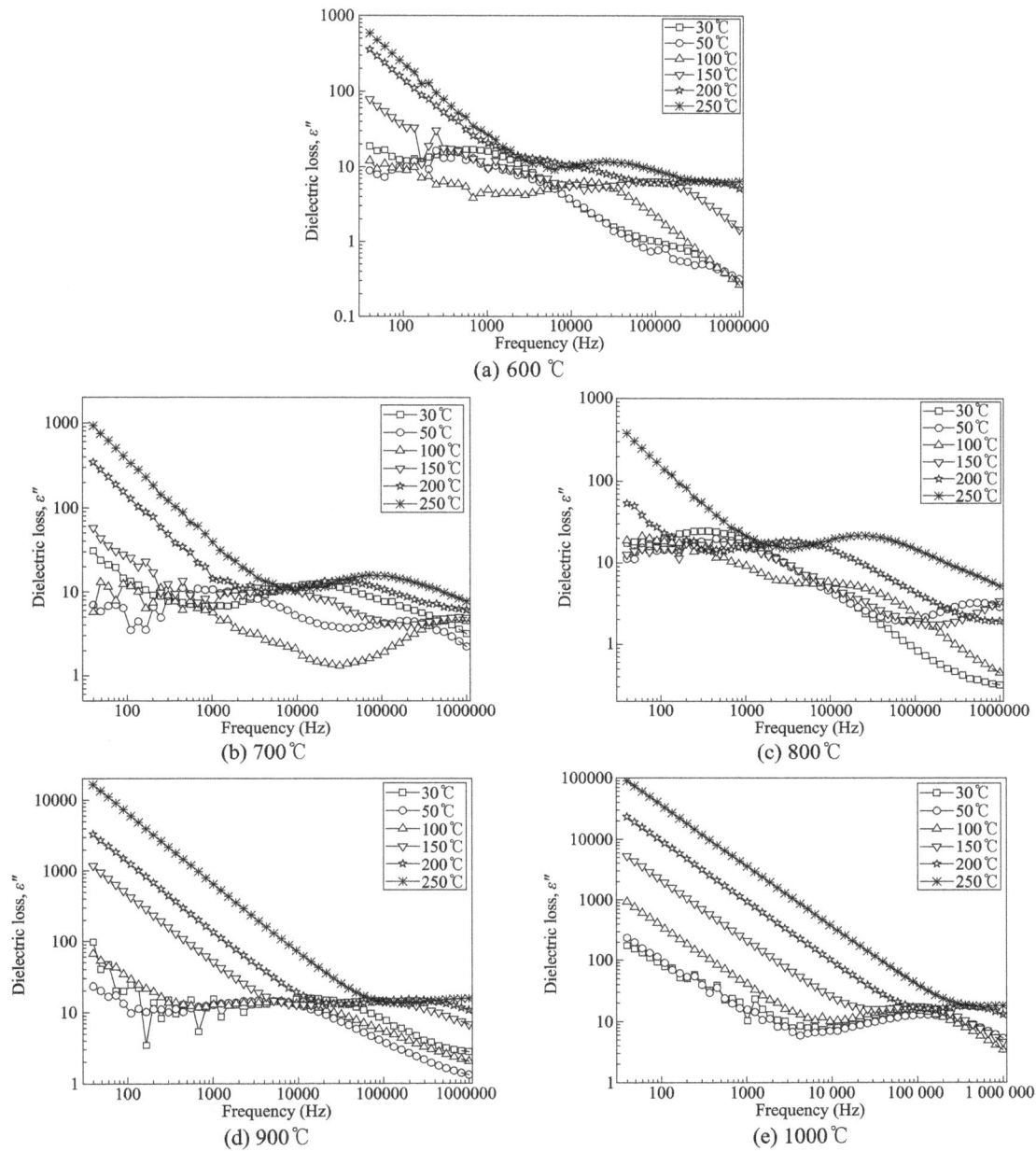

Fig. 6   Dielectric loss factor of $Ni_{0.5}Zn_{0.5}Fe_2O_4$ sintered at different temperatures: (a) 600 ℃, (b) 700 ℃, (c) 800 ℃, (d) 900 ℃ and (e) 1000 ℃.

respect to the applied alternating electric field. When the frequency of the external AC field is equal to the hopping frequency of the charge carriers, the maximum electrical energy is transferred to the oscillating ions and we observe that there are broad peaks in power loss. A broad peak of ε'' indicates the occurrence of distribution of relaxation time rather than a single relaxation time [16]. The increment of ε'' is low from 30 ℃ to 250 ℃ especially at higher frequency. The atoms or molecules in the sample

cannot in most cases orient themselves in the low temperature region. As the temperature rises, the orientation of the dipoles is facilitated and this increases the permittivity. When the frequency of applied AC field is much larger than the hopping frequency of electrons, the electrons do not have an opportunity to jump at all and the energy loss is small. In general, the local displacements of electronic charge carriers cause the dielectric polarization in materials. Therefore the marked decrease in ε'' is due to the decreasing ability of the jumping electrons to follow the alternating frequency of AC electric field beyond certain critical frequency. This explains the decreasing trends for ε'' with increasing frequency. As observed from the graphs, the decrement of ε'' is very rapid at higher temperature in low frequency region, but at frequency 1 kHz and above it decays with slow rate reaching a constant value at high frequencies and low temperature. This proposes a thermally activated hopping process with increasing temperature at low frequencies.

$Ni_{0.5}Zn_{0.5}Fe_2O_4$ sintered at low temperature has the lowest dielectric constant. This may be due to the incomplete phase formation and low crystallization of $Ni_{0.5}Zn_{0.5}Fe_2O_4$. Another reason may be the agglomerations of powders bonded with grains as mentioned above. Agglomerations of powders become a resistant for ions to polarize between grains and grain boundaries [17]. ε' and ε'' increase with increasing temperature which is normally an expected behavior that has been observed in most of the ferrites. The hopping of charge carriers is thermally activated with the temperature rise, and hence the dielectric polarization increases causing an increase in ε' and ε'' with temperature. High temperature sintering also leads to the escape of Zn^{2+} ions from the lattice, which results in greater structural imperfections and high dielectric loss. From the SEM images, the neck growth among the grains increases as sintering temperature increases. On the other hand, dielectric constant of the sample also possesses the phenomena. Polarization of ions in grains may increase with grain necking growth. Polarization of ions in grains is the main factor that influences the dielectric constant. The grain size increases, but the porosity decreases with increasing sintering temperature. The increase in the grain size decreases the grain boundary density between them. The sample becomes denser and more homogenous. Low porosity, high density and increasing grain size increase the dielectric permittivity. It is found that higher temperature which reduces pores can cause the larger dielectric constant ε' at low frequency which is due to the blocking of charge carriers at the electrodes. Space charge polarization induces the dispersion of ε' because the enhanced mobility of charge carriers increases as the temperature increases which is thermally generated from the beginning resulting in an increase of ε'.

4 Conclusions

A single sample of $Ni_{0.5}Zn_{0.5}Fe_2O_4$ ferrite nanoparticles was successfully synthesized via high energy ball milling and the parallel evolutions of dielectric and microstructural properties were obtained. High energy ball milling is an effective technique to improve the preparation of Ni–Zn ferrite nanoparticles at lower temperatures. The particles are in the nanometer size of around 22 nm in average. The effect of increasing sintering temperature changes the microstructure of the material, leading to the changes in the dielectric properties of the material. The results show that single phase $Ni_{0.5}Zn_{0.5}Fe_2O_4$ could not be formed during milling alone and therefore requires sintering. The crystallization of the ferrite increases with increasing sintering temperature, which decreases the porosity and increases the density and crystallite size. The activation energies of grain growth (Q) are 4.56 kJ/mol, 27.94 kJ/mol and 109.139 kJ/mol where the rate of grain growth is slower initially from 600 ℃ to 850 ℃ but grows rapidly at higher sintering temperature above 850 ℃. The sintered density of the sample was between 4.575 g/cm^3 and 5.179 g/cm^3 or about 86%–97% of the theoretical density of 5.3573 g/cm^3. Both ε' and ε'' decrease with increasing frequency, which is a normal dielectric behavior in ferrites. The optimum sintering temperature having the highest dielectric constant and the lowest dielectric loss factor, which are 14.22 and 2.762 respectively, is 900 ℃.

Acknowledgements

The authors acknowledge the Fundamental Research Grant Scheme (FRGS) Project No. 01-04-10-862 FR, Graduate Research Fellowship given to the graduate student (Rafidah Hassan) and the Physics Department, Faculty of Science, UPM.

References

[1] Agilent Technologies. Agilent solutions for measuring permittivity and permeability with LCR

meters and impedance analyzers. Application Note 1369-1. 2006. Available at http://cp.literature.agilent.com/litweb/pdf/5980-2862EN.pdf.

[2] Matsushita N, Kondo K, Yoshida S, *et al.* Ni–Zn ferrite films synthesized from aqueous solution usable for sheet-type conducted noise suppressors in GHz range. *J Electroceram* 2006, **16**: 557–560.

[3] Hassan J, Yen, FM, Hashim M, *et al.* Dielectric permittivity of nickel ferrites at microwave frequencies 1 MHz to 1.8 GHz. *Ionics* 2007, **13**: 219–222.

[4] Verma A, Chatterjee R. Effect of zinc concentration on the structural, electrical and magnetic properties of mixed Mn–Zn and Ni–Zn ferrites synthesized by the citrate precursor technique. *J Magn Magn Mater* 2006, **306**: 313–320.

[5] Mangalaraja RV, Anathakumar S, Manohar P, *et al.* Magnetic, electrical and dielectric behaviour of $Ni_{0.8}Zn_{0.2}Fe_2O_4$ prepared through flash combustion technique. *J Magn Magn Mater* 2002, **253**: 56–64.

[6] Ismayadi I, Hashim M, Khamirul AM, *et al.* The effect of milling time on $Ni_{0.5}Zn_{0.5}Fe_2O_4$ compositional evolution and particle size distribution. *Am J Applied Sci* 2009, **6**: 1548–1552.

[7] El-Eskandarany MS, Sumiyama K, Suzuki K. Crystalline-to-amorphous phase transformation in mechanically alloyed $Fe_{50}W_{50}$ powders. *Acta Mater* 1997, **45**: 1175–1187.

[8] Rahman MN. *Ceramic Processing and Sintering*, 2nd edn. CRC Press, 2003.

[9] German RM. *Sintering Theory and Practice*. John Wiley & Sons Inc., 1996.

[10] Coble RL. Sintering crystalline solids. I. Intermediate and final state diffusion models. *J Appl Phys* 1961, **32**: 787–792.

[11] Ismail I, Hashim M. Structural and magnetic properties evolution of fine-grained $Ni_{0.5}Zn_{0.5}Fe_2O_4$ series synthesized via mechanical alloying. *Australian Journal of Basic and Applied Sciences* 2011, **5**: 1865–1877.

[12] Rao BP, Rao PSVS, Rao KH. Densification, grain growth and microstructure of Ni–Zn ferrites. *J Phys IV France* 1997, **07**: C1-241–C1-242.

[13] Koops CG. On the dispersion of resistivity and dielectric constant of some semiconductors at audiofrequencies. *Phys Rev* 1951, **83**: 121.

[14] Wagner KW. Zur theorie der unvollkommenen dielektrika. *Annalen der Physik* 1913, **345**: 817–855.

[15] Maxwell JC. *A Treatise on Electricity and Magnetism, Vol. 1*. New York: Oxford University Press, 1973: 828.

[16] Olofa SA. Oscillographic study of the dielectric polarization of Cu-doped NiZn ferrite. *J Magn Magn Mater* 1994, **131**: 103–106.

[17] Jonscher AK. *Dielectric Relaxation in Solids*. London: Chelsea Dielectrics Press, 1983: 231.

13

Abrupt change of dielectric properties in mullite due to titanium and strontium incorporation by sol–gel method

Biplab Kumar PAUL, Kumaresh HALDAR, Debasis ROY, Biswajoy BAGCHI, Alakananda BHATTACHARYA, Sukhen DAS[*]

Physics Department, Jadavpur University, Kolkata-700 032, India

Abstract: Highly crystallized mullite has been achieved at temperatures of 1100 ℃ and 1400 ℃ by sol–gel technique in presence of titanium and strontium ions of different concentrations: $G_0 = 0$ M, $G_1 = 0.002$ M, $G_2 = 0.01$ M, $G_3 = 0.02$ M, $G_4 = 0.1$ M, $G_5 = 0.2$ M and $G_6 = 0.5$ M. X-ray diffraction (XRD), Fourier transform infrared spectroscopy (FTIR), field emission scanning electron microscopy (FESEM), LCR meter characterized the samples. Mullite formation was found to depend on the concentration of the ions. The dielectric properties (dielectric constant, loss tangent and AC conductivity) of the composites have been measured, and their variation with increasing frequency and concentration of the doped metals was investigated. All the experiments were performed at room temperature. The composites showed maximum dielectric constants of 24.42 and 37.6 at 1400 ℃ of 0.01 M concentration for titanium and strontium ions at 2 MHz, respectively. Due to the perfect nature of the doped mullite, it can be used for the fabrication of high charge storing capacitors and also as ceramic capacitors in the pico range.

Keywords: mullite; sol–gel technique; X-ray diffraction (XRD); dielectric properties; field emission scanning electron microscopy (FESEM)

1 Introduction

Mullite is a promising engineering ceramic material for use in optical, dielectric and structural applications. Mullite has a unique combination of properties, such as high melting point (1830 ℃), good electrical resistance, good mechanical strength, low thermal expansion coefficient, high strength and high creep resistance at any temperature range. Mullite is also a leading candidate material for high transmitting IR windows, electronic substrates, humidity sensors, protective coatings, electrical insulators and turbine engine components, etc.

Electronic industry is continuously trying to develop processes that are more advanced and lead to forecasting transistor density and chip complexity, and operating speed or frequency for future technological developments [1,2]. The main challenge is to carry electric power and distribute clock signals that control the timing and synchronize the operation. This challenge extends beyond the material properties and technology and also involves system architecture [3–7]. Controlled chip connections used in high package density logic devices require a good compatibility of thermal expansion coefficient (TEC) between substrate and Si chip. The mullite composite system has required

* Corresponding author.
E-mail: sdasphysics@gmail.com

strength and much closer match of TEC with Si chip than alumina.

Several reports have been published dealing with synthesis of mullite composites in presence of various mineralizing agents to improve the mechanical and chemical properties, and also there are some recent publications related to the dielectric properties of these modified mullite composites [8–14]. We have studied the dielectric constant, loss tangent, AC conductivity of the mullite composites doped with varying concentrations of tungsten ions with different frequencies at room temperature. We also have studied the dielectric constant, loss tangent, AC conductivity and magnetization of the mullite composites doped with varying concentrations of iron ions with different frequencies at room temperature. The results indicate that the sample of 0.01 M concentration has the highest dielectric constant 24.42 at frequency 2 MHz [15,16].

2   Experimental

2. 1   Sample materials

Chemicals used in the preparation of mullite precursor gels were aluminium nitrate nonahydrate $(Al(NO_3)_3 \cdot 9H_2O$, MERCK, India, 99.9%), aluminium isopropoxide $(Al(-O-i-Pr)_3$, puriss, Spectrochem Pvt. Ltd., India), tetraethyl orthosilicate $(Si(OC_2H_5)_4$, TEOS, MERCK, Germany), titanium isopropoxide $(Ti\{OCH(CH_3)_2\}_4$, MERCK, Germany, 99.9%) and strontium chloride hexahydrate $(SrCl_2 \cdot 6H_2O$, MERCK, India, 99.9%).

2. 2   Sample preparation and characterization

Mullite precursor gel powder was synthesized by dissolving stoichiometric amounts of $Al(-O-i-Pr)_3$ and TEOS in 0.5 M solution of $Al(NO_3)_3 \cdot 9H_2O$ [12,13]. The molar ratio of $Al(-O-i-Pr)_3 : Al(NO_3)_3 \cdot 9H_2O$ was kept at 3.5:1. The molar ratio of $Al:Si$ was 3:1 [12,13].

For preparation of the doped gels, the titanium and strontium salts were added to the original solution in the ratio of $Al:Si:x$, where x is the concentration of the metal salts in molarity. The titanium and strontium salts were added such that in the final solution $x = 0.002$ M (G_1), 0.01 M (G_2), 0.02 M (G_3), 0.1 M (G_4), 0.2 M (G_5), 0.5 M (G_6) [12,13]. The sol would be in the gel form after vigorous stirring for 5 h, and the sol was

maintained overnight at 70 ℃. Finally, the gel was dried at 120 ℃. The samples were then pelletized and sintered at 1100 ℃ and 1400 ℃ or 3 h in a muffle furnace under air atmosphere (heating rate 10 ℃/min) [12,13].

2. 3   Instruments used

The crystalline phases developed in the samples sintered at 1100 ℃ and 1400 ℃ were analyzed by X-ray powder diffractometer (XRD, model-D8, Bruker AXS, Wisconsin, USA) using Cu Kα radiation at 1.5418 Å and operating at 40 kV with a scan speed of 1 s/step.

The characteristic stretching and bending modes of vibration of chemical bonds of a sample can be effectively evaluated by spectroscopic methods. 1% of the sample was mixed with spectroscopy grade KBr, pelletized to form disc and analyzed by Fourier transform infrared (FTIR) spectroscopy (FTIR-8400S, Shimadzu).

AC parameters such as capacitance (C) and dissipation factor $(\tan\delta)$ of the samples were measured in the frequency range of 20 Hz to 2 MHz using LCR meter (HP Model 4274 A, Hewlett-Packard, USA). The variation of dielectric constant and loss tangent was studied by recording these parameters using sample pellets of uniform thickness at ten different frequencies from 20 Hz to 2 MHz.

Morphology of the sintered gels was observed by field emission scanning electron microscopy (FESEM, JSM 6700F, JEOL Ltd., Tokyo, Japan). Samples were etched with 25% HF solution. About 2 mg of each sample was dispersed in ethanol, and a single drop was placed on copper grid for sample preparation.

3   Results and discussion

From the X-ray diffractograms, it can be seen that the undoped sample shows considerable mullite phase at 1100 ℃ and 1400 ℃, while for the doped samples, prominent mullite peaks are also obtained and changing with the concentrations. Mullite phase (JCPDS No. 150776) increases with increasing concentration of metal ions (Figs. 1 and 2). From the diffractograms, it has been observed that with increase in the concentration of doped titanium and strontium ions, mullite phase in the composites increases. The rutile phase (JCPDS No. 211276) is obtained at 1100 ℃ and 1400 ℃ for titanium doped mullite. The

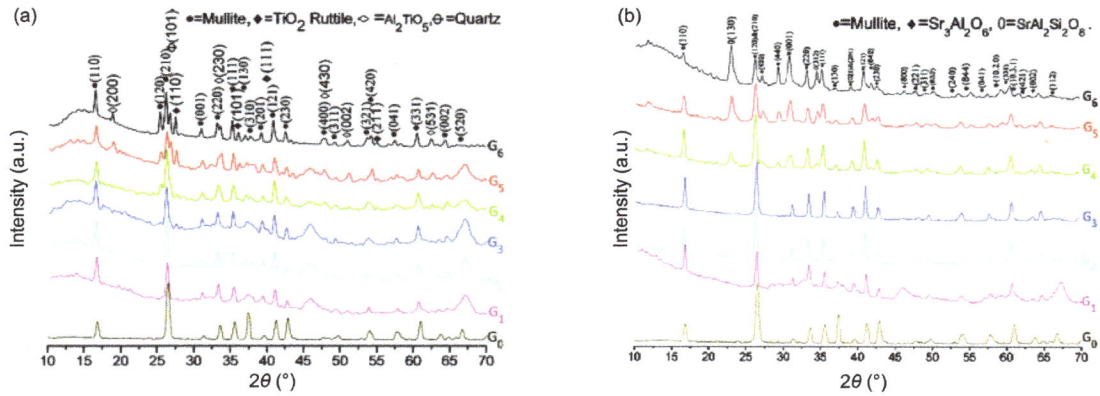

Fig. 1   XRD patterns of (a) titanium and (b) strontium doped mullite precursor gels sintered at 1100 ℃ with increasing doping concentration.

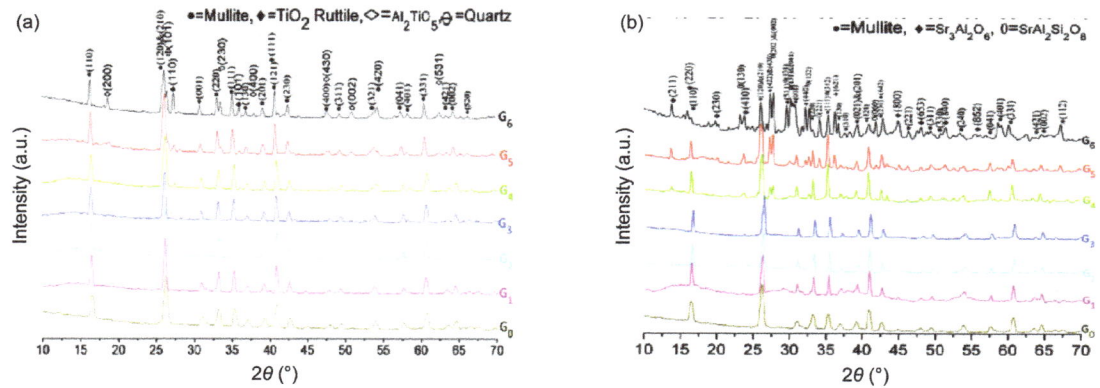

Fig. 2   XRD patterns of (a) titanium and (b) strontium doped mullite precursor gels sintered at 1400 ℃ with increasing doping concentration.

diffractograms show that the amount of mullite formation is more in the case of strontium for all concentrations. The hump at the lower scattering angle shows that glass phase increases with the increase of metal ion concentration. Aluminates and oxide phases are also observed in the diffractograms. The phase variation of the composite is due to the changed concentration of each metal ion and John–Teller distortion [12,13]. Interaction of the metal ions with the alumina and silica components of the gels is responsible for the accelerated transformation to mullite phase. The content of crystalline mullite decreases and the background increases due to the increase of metal ion concentration and the formation of metal silicate and aluminate phases of the samples. The "mineralizing" effect continues for samples G_4 to G_6 with respect to control sample G_0. The "mineralizing" effect of transition metals on phase transformation of mullite is well documented in Refs. [8–14]. Probably there are two possibilities, either complete incorporation of metal ions in mullite structure or dissolution of metal ions in the Si-rich

glassy phase. Such amorphous phase increases with the increased metal ions, as an account of decrease of the crystalline phases. The densification of the composite may be due to increased consolidation of the composite because of the molten state of the sintered gel at 1400 ℃ ((Fig. 11(d)) and for the increasing metal ion concentration.

By spectroscopic method, the characteristic stretching and bending modes of vibration of chemical bonds of a sample can be effectively evaluated. 1% of the sample is mixed with spectroscopy grade KBr, pelletized to form disc and analyzed by FTIR spectroscopy. Figure 3 shows the FTIR spectra of titanium and strontium doped sintered gels at 1100 ℃. Mullite gives characteristic bands at wave numbers around 463 cm^{-1}, 542 cm^{-1} (AlO_6), 731 cm^{-1} (AlO_4), 828 cm^{-1} (AlO_4), 1074 cm^{-1} (Si phase), 1130 cm^{-1} (Si–O stretching mode), 1176 cm^{-1} and 1368 cm^{-1} [13,14], and 1226 cm^{-1} and 1450 cm^{-1} for rutile at 1100 ℃ [15]. Strontium doped mullite gives characteristic bands at wave numbers around 465 cm^{-1},

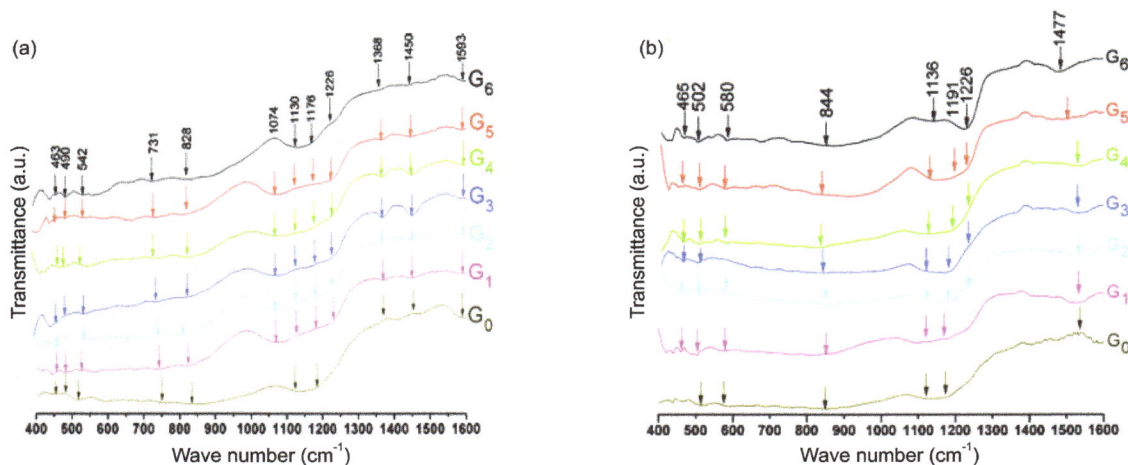

Fig. 3   FTIR patterns in transmittance mode of (a) titanium and (b) strontium doped mullite precursor gels sintered at 1100 ℃ with increasing doping concentration.

502 cm^{-1}, 580 cm^{-1} (AlO$_6$), 844 cm^{-1} (AlO$_4$) and 1136 cm^{-1} (Si–O stretching mode), and 1191 cm^{-1} and 1226 cm^{-1} for mullite, and for strontium aluminates (Sr$_3$Al$_2$O$_6$) at 1477 cm^{-1} at 1100 ℃ [16].

Figure 4 shows the FTIR spectra of titanium and strontium doped mullite at 1400 ℃. Sanad *et al.* [10,14,17] showed that all the IR spectra obtained for the Y^{3+}, Gd^{3+} doped samples are almost similar to the characteristic peaks of pure o-mullite sample in their overall appearance. For titanium at 1400 ℃, characteristic bands are obtained at wave numbers around 463 cm^{-1}, 509 cm^{-1}, 570 cm^{-1} (AlO$_6$), 740 cm^{-1} (AlO$_4$), 820 cm^{-1} (AlO$_4$), 1114 cm^{-1} (Si–O stretching mode) and 1173 cm^{-1} [16], and 609 cm^{-1}, 1405 cm^{-1} and 1593 cm^{-1} for rutile at 1400 ℃ [15,16]. For strontium at 1400 ℃, characteristic bands are obtained

at wave numbers 455 cm^{-1}, 613 cm^{-1} (AlO$_6$), 700 cm^{-1} (AlO$_4$), 850 cm^{-1} (AlO$_4$), 950 cm^{-1} (Si–OH), 1092 cm^{-1} (Si–O phase), 1152 cm^{-1} (Si–O stretching mode) and 1526 cm^{-1} (Sr$_3$Al$_2$O$_6$).

The dielectric constant or relative permittivity (ε_r) of each sample is calculated from the capacitance using the formula:

$$\varepsilon_r = (C \times d)/(A\varepsilon_0) \tag{1}$$

where C is the capacitance of the material; d is the thickness of the pellet; A is the area of cross section; and ε_r and ε_0 are the dielectric constant and permittivity of free space, respectively [18,19].

The dielectric properties of materials are used to describe electrical energy storage, dissipation and energy transfer. Electrical storage is the manifestation of dielectric polarization. The variation of dielectric

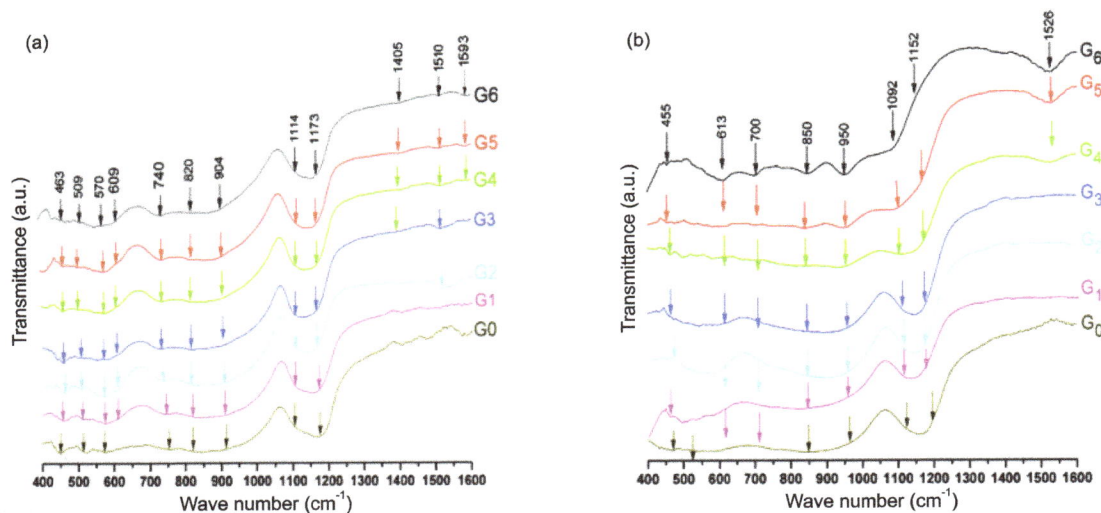

Fig. 4   FTIR patterns in transmittance mode of (a) titanium and (b) strontium doped mullite precursor gels sintered at 1400 ℃ with increasing doping concentration.

constant with frequency of titanium and strontium doped mullite composites at 1100 ℃ are shown in Fig. 5 and similarly at 1400 ℃ in Fig. 6. From the plots, it is clear that in all the cases, dielectric constant decreases with increase in frequency and attains a saturation tendency at 2 MHz for each concentration of doped metal. This behavior of dielectric may be explained qualitatively by the supposition of the mechanism of the polarization process in mullite–titanium/strontium. The electron-hopping model [13] can explain the electrical conduction mechanism. It is known that the effect of polarization is to reduce the field inside the medium. Therefore, the dielectric constant of a substance may be decreased substantially as the frequency is increased [20]. The electronic polarizations can orient themselves with the electric field at the lower frequency range, but at higher

frequency the internal individual dipoles contributing to the dielectric constant cannot move instantly. So as frequency of an applied voltage increases, the dipole response is limited and the dielectric constant diminishes [21,22]. The drop in resistivity with frequency predicts the presence of glassy phase /amorphous phase in the mullite structure that may increase the mobility of the ions such as Ti^{4+}/Sr^{2+} and Al^{3+} which find an easy path to move and hence increase the electrical conductivity. Moreover, it is known that the incorporation of transition metals in the periodic lattice of the mullite crystal structure helps in attending a lower band structure [11,14]. The dielectric results of Sanad et al.'s data are in good agreements with the data shown at 1100 ℃ and 1400 ℃ of 0.01 M concentration for titanium and strontium ions in the MHz frequency region [10,14,17].

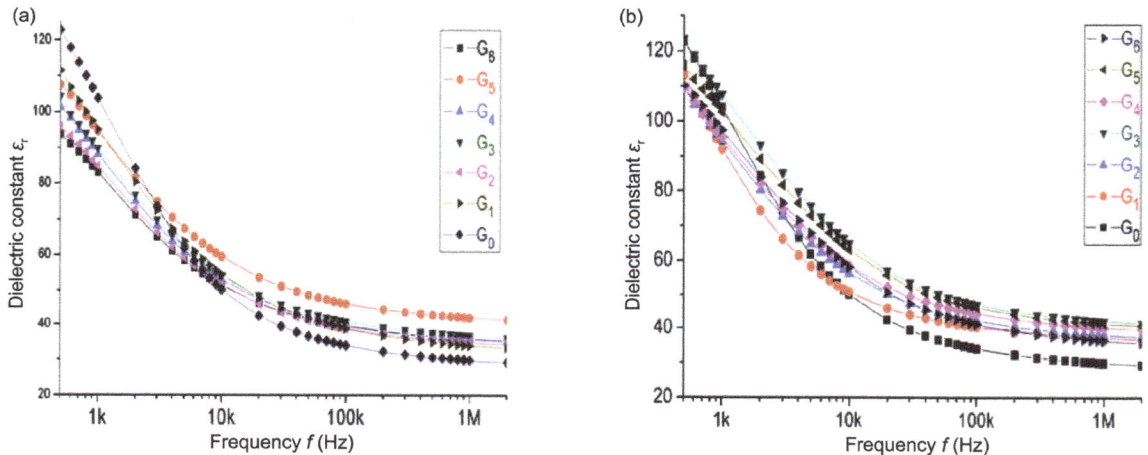

Fig. 5   Frequency response dielectric constant behavior of (a) titanium and (b) strontium doped mullite precursor gels sintered at 1100 ℃ with increasing doping concentration.

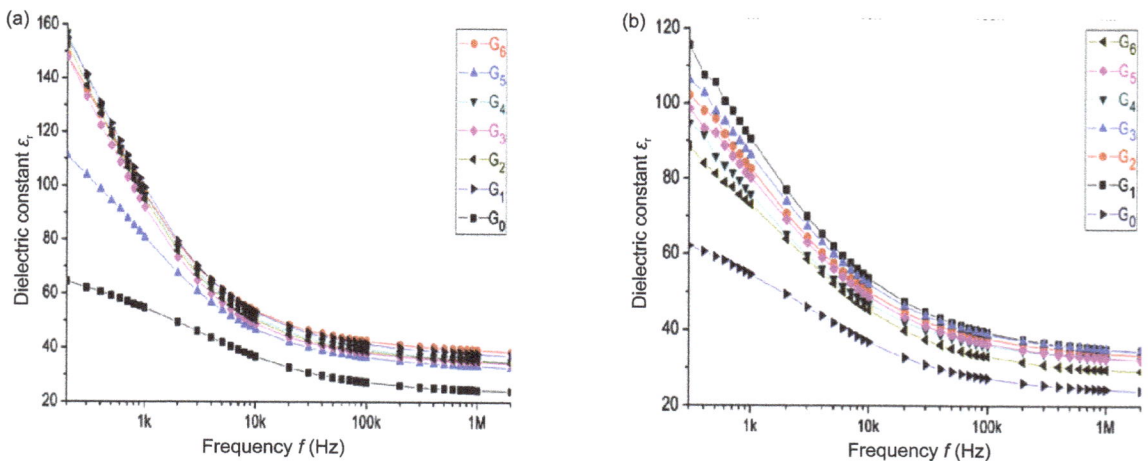

Fig. 6   Frequency response dielectric constant behavior of (a) titanium and (b) strontium doped mullite precursor gels sintered at 1400 ℃ with increasing doping concentration.

The dielectric loss (tanδ) of all samples is measured in the frequency range of 20 Hz to 2 MHz in the room temperature and is graphically shown in Figs. 7 and 8 for 1100 ℃ and 1400 ℃, respectively. It is found that for all samples, tanδ decreases with increasing frequency and reaches constant value at 2 MHz, but initially it increases until 10 kHz. Dissipation factor initially increases maybe due to the greater electronic polarization of the metal ions and also due to the formation of the metal aluminates and oxides. After 10 kHz, the internal electric field will be responsible for the decrement of the dissipation factor. The electronic polarizations can orient themselves with the electric field at the lower frequency range, but at higher frequency, the internal individual dipoles contributing to the dielectric constant cannot move instantly. So as the frequency of an applied voltage

increases, the dipole response is limited and the dielectric constant diminishes, and similarly the dielectric loss tanδ decreases.

AC conductivity of the samples is then calculated using the formula:

$$\sigma_{AC} = 2\pi f \tan\delta \varepsilon_r \varepsilon_0 \qquad (2)$$

where f is the frequency in Hz; tanδ is the dielectric loss factor; and ε_r and ε_0 are the dielectric constant of the material and permittivity of free space, respectively [23,24].

In σ_{AC} vs. f graphs, a linear increment of AC conductivity with frequency for all doping concentrations is observed for 1100 ℃ (Fig. 9) and 1400 ℃ (Fig. 10). It has been observed that the increment of AC conductivity suddenly jumps from G_4 to G_5. The linearity of the plots follows the frequency

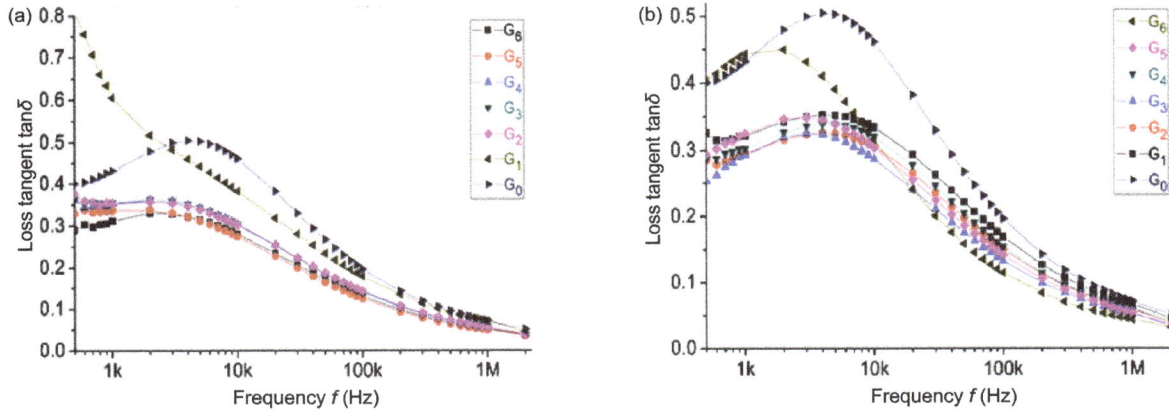

Fig. 7 Frequency response loss tangent behavior of (a) titanium and (b) strontium doped mullite precursor gels sintered at 1100 ℃ with increasing doping concentration.

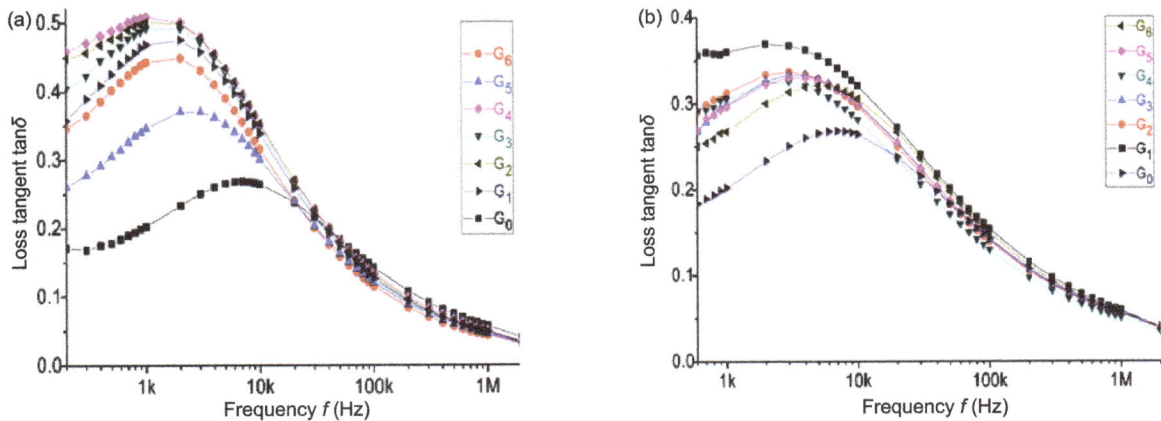

Fig. 8 Frequency response loss tangent behavior of (a) titanium and (b) strontium doped mullite precursor gels sintered at 1400 ℃ with increasing doping concentration.

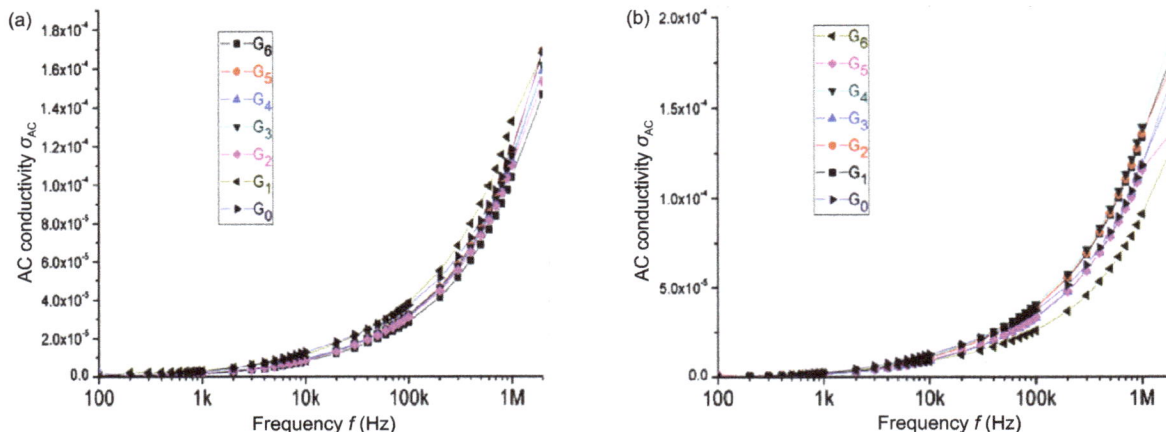

Fig. 9   Frequency response AC conductivity behavior of (a) titanium and (b) strontium doped mullite precursor gels sintered at 1100 ℃ with increasing doping concentration.

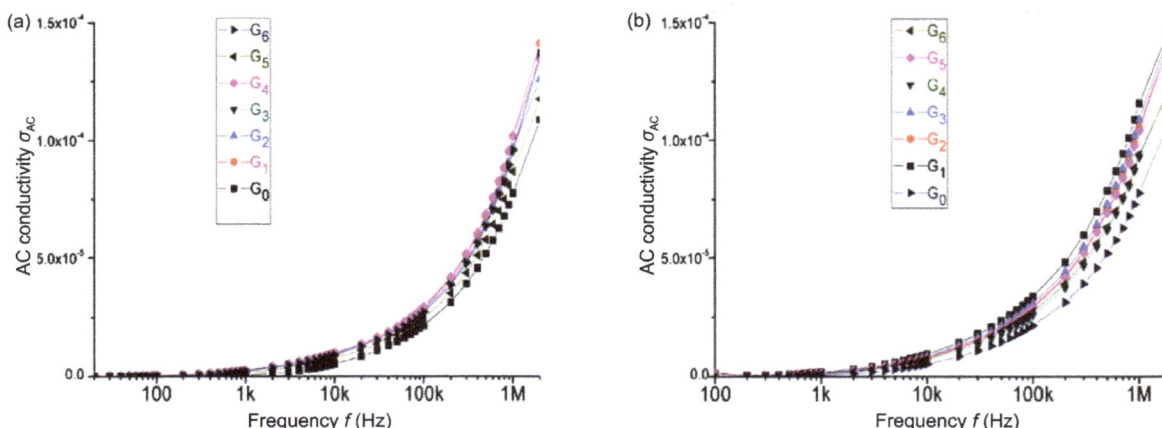

Fig. 10   Frequency response AC conductivity behavior of (a) titanium and (b) strontium doped mullite precursor gels sintered at 1400 ℃ with increasing doping concentration.

dependent part of Jonscher's universal power law which can be represented by the equation:

$$\sigma(\omega) = \sigma_{DC} + \sigma_0 \omega^s \qquad (3)$$

where σ_{DC} is the DC (or frequency independent) conductivity; σ_0 is a temperature dependent parameter; and s lies in the range of $0 < s < 1$ [20–25].

The morphology of the mullite particles with G_3 concentration of the doped metal was investigated by FESEM. G_3 sample shows titanium doped mullite nanoparticles of size 50 nm at 1100 ℃ and 100 nm at 1400 ℃ [25]. Similarly for strontium doped mullite, distinct elongated morphology of mullite particles of size 3 μm is embedded in the matrix (Fig. 11). The change in morphology of the sample is due to the combined effect of the doped metals and the sintering temperature. The particles in the case of strontium doped mullite take the shape of whiskers at 1400 ℃

due to the "mineralizing" effect of Sr^{2+} ions.

4   Conclusions

Titanium and strontium doped mullite composites have been synthesized by sol–gel technique, and their phase evolution and dielectric properties have been investigated. The dielectric constant decreased with frequency for all the samples attaining constancy at high frequency, which is a normal behavior for dielectric ceramics. Reported maximum dielectric constants are 24.42 and 37.6 at 1400 ℃ for 0.01 M concentration for titanium and strontium ions at 2 MHz, respectively.

AC conductivity increased with frequency following Jonscher's power law and was found to depend on the amount of glassy phase and concentration of mobile ions present in the composites. The doped mullite may

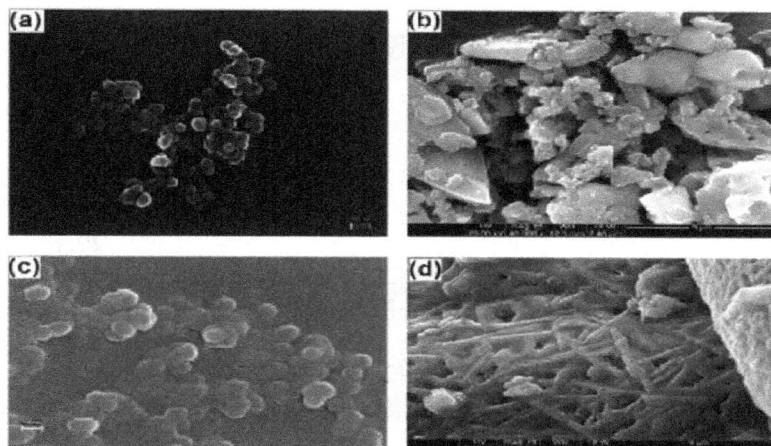

Fig. 11   SEM images of G_3 samples: (a) titanium doped mullite fired at 1100 ℃, (b) strontium doped mullite fired at 1100 ℃, (c) titanium doped mullite fired at 1400 ℃ and (d) strontium doped mullite fired at 1400 ℃.

be used for the fabrication of high charge storing capacitors and also as ceramic capacitors in the pico range.

Acknowledgements

We are grateful to DST and UGC (PURSE program), Government of India, for the financial assistance.

References

[1] Maex K, Baklanov MR, Shamiryan D, *et al*. Low dielectric constant materials for microelectronics. *J Appl Phys* 2003, **93**: 8793.

[2] Ebadzadeh T, Lee WE. Processing–microstructure–property relations in mullite–cordierite composites. *J Eur Ceram Soc* 1998, **18**: 837–848.

[3] Kurihara T, Horiuchi M, Takeuchi Y, *et al*. Mullite ceramic substrate for thin film application. Proceedings of the 40th Electronic Components and Technology Conference, Las Vegas, NV, 1990: 68–75.

[4] Ramakrishnan V, Goo E, Roldan JM, *et al*. Microstructure of mullite ceramics used for substrate and packaging applications. *J Mater Sci* 1992, **27**: 6127–6130.

[5] Viswabaskaran V, Gnanam FD, Balasubramanian M. Mullite from clay–reactive alumina for insulating substrate application. *Appl Clay Sci* 2004, **25**: 29–35.

[6] Camerucci MA, Urretavizcaya G, Castro MS, *et al*. Electrical properties and thermal expansion of cordierite and cordierite–mullite materials. *J Eur Ceram Soc* 2001, **21**: 2917–2923.

[7] Kanka B, Schneider H. Sintering mechanisms and microstructural development of coprecipitated mullite. *J Mater Sci* 1994, **29**: 1239–1249.

[8] Sanad MMS, Rashad MM, Abdel-Aal EA, *et al*. Synthesis and characterization of nanocrystalline mullite powders at low annealing temperature using a new technique. *J Eur Ceram Soc* 2012, **32**: 4249–4255.

[9] Esharghawi A, Penot C, Nardou F. Contribution to porous mullite synthesis from clays by adding Al and Mg powders. *J Eur Ceram Soc* 2009, **29**: 31–38.

[10] Sanad MMS, Rashad MM, Abdel-Aal EA, *et al*. Effect of Y^{3+}, Gd^{3+} and La^{3+} dopant ions on structural, optical and electrical properties of o-mullite nanoparticles. *J Rare Earth* 2014, **32**: 37–42.

[11] Sanad MMS, Rashad MM, Abdel-Aal EA, *et al*. Mechanical, morphological and dielectric properties of sintered mullite ceramics at two different heating rates prepared from alkaline monophasic salts. *Ceram Int* 2013, **39**: 1547–1554.

[12] Roy DS, Bagchi BJ, Bhattacharya AN, *et al*. A comparative study of densification of sol–gel-derived nano-mullite due to the influence of iron, nickel and copper ions. *Int J Appl Ceram Tec* 2013, DOI: 10.1111/ijac.12114.

[13] Roy DS, Bagchi BJ, Das SK, *et al*. Electrical and dielectric properties of sol–gel derived mullite doped with transition metals. *Mater Chem Phys* 2013, **138**: 375–383.

[14] Sanad MMS, Rashad MM, Abdel-Aal EA, *et al*. Optical and electrical properties of Y^{3+} ion

substituted orthorhombic mullite $Y_{(x)}Al_{(6-x)}Si_2O_{13}$ nanoparticles. *J Mater Sci: Mater Electron* 2014, **25**: 2487–2493.

[15] Archana J, Navaneethan M, Hayakawa Y. Hydrothermal growth of monodispersed rutile TiO_2 nanorods and functional propertie. *Mater Lett* 2013, **98**: 38–41.

[16] Misevicius M, Scit O, Grigoraviciute-Puroniene I, *et al*. Sol–gel synthesis and investigation of un-doped and Ce-doped strontium aluminates. *Ceram Int* 2012, **38**: 5915–5924.

[17] Sanad MMS, Rashad MM, Abdel-Aal EA, *et al*. Effect of Gd^{3+} ion insertion on the crystal structure, photoluminescence, and dielectric properties of o-mullite nanoparticles. *J Electron Mater* 2014, **43**: 3559–3566.

[18] Patil DR, Lokare SA, Devan RS, *et al*. Studies on electrical and dielectric properties of $Ba_{1-x}Sr_xTiO_3$. *Mater Chem Phys* 2007, **104**: 254–257.

[19] See A, Hassan J, Hashim M, *et al*. Dielectric variations of barium titanate additions on mullite–kaolinite sample. *Solid State Sci Tech* 2008, **16**: 197–204.

[20] Ravinder D, Mohan GR, Prankishan, *et al*. High frequency dielectric behavior of aluminum-substituted lithium ferrites. *Mater Lett* 2000, **44**: 256–260.

[21] Chakraborty AK. Role of hydrolysis water–alcohol mixture on mullitization of Al_2O_3–SiO_2 monophasic gels. *J Mater Sci* 1994, **29**: 6131–6138.

[22] Chakraborty AK. Effect of pH on 980 ℃ spinel phase-mullite formation of Al_2O_3–SiO_2 gels. *J Mater Sci* 1994, **29**: 1558–1568.

[23] Zhang Y, Wu Z, Wang S, *et al*. Density of Li_2TiO_3 solid tritium breeding ceramic pebbles. *Adv Mat Res* 2011, **177**: 310–313.

[24] Oréfice RL, Vasconcelos WL. Sol–gel transition and structural evolution on multicomponent gels derived from the alumina–silica system. *J Sol–Gel Sci Technol* 1997, **9**: 239–249.

[25] Roy DS, Bagchi BJ, Das SK, *et al*. Dielectric and magnetic properties of sol–gel derived mullite–iron nanocomposite. *J Electroceram* 2012, **28**: 261–267.

14

Evaluation of SiC–porcelain ceramics as the material for monolithic catalyst supports

Oleg SMORYGO[a,*], Alexander MARUKOVICH[a],
Vitali MIKUTSKI[a], Vladislav SADYKOV[b]

[a]*Powder Metallurgy Institute, 41, Platonov Str., 220005, Minsk, Belarus*
[b]*Boreskov Institute of Catalysis, 5, Lavrentiev Ave., 630090, Novosibirsk, Russia*

Abstract: Mechanical and thermal properties of SiC–porcelain ceramics were studied in the wide SiC content range of 0–95%. Microstructure evolution, shrinkage at sintering, porosity, mechanical strength, elastic modulus, coefficient of thermal expansion (CTE) and thermal conductivity were studied depending on SiC content. The optimal sintering temperature was 1200 ℃, and the maximum mechanical strength corresponded to SiC content of 90%. Parametric evaluation of the ceramic thermal shock resistance revealed its great potential for thermal cycling applications. It was demonstrated that the open-cell foam catalyst supports can be manufactured from SiC–porcelain ceramics by the polyurethane foam replication process.

Keywords: SiC; porcelain; mechanical properties; thermal properties; foam

1 Introduction

Efficiency of catalytic process with strong heat flux (e.g., hydrocarbon steam reforming) is strongly dependent on the catalyst thermal conductivity. Catalysts supported on the metal alloy monolithic supports ensure effective heat transfer in reactors thus providing a uniform temperature distribution (both axial and radial), and this affects overall catalyst performance [1–3]. Low robustness of metallic catalyst supports due to corrosive degradation in the reaction media (600–900 ℃, water vapor, decomposition products of hydrocarbon, aggressive admixtures) limits their practical application [1,4]. This problem does not arise when the catalyst supports are made from various oxide ceramics exhibiting excellent corrosive resistance [5–7]. Conventional oxide ceramics,

however, have low thermal conductivity which can be the cause of strong temperature gradients within catalytic reactor, and local hot zones can appear resulting in the catalyst sintering and its fast deactivation [8]. The use of the silicon carbide catalyst supports in the hydrocarbon steam reforming and other strongly endothermic or exothermic process [9] looks promising taking into account their attractive combination of high thermal conductivity and excellent corrosive resistance. The data in Ref. [10] demonstrated that the substitution of the alumina foam catalyst supports by the silicon carbide ones results in the increase of the cobalt based catalyst selectivity from 54% to 80% due to more uniform temperature distribution in the reactor.

Extensive application of SiC catalyst supports is restrained by the considerable manufacturing cost resulting from high sintering temperature of above 2100 ℃ [11]. Ceramic materials with high SiC content (up to 90%–95%) and high mechanical strength

---
* Corresponding author.
E-mail: olegsmorygo@yahoo.com

(450–500 MPa) as well as SiC foam catalyst supports can be synthesized via liquid phase sintering with sintering aids like Al_2O_3–Y_2O_3 [12,13]. This process is often classified as "low temperature sintering", but actually the sintering occurs at rather high temperatures of 1700–1800 ℃. Besides, the process implies protective sintering atmospheres in order to prevent SiC oxidation, which also contributes to the product manufacturing cost. That is why many efforts were undertaken during the last decade to develop compositions ensuring SiC based ceramics at much lower sintering temperatures. It was found that ceramic materials with high SiC content and reasonable mechanical properties can be synthesized by sintering in air with alkali or alkali-earth silicate sintering aids at temperatures as low as 1100–1200 ℃ [14–19]. Various sintering aids containing alkali oxides were successfully applied in these researches: porcelain and its polishing residues, bentonite, and art glass. Potentially, if a reasonable combination of mechanical and thermal properties can be attained, this type of ceramics can have a great commercial potential as the material for monolithic catalyst supports: its low manufacturing cost is predetermined by cheap initial materials and low sintering temperature without special protective atmosphere. However, no study performed complex analysis of mechanical and thermal properties of this type of ceramics before, and no study estimated this type of ceramics as the material for monolithic catalyst supports. Besides, the referred papers presented experimental data on different and rather narrow SiC content ranges, and hence the authors did not report on the evolution of microstructure and properties in the whole SiC content range.

In this paper, we studied sintering regimes, microstructure evolution, and mechanical and thermal properties of SiC–porcelain ceramics in the wide SiC content range of 0–95%. Complex analysis of properties that influence the performance in thermal cycling applications was performed, and SiC–porcelain ceramics' potential as the catalyst support material was analyzed via parametric evaluation. It was demonstrated that open-cell ceramic foam catalyst supports can be manufactured from this type of ceramics by the polyurethane foam replication process.

2 Experimental procedures

Industrial porcelain slip and technical-grade silicon

carbide powder ($d_{50} = 10.5$ µm, purity > 98%) were used in this study. The slip was received from PJSC "Dobrush Porcelain Factory", Belarus; its chemical composition is given in Table 1. The compositions were prepared as follows. The porcelain slip was dried at 110 ℃ and then mixed with the silicon carbide powder in aqueous media using a high-velocity stirrer (250 rpm, 1 h). Then the mixture was dried at 110 ℃, and 10% polyvinyl alcohol (PVA) aqueous solution (10 wt%) was added. Ceramics and PVA solution were mixed in porcelain mortar, and after rubbing through the sieve with the opening size of 100 µm, experimental samples were prepared by uniaxial die pressing (100 MPa). SiC content in the mixture varied within 0–95 wt% (related to the porcelain's solid) in this study. Dimensions of the samples were dependent on the characterization procedure; the details will be given below. After drying at room temperature for 24 h and at 110 ℃ for 4 h, the samples were heated in an electric furnace to the sintering temperature (1100–1250 ℃, 10 ℃/min) and sintered in air atmosphere for 1 h.

Mechanical properties (three-point bending strength and elastic modulus) were studied using Tinius Olsen H150KU testing machine, and the sample dimensions were 55 mm × 10 mm × 5 mm. The same samples were used for measuring open porosity by hydrostatic weighing in ethanol. Scanning electron microscope (SEM) Mira, Tescan, with energy dispersive X-ray (EDX) analyzer INCA-350, Oxford Instruments, was used for the fracture surface and microstructure examinations. Prior to the microstructure examination, the samples were mounted with the epoxy resin, ground following usual metallographic routines and finally etched at room temperature in 5% HF aqueous solution for 30 s. Coefficient of thermal expansion (CTE) was measured in vacuum within the temperature range of 20–500 ℃ using Netzsch 402-E horizontal pushrod dilatometer; the samples with diameter of 7 mm and length of 25 mm were examined. Thermal conductivity coefficient was measured at room temperature under the steady-state thermal conditions using IT-2 thermal conductivity meter, RIAPP, Belarus;

Table 1 Chemical composition of porcelain and its loss of ignition (LOI) at 1200 ℃

Chemical composition (wt%)							LOI
Al_2O_3	SiO_2	K_2O	Na_2O	Fe_2O_3	CaO	TiO_2	(wt%)
25.07	70.59	2.04	0.90	0.52	0.46	0.42	7.1

the sample dimensions were 15 mm × 15 mm × 0.5 mm.

Open-cell foam catalyst supports were manufactured from SiC–porcelain composition with optimal combination of mechanical and thermal properties via the polyurethane foam replication route. Details of the process applied in this study are described in Ref. [20].

3 Results and discussion

3. 1 Processes during the sintering

The sintering regimes of SiC–porcelain ceramics are pre-defined by interactions that occur at temperatures of above 1000 ℃; major interactions and the accompanying microstructure changes in this system were described before in Refs. [21–23]. In brief, SiC particles are protected by a thin vitreous silica film at normal conditions, which ensures excellent corrosive resistance of SiC ceramics in air. However, when temperature exceeds the melting point of the alkaline salts in porcelain, the silica film reacts with the molten salt, and it results in the formation of the alkaline silicate liquid. Thus, silica film is broken due to corrosion; oxygen diffuses easily to SiC and reacts with it. Silica and gaseous carbon oxides are the products of this reaction. If outward release of the gaseous products is hindered, they are entrapped by the silicate melt (liquid phase), and intensive generation of the secondary porosity occurs. The increase of sintering temperature favors accelerated corrosion of the protective layer, and it results in the increased final porosity and coarser pores. If the temperature is too high, the surface swelling and even formation of a highly-porous cellular structure in the sample's volume can be observed. Hence, optimization of sintering conditions for the studied compositions implies the compromise between maximum possible densification due to the liquid phase sintering, on one hand, and the absence of bubbling due to the gaseous product releasing, on the other hand. Different authors who studied SiC interaction with the alkali silicate melt sintering aids stated rather close optimal sintering temperatures; some non-principal deviations can be referred to differences in the sintering aid compositions and different targeted SiC content ranges in the ceramics. The reported sintering aids and corresponding optimal temperatures are as follows: $Li_2O–Al_2O_3–SiO_2$, 1150 ℃ [17]; bentonite, 1100 ℃ [14];

porcelain tile residues, 1200 ℃ [23]; porcelain, 1180 ℃ [22].

In order to determine preferable sintering temperature for the studied SiC–porcelain system, samples with SiC content of 85% were sintered in air for 1 h at temperatures ranging from 1100 ℃ to 1250 ℃. Dependence of the linear shrinkage on sintering temperature is demonstrated in Fig. 1. The shrinkage is not strongly dependent on temperature and just slightly decreased from 1.65% to 1.2% with the sintering temperature increase. SiC is the reactive filler towards porcelain, and hence the ceramic shrinkage is the resultant of the following phenomena: (I) densification due to the liquid phase formation; (II) formation of less dense phase due to SiC oxidation to SiO_2 during sintering ($\rho = 3.21$ g/cm^3 and $\rho = 2.2$ g/cm^3 for SiC and vitreous silica, respectively); (III) expansion due to the gas generation and consequent bubbling the liquid phase (resulting from SiC reaction with the alkali silicate melt). For the studied system, the driving forces of the volumetric expansion compensate the driving forces of the ceramic shrinkage: higher sintering temperatures result in more intensive SiC oxidation, and hence more intensive SiO_2 formation and gas generation. That is why obvious increase of the shrinkage with the sintering temperature increase is not observed. Furthermore, some shrinkage decrease is stated. Visible surface glass swelling and large bubbles at the sample surface are observed after sintering at 1250 ℃ (Fig. 2). Besides, coarse and isolated round pores are detected inside the samples, which is resulted evidently from the gas generation and its entrapment during the sintering. Samples sintered at 1200 ℃ retain their original shape without any change of the surface condition, and this sintering temperature is used in all further studies.

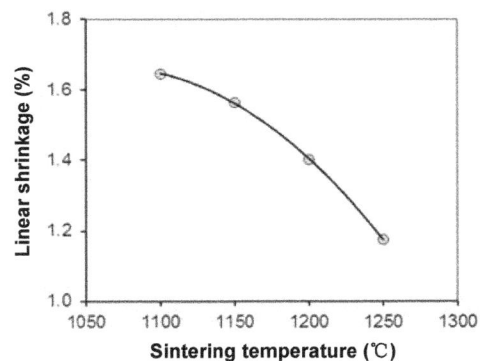

Fig. 1 Effect of sintering temperature on linear shrinkage (85% SiC).

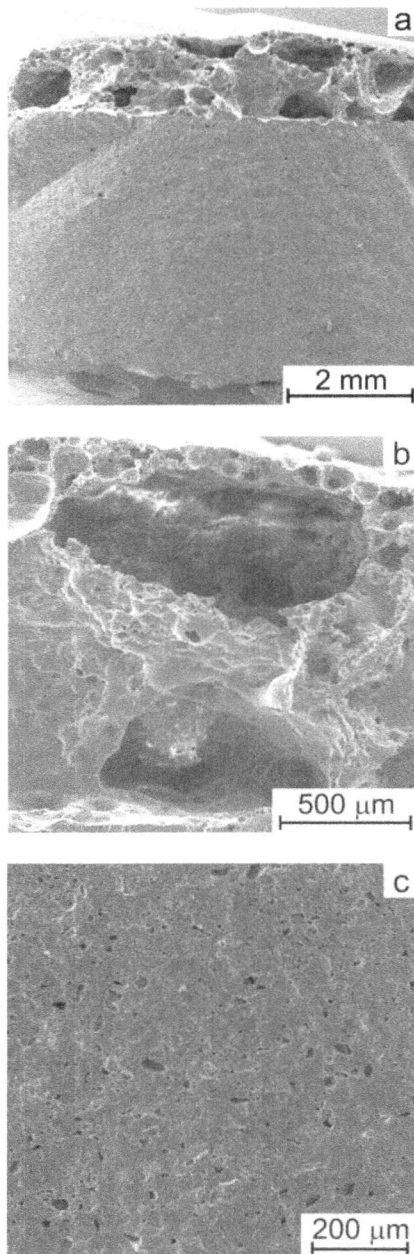

Fig. 2 SEM images of the SiC–porcelain sample with the surface glass swelling (85% SiC, 1250 °C): (a) fracture surface; (b) bubbles on the top surface; (c) bulk microstructure.

Microstructures of the porcelain and SiC–porcelain ceramics sintered at 1200 °C are presented in Fig. 3. As it is expected, remarkable residual porosity with the pooled glass inclusions can be observed in the porcelain sample; regular sintering temperature of this porcelain is 1340–1360 °C according to the supplier. At low and medium SiC contents, both SiC grains and pooled glass inclusions are observed, and the intergrain

phase has evidently higher residual porosity compared to pure porcelain. At the highest SiC content, SiC grains are bound by a highly porous glassy phase. Compositional and resultant microstructural differences among the different ceramics in this study affect their properties, which will be discussed in the following.

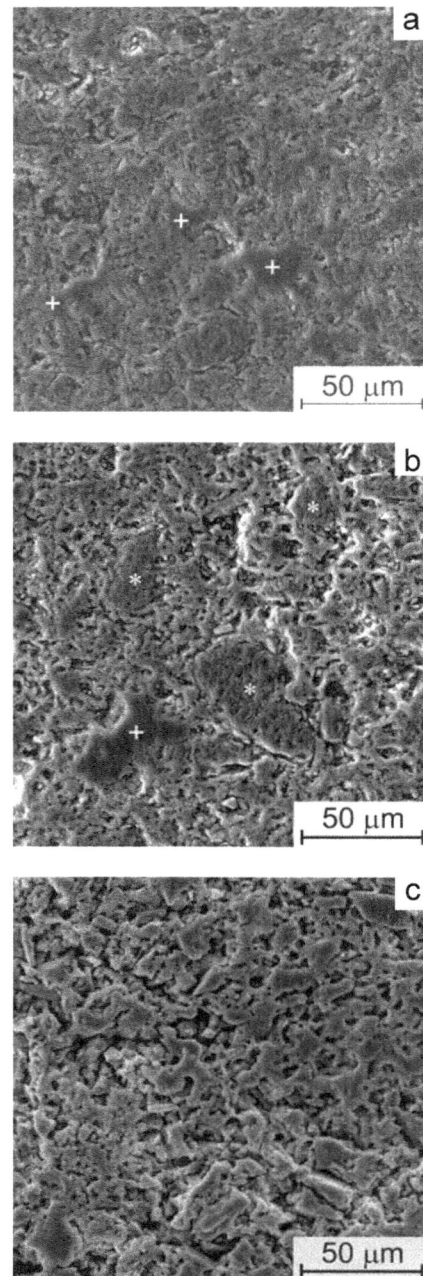

Fig. 3 SEM images of the microstructure of SiC–porcelain ceramics at different SiC contents: (a) porcelain; (b) 40% SiC; (c) 85% SiC. Pooled glass is marked with "+"; SiC grains are marked with " * ".

Figures 4, 5 and 6 enable to analyze the evolutions of ceramic shrinkage at sintering, apparent density and open porosity depending on SiC content, respectively. Linear shrinkage decreases from 6.8% to 0.8% with SiC content increase. At the same time, there is no strict correlating dependence for the sample apparent density that decreases from 2.27 g/cm^3 to 2.01 g/cm^3 only in the same SiC content range. Taking into account large difference among densities of the ceramic constituents ($\rho = 2.27$ g/cm^3 and $\rho = 3.21$ g/cm^3 for the studied porcelain and SiC, respectively), one can state that the expected density increase is compensated by the residual porosity increase. The comparative analysis of Figs. 5 and 6 reveals that the apparent density decrease by 11.5% is accompanied by the open porosity increase from 15% to 46%, which is equivalent to 1.5-fold relative density decrease. Gradual elimination of closed porosity that is not measurable by the applied method and simultaneous formation of open porosity at higher SiC contents are the most evident reasons of this mismatch.

Formation of the predominantly open porosity is stated to favor faster SiC oxidation during the sintering in air. The effect of SiC content on SiC oxidation rate during the sintering can be understood from Figs. 7 and 8. The weight loss after sintering of compositions with the lowest SiC contents changes to the weight gain at its higher contents. In Fig. 7, the solid line is the measured weight change after sintering at 1200 ℃ for 1 h. This dependence results from two simultaneous phenomena: the porcelain weight loss of ignition (LOI) and the weight gain due to SiC oxidation. Contribution of the porcelain weight loss can be easily estimated from its portion in the initial mixture and the measured LOI is 7.1%; the dashed line in the graph corresponds to this dependence. Subtraction of the dashed line from the solid line gives the dashed-dot line demonstrating the weight gain due to SiC oxidation. Assuming that all the oxidized SiC is transformed to SiO_2, one can estimate the rate of the oxidized SiC in the ceramics depending on its composition (Fig. 8). Larger open porosity seems to be the only reason of the more intensive SiC oxidation: the open porosity favors easier oxygen supply to the non-protected SiC particle surfaces as well as much

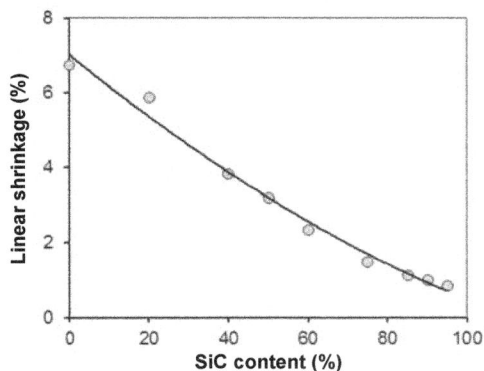

Fig. 4 Effect of SiC content on linear shrinkage at sintering.

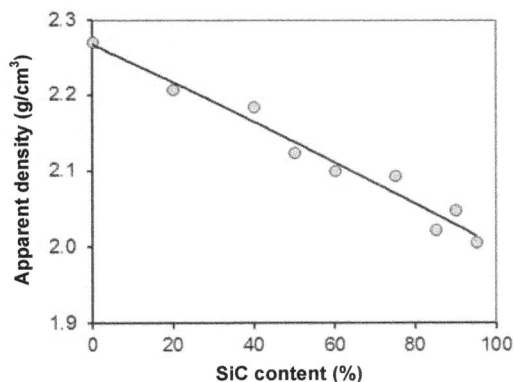

Fig. 5 Effect of SiC content on apparent density.

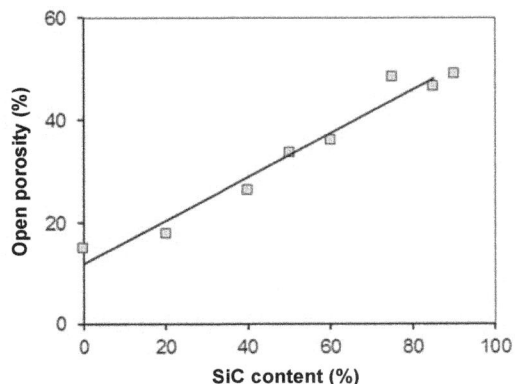

Fig. 6 Effect of SiC content on open porosity.

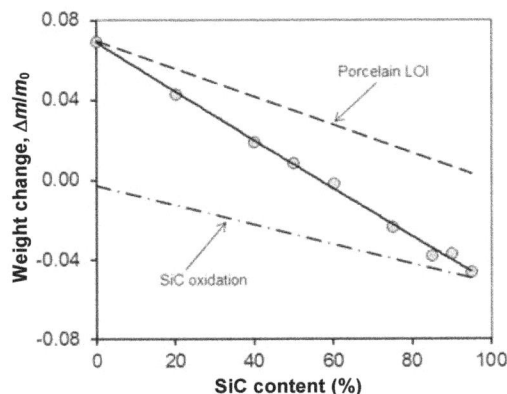

Fig. 7 Effect of SiC content on the ceramic weight change after sintering.

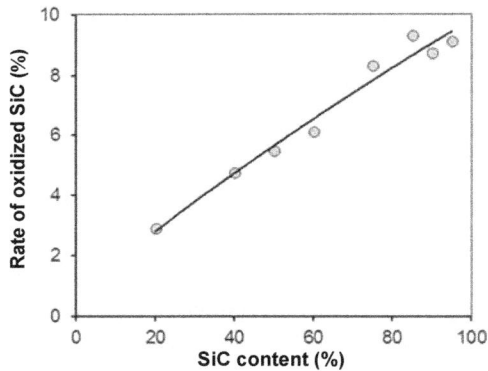

Fig. 8 Effect of SiC content on estimated rate of oxidized SiC.

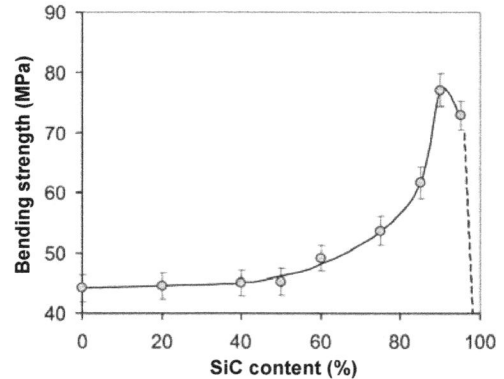

Fig. 9 Effect of SiC content on bending strength.

easier release of the gaseous SiC oxidation products. According to this estimation, up to 9% of total SiC amount oxidizes during the sintering. This fact must be taken into account when monolithic catalyst supports with high surface-to-volume ratio are manufactured. The surface-to-volume ratio of the bar-type samples used in this study is ~640 m^2/m^3. Specific surface area of the open-cell foam catalyst supports with typical 10–30 ppi cell size grades (this corresponds to the cell diameter range of ~2–6 mm) and open porosity of 0.85 is roughly equal to ~330–1000 m^2/m^3 [24]. This is very close to the same parameters of the experimental samples, and similar oxidation rate can be expected.

3. 2   Mechanical properties

As it can be seen in Fig. 9, the bending strength is almost independent on SiC content until 60% of SiC (55–59 MPa). Further SiC content increase results in the fast strength increase that attains the maximum value at 90% of SiC (77 MPa). Hypothetic continuation of experimental curve is given as the dashed line in the graph. Thus, direct correlation between the ceramic residual porosity and its mechanical strength is not stated. Similar strength maxima are observed in the SiC–art glass system at 90% of SiC [18] and SiC–bentonite system at 90%–95% of SiC [14]. In Ref. [18], such maximum is referred to the best dispersion of the binding glass phase throughout the composite; at higher glass loadings it forms continuous phase with low toughness and significant pooling, which facilitates the crack initiation and propagation. Fracture surface images of different compositions synthesized in this research are demonstrated in Fig. 10. The fracture surface views are similar in porcelain and SiC–porcelain ceramics at low

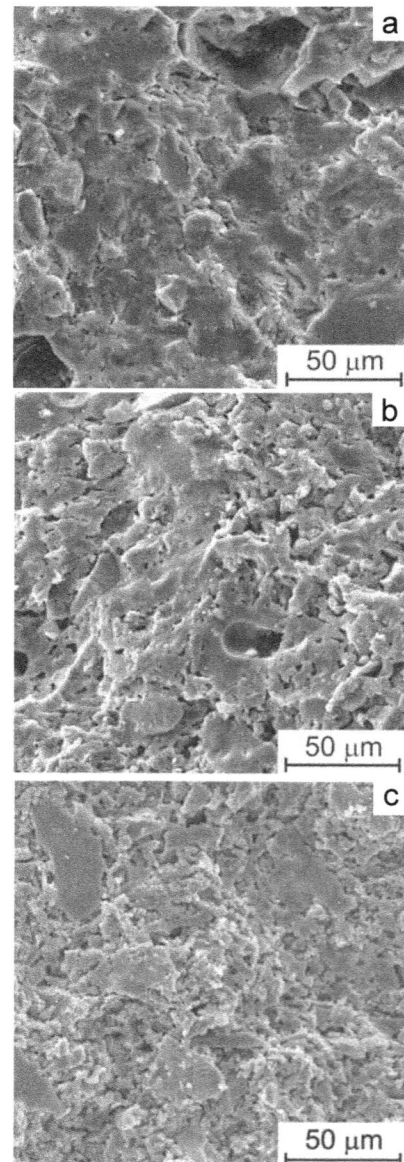

Fig. 10 SEM images of the SiC–porcelain ceramic fracture surfaces at different SiC contents: (a) porcelain; (b) 20% SiC; (c) 85% SiC.

and medium SiC contents, because the fracture runs through the porcelain phase in all cases. That is why SiC grains can not be differentiated in Figs. 10(a) and 10(b). The presence of SiC in the ceramics could be only detected indirectly due to much higher residual porosity generated because of SiC oxidation. At the highest SiC content, the fracture occurs predominantly along the SiC–porcelain interface, which presumably must be responsible for the sharp mechanical strength increase.

Thus, based on the mechanical strength and microstructure investigations, one can distinguish three characteristic SiC content areas with different fracture behaviors. At low and medium SiC contents (less than 50%–60%), the measured strength values are very close (i.e., the deviations are within the experimental error). In this content range, SiC grains are distributed loosely in the binding porcelain phase. That is why the crack runs through this phase, the crack path is linear, and the bending strength is controlled by this phase. Sharp strength increase with the SiC content increase from 60% to 90% can be explained by gradual microstructure changes. In this range, the space among SiC particles is still filled by the porcelain phase, but the crack path becomes tortuous, and deflection of the crack path becomes more and more significant since the packing density of strong SiC grains is increasing [25]. Further SiC content increase results in the strength decrease because the porcelain binding phase does not fill the intergranular space, and, according to the minimum solid area model for porous materials [26], mechanical strength of porous ceramics σ is related to its porosity ε by the equation $\sigma = \sigma_0\exp(-b\varepsilon)$, where σ_0 is the strength of the corresponding fully dense material and b is a constant depending on the pore shape and alignment. Finally, the curve on Fig. 9 must tend to zero as far as SiC content approaches 100% since direct sintering of SiC is impossible at such low temperatures [27].

Similar but not so pronounced dependence is stated for elastic modulus (Fig. 11): it is initially invariable, but increases at the highest SiC contents. Unfortunately, exact location of the modulus maximum is not stated in this study; hypothetical continuation of the curve is given as the dashed line in the graph. It should be emphasized that the elastic modulus values are rather low in all the studied range due to high residual porosity (~9–11 GPa). It also should be noted that,

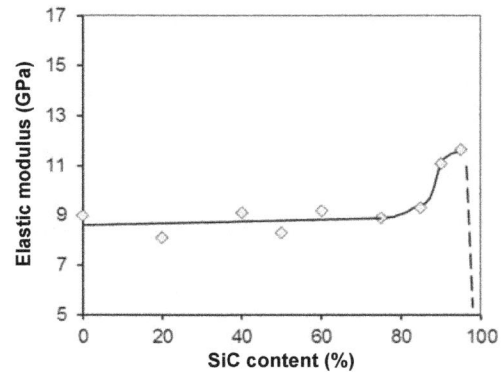

Fig. 11   Effect of SiC content on elastic modulus.

according to the additivity law [28], elastic modulus value is expected to increase smoothly with the SiC fraction increase in all the SiC content range that ensures continuous filling of the intergranular space between SiC grains with the porcelain phase (i.e., until 85%–90% of SiC). However, the expected increase of the modulus seems to be compensated by the increasing residual porosity. The increase of the modulus value in Fig. 11 at SiC content of 90%–95% presumably results from the tightest packing of SiC grains within the porcelain matrix. Further sharp drop of elastic modulus value to zero can be expected taking into account the minimum solid area model [26] and the fact that SiC grains cannot be sintered directly at 1200 ℃.

3. 3   Thermal properties

Evolutions of the ceramic thermal properties (CTE and thermal conductivity) depending on SiC content are illustrated in Figs. 12 and 13. As it is expected, CTE gradually decreases and thermal conductivity increases with the SiC content increase; near-linear dependencies are observed in both cases in the studied range. Thermal conductivity coefficient is equal to ~10–12 W/(m·K) at the highest SiC contents, which is comparable by value of magnitude to the same parameter of stainless steels and nickel-chromium alloys. The measured thermal properties in combination with mechanical properties provide the necessary data to estimate thermal shock resistance of the synthesized ceramics. The strength controlled thermal fracture resistance of the engineering ceramics can be roughly estimated using statistical parameters $R = \sigma_f/(\alpha E)$ (high heat transfer) and $R' = \lambda\sigma_f/(\alpha E)$ (low heat transfer) [29], where λ is the thermal conductivity coefficient; σ_f is the stress to failure; α is the CTE; and E is the elastic modulus. Dependencies calculated from experimental data are presented in Fig. 14; both R and R' values increase with the SiC content increase. Analysis of these experimental curves and the data on many usual engineering ceramics [29] reveals superior R and R' values of the studied SiC–porcelain ceramics with SiC content of 85%–90% compared to conventional cordierite, mullite and

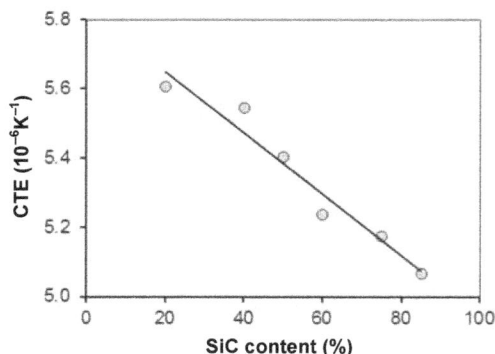

Fig. 12   Effect of SiC content on coefficient of thermal expansion.

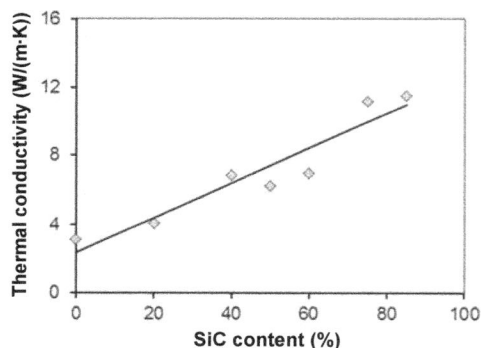

Fig. 13   Effect of SiC content on thermal conductivity coefficient.

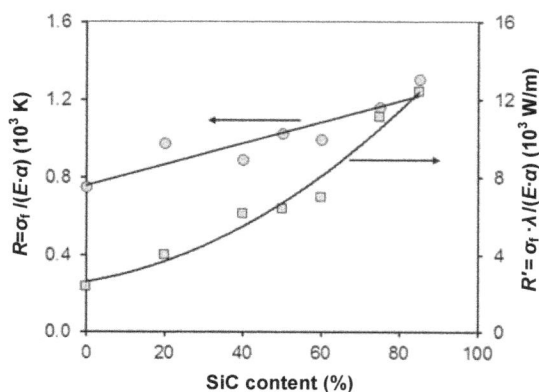

Fig. 14   Effect of SiC content on statistical parameters R and R'.

Y-PSZ. This makes the alkali silicate bound SiC ceramics a rather attractive object of detailed studies in the thermal cycling and thermal shock applications (e.g., as catalyst supports), because the production cost benefits can be expected due to comparatively low sintering temperature, simple sintering regimes and cheap raw materials. However, practical operating temperature range, durability and thermal shock behavior of this type of ceramics should be studied in more detail in further researches.

3. 4   Ceramic foam manufacture

Samples of open-cell ceramic foam that can be used as the catalyst support were manufactured from SiC–porcelain ceramics by the polyurethane foam replication process. Details of the process applied in this study are described in Ref. [20]. In brief, reticulated polyurethane foam samples with the nominal cell density of 30 ppi were subjected to multiple impregnation with aqueous suspensions comprising SiC powder and industrial porcelain at the solid residue ratio of 85/15, and carboxymethyl cellulose as the rheological additive (0.1 wt%). After each impregnation using suspensions with the controlled moisture content [20], the samples were subjected to centrifuging to remove excess suspension and drying at 110 °C. When the targeted porosity of 83%–85% was attained, the samples were sintered in air at 1200 °C for 1 h. No visible surface flaw on the foam strut surfaces is observed. Porcelain contains clay and bentonite, which is essential for the sample green density after the start of the polyurethane decomposition. Pore structure of the SiC–porcelain foam synthesized in this study is shown in Fig. 15. Thermal and mechanical properties of the foam were

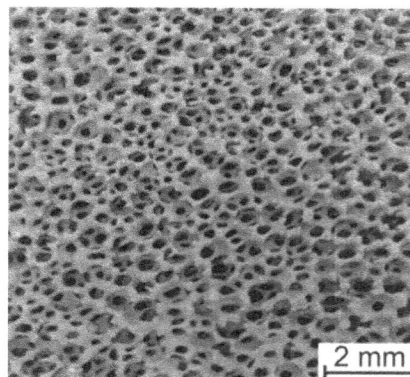

Fig. 15   Pore structure of the SiC–porcelain ceramic foam.

not studied in this research, but they can be estimated from the experimental data on SiC–porcelain ceramic properties using the extensively recognized Ashby's equations [30].

4   Conclusions

Mechanical and thermal properties of SiC–porcelain ceramics were studied in the wide SiC content range of 0–95%. Microstructure evolution, shrinkage at sintering, porosity, mechanical strength, elastic modulus, CTE and thermal conductivity were studied depending on SiC content. The optimal sintering temperature was 1200 ℃, and the maximum three-point bending strength (77 MPa) corresponded to SiC content of 90%. This strength was attained at rather low linear shrinkage at sintering of ~1%. Introducing SiC resulted in the residual porosity increase, including formation of secondary pores due to SiC oxidation and gaseous carbon oxide generation. Higher residual porosity resulted in the more intensive SiC oxidation during the sintering; up to 9% of the introduced SiC oxidized during the sintering. Compared to porcelain ceramics, CTE of this composite ceramics decreased from $5.6\times10^{-6}\ K^{-1}$ to $5\times10^{-6}\ K^{-1}$, and its thermal conductivity increased from 3 W/(m·K) to 12 W/(m·K). Parametric evaluation of the ceramic thermal shock resistance revealed its great potential as the material for thermal cycling applications: higher values of statistical parameters compared to the most conventional engineering ceramics were stated. It was demonstrated that the open-cell foam catalyst supports can be manufactured from SiC–porcelain ceramics by polyurethane foam replication process.

Acknowledgements

The research was supported by Integrated Project T12CO-020 of the National Academy of Sciences of Belarus and Siberian Branch of the Russian Academy of Sciences.

References

[1] Li Y, Wang Y, Zhang Z, et al. Oxidative reformings of methane to syngas with steam and CO_2 catalyzed by metallic Ni based monolithic catalysts. *Catal Commun* 2008, **9**: 1040–1044.

[2] Visconti CG, Tronconi E, Groppi G, et al. Monolithic catalysts with high thermal conductivity for the Fischer–Tropsch synthesis in tubular reactors. *Chem Eng J* 2011, **171**: 1294–1307.

[3] Sadykov V, Sobyanin V, Mezentseva N, et al. Transformation of CH_4 and liquid fuels into syngas on monolithic catalysts. *Fuel* 2010, **89**: 1230–1240.

[4] Smorygo O, Mikutski V, Marukovich A, et al. Structured catalyst supports and catalysts for the methane indirect internal steam reforming in the intermediate temperature SOFC. *Int J Hydrogen Energ* 2009, **34**: 9505–9514.

[5] Faure R, Rossignol F, Chartier T, et al. Alumina foam catalyst supports for industrial steam reforming processes. *J Eur Ceram Soc* 2011, **31**: 303–312.

[6] Rennard D, French R, Czernik S, et al. Production of synthesis gas by partial oxidation and steam reforming of biomass pyrolysis oils. *Int J Hydrogen Energ* 2010, **35**: 4048–4059.

[7] Manfro RL, Ribeiro NFP, Souza MMVM. Production of hydrogen from steam reforming of glycerol using nickel catalysts supported on Al_2O_3, CeO_2 and ZrO_2. *Catalysis for Sustainable Energy* 2013, **1**: 60–70.

[8] Wang C, Wang T, Ma L, et al. Steam reforming of biomass raw fuel gas over NiO–MgO solid solution cordierite monolith catalyst. *Energ Convers Manage* 2010, **51**: 446–451.

[9] Twigg MV, Richardson JT. Fundamentals and applications of structured ceramic foam catalysts. *Ind Eng Chem Res* 2007, **46**: 4166–4177.

[10] Lacroix M, Dreibine L, de Tymowski B, et al. Silicon carbide foam composite containing cobalt as a highly selective and re-usable Fischer–Tropsch synthesis catalyst. *Appl Catal A: Gen* 2011, **397**: 62–72.

[11] Guo X, Cai X, Zhu L, et al. Preparation and properties of SiC honeycomb ceramics by pressureless sintering technology. *J Adv Ceram* 2014, **3**: 83–88.

[12] Magnani G, Minoccari GL, Pilotti L. Flexural strength and toughness of liquid phase sintered silicon carbide. *Ceram Int* 2000, **26**: 495–500.

[13] Chen F, Yang Y, Shen Q, et al. Macro/micro structure dependence of mechanical strength of low temperature sintered silicon carbide ceramic foams. *Ceram Int* 2012, **38**: 5223–5229.

[14] Soy U, Demir A, Caliskan F. Effect of bentonite addition on fabrication of reticulated porous SiC ceramics for liquid metal infiltration. *Ceram Int* 2011, **37**: 15–19.

[15] Medri V, Fabbri S, Ruffini A, *et al.* SiC-based refractory paints prepared with alkali aluminosilicate binders. *J Eur Ceram Soc* 2011, **31**: 2155–2165.

[16] Medri V, Ruffini A. Alkali-bonded SiC based foams. *J Eur Ceram Soc* 2012, **32**: 1907–1913.

[17] Pan Y, Baptista JL. Low-temperature sintering of silicon carbide with $Li_2O–Al_2O_3–SiO_2$ melts as sintering aids. *J Eur Ceram Soc* 1996, **16**: 745–752.

[18] Tucker MC, Tu J. Ceramic coatings and glass additives for improved SiC-based filters for molten iron filtration. *Int J Appl Ceram Tec* 2014, **11**: 118–124.

[19] Yao X, Tan S, Zhang X, *et al.* Low-temperature sintering of SiC reticulated porous ceramics with $MgO–Al_2O_3–SiO_2$ additives as sintering aids. *J Mater Sci* 2007, **42**: 4960–4966.

[20] Leonov AN, Smorygo OL, Sheleg VK. Monolithic catalyst supports with foam structure. *React Kinet Catal L* 1997, **60**: 259–267.

[21] Shui A, Xi X, Wang Y, *et al.* Effect of silicon carbide additive on microstructure and properties of porcelain ceramics. *Ceram Int* 2011, **37**: 1557–1562.

[22] García-Ten J, Saburit A, Bernardo E, *et al.* Development of lightweight porcelain stoneware tiles using foaming agents. *J Eur Ceram Soc* 2012, **32**: 745–752.

[23] Bernardin AM, da Silva MJ, Riella HG. Characterization of cellular ceramics made by porcelain tile residues. *Mat Sci Eng A* 2006, **437**: 222–225.

[24] Smorygo O, Mikutski V, Marukovich A, *et al.* An inverted spherical model of open-cell foam structure. *Acta Mater* 2011, **59**: 2669–2678.

[25] Ritchie RO. Mechanisms of fatigue crack propagation in metals, ceramics and composites: Role of crack tip shielding. *Mat Sci Eng A* 1988, **103**: 15–28.

[26] Kayal N, Dey A, Chakrabarti O. Synthesis of mullite bonded porous SiC ceramics by a liquid precursor infiltration method: Effect of sintering temperature on material and mechanical properties. *Mat Sci Eng A* 2012, **556**: 789–795.

[27] Riedel R, Passing G, Schönfelder H, *et al.* Synthesis of dense silicon-based ceramics at low temperatures. *Nature* 1992, **355**: 714–717.

[28] Gunyaev GM. Polycomponent high-modulus composites. *Polymer Mechanics* 1977, **13**: 685–692.

[29] Lu TJ, Fleck NA. The thermal shock resistance of solids. *Acta Mater* 1998, **46**: 4755–4768.

[30] Ashby MF. The properties of foams and lattices. *Phil Trans R Soc A* 2006, **364**: 15–30.

15

Microhardness, microstructure and electrical properties of ZVM ceramics

Abdel-Mageed H. KHAFAGY[a], Sanaa M. EL-RABAIE[b], Mohamed T. DAWOUD[b,*], M. T. ATTIA[b]

[a]Physics Department, Faculty of Science, Menufiya University, Shebin El-Koom 32511, Egypt
[b]Physics and Engineering Mathematics Department, Faculty of Electronic Engineering, Menufiya University, Menouf 32952, Egypt

Abstract: The effect of Mn_3O_4 addition on microhardness, microstructure and electrical properties of vanadium oxide doped zinc oxide varistor ceramics is systematically investigated. The Vicker's microhardness H_V has decreased with increasing the amount of Mn_3O_4. Also, the average grain size has decreased from 27.51 μm to 19.55 μm with increasing the amount of Mn_3O_4 up to 0.50 mol%, whereas an increase in Mn_3O_4 up to 0.75 mol% has caused the average grain size to increase and then it decreases with increasing Mn_3O_4 from 0.75 mol% to 1.00 mol%. The sintered density has decreased from 5.38 g/cm^3 to 5.31 g/cm^3 with increasing the amount of Mn_3O_4. The varistor ceramic modified with 0.50 mol% Mn_3O_4 has exhibited excellent nonlinear properties, with 16.29 for the nonlinear coefficient and 441.9 μA/cm^2 for the leakage current density. Furthermore, the sample doped with 0.50 mol% Mn_3O_4 has been found to possess donor density as 0.77×10^{18} cm^{-3} and 0.916 eV barrier height.

Keywords: ceramics; electrical properties; microstructure; varistor; Mn_3O_4 doped ZnO–V_2O_5 varistor

1 Introduction

ZnO based varistors are semiconducting ceramic devices exhibiting nonlinear symmetrical current–voltage (I–V) or current density–electric field (J–E) characteristics for both voltage polarities [1]. This is due to the presence of inter-granular layers with high resistance between ZnO grains having low resistance and Schottky barrier arising due to the surface states caused by the dopants. Such behaviors enable them to absorb large amount of energy. Consequently, they have been extensively used for protecting electronic circuits, against voltage surges and for the voltage stabilization of electrical power lines.

The I–V characteristic relation is expressed as [1]:

$$I = KV^\alpha$$

where K is a constant and α is the nonlinear coefficient related to the material's microstructure. Besides the above coefficient, there are two other parameters, the breakdown voltage V_B and the leakage current I_L. The breakdown voltage is calculated by the simple equation [2,3]:

$$V_B = N_{gb} \times V_{gb} = (t/D)V_{gb}$$

where N_{gb} is the number of grain boundaries across the varistor's thickness t; D is the average grain size; and

* Corresponding author.
E-mail: dawoud.mohamed73@yahoo.com

V_{gb} is the average voltage per grain boundary (≈ 3 V).

Also, in terms of the current density J and the applied electric field E, the above I–V relation is rewritten as

$$J = (E/C)^{\alpha}$$

where C is a constant. Then the breakdown field $E_B = V_{gb}/D$ [3] is usually taken as the field applied when the current flowing through the varistor is 1 mA [4]. Since the grain boundary barriers in the ZnO based varistors are of Schottky type, the current density J in the first ohmic region of varistor characteristics is related to the electric field E at a given absolute temperature T by the equation [5,6]:

$$J = AT^2 \exp[(\beta E^{1/2} - \varphi_B)/(kT)]$$

where $A = 4\rho emk^2/h^3$ is the Richardson's constant; ρ is the varistor density; e is the electron charge; m is the electron mass; k is the Boltzmann constant; h is the Plank's constant; β is a constant related to the grain number per unit length and the barrier width ω; and φ_B is the Schottky barrier height formed on the grain surfaces and arises due to the surface states caused by the dopants. Therefore, the electrical conductivity σ of ZnO varistors could be calculated from the slopes of the J–E curves in the first and third ohmic regions, while in the second nonlinear region where the current strongly increases due to the decrease of φ_B, the conductivity is given by [7]:

$$\sigma_2 = \sigma_1 \exp[(\alpha-1)(E_2-E_1)/E_2]$$

where σ_1 is the conductivity corresponding to the first ohmic region; σ_2 is the conductivity of the second region; and E_1 and E_2 are two values on the linear part of the breakdown region.

The effects of doping ZnO based ceramics with few mole percentages of different oxides of some metals, e.g., Bi, Sb, Co, Mn, Ni, Cr, Al and Fe, have been investigated by several authors [8–16]. Reported results of those studies show that the important physical properties mentioned above are closely related to the compositions and microstructures, i.e., they are dependent on density, grain size, morphology, distribution of any secondary phases and porosity. Therefore, by controlling these parameters, the nonlinear coefficient and breakdown field values could be improved in order to obtain the required electrical performance. Also, each of the above dopants plays a distinctive role in the subtle tuning of the final nonlinear characteristics of the varistor ceramics, while the overall electrical characteristics are resulted from the collective effect of all the microstructural features.

Multilayer chip ceramic products require that the inner-electrode materials should have their melting points (MP) relatively higher than the sintering temperature of their components [17–19]. Therefore, new ceramics are developed where a silver inner-electrode (MP = 961 ℃) can be used for making multilayer components which have their sintering temperatures around 900 ℃. Among those products, the binary system ZnO–V$_2$O$_5$ has become a good candidate [17,20–24], since it can be sintered at a temperature in the vicinity of about 900 ℃ which is important for multilayer chip applications due to its ease of co-firing (sintering) with silver inner-electrode without using expensive palladium or platinum metals as electrode materials. However, effects of adding some different metal oxides and rare earths to the binary system ZnO–V$_2$O$_5$ on the electrical properties have been investigated by several authors, e.g., Mn ions [25], Mn$_3$O$_4$ and Nb$_2$O$_5$ [26], Mn$_3$O$_4$ and Er$_2$O$_3$ [27], MnO$_2$–Er$_2$O$_3$ and Nb$_2$O$_5$ [28,29], MnO$_2$–Nb$_2$O$_5$ and Bi$_2$O$_3$ [30], and MnO$_2$ [31,32]. The reported results of these works are dealt with the effects of different doping or the effects of sintering temperature on the microstructure, nonlinear and conduction behaviors of tested ceramics.

The present work is aimed to investigate and throw more lights on the microstructure and electrical properties of xMn$_3$O$_4$–0.5V$_2$O$_5$–(99.5–x)ZnO (ZVM) ceramics, where $x = 0$, 0.15, 0.30, 0.50, 0.75 and 1.00 mol%, as prepared by the conventional ceramic method and sintered at 900 ℃ for 5 h. Besides, X-ray diffraction, scanning electron microscopy and energy dispersive X-ray analyses have been used to characterize the prepared ceramics.

2   Experimental

Reagent-grade raw materials of ZnO, V$_2$O$_5$ and Mn$_3$O$_4$ were used for preparing ZnO based ceramics by solid state reaction technique according to the formula xMn$_3$O$_4$–0.5V$_2$O$_5$–(99.5–x)ZnO, where $x = 0$, 0.15, 0.30, 0.50, 0.75 and 1.00 mol%. Weighed raw materials were mixed by ball-milling in exist of acetone for 28 h. The mixture was dried at 120 ℃ for 12 h. The dried mixture was milled using an agate mortar and after 0.8 wt% polyvinyl alcohol binder addition for 3 h. The powder was uniaxially pressed into discs of 12 mm in diameter and 1.5 mm in thickness at a pressure of 9500 psi. The discs were put in porcelain crucible and sintered at 900 ℃ in air for 5 h, and then the furnace was cooled to room temperature. The sintered samples were lapped and polished to 1.0 mm thickness. Silver paste was coated on both surfaces of the samples, and the ohmic contacts were formed by heating at 300 ℃ for 5 min.

At first, both the sintered density and the Vicker's microhardness H_V were measured at room temperature 27 ℃. The density ρ was determined by the Archimedes method with toluene as immersing liquid, while H_V was found by using a microhardness tester (HMV-2, Ver. 1.28, Shimatzu, Japan) where each sample was loaded at force equal to 4.902 N for 30 s and then H_V was estimated according to the relation $H_V = 0.1891 P/d^2$, where P is the load applied and d is the diagonal length of indenter impression.

The current density–electric field $(J–E)$ characteristics of the samples were recorded at room temperature. An electric circuit composed from a 500 V DC Gelman power supply (Model: 38207 ANN ARBOR, MICH., USA 48106), a digital multi-meter (GW Digital Multimeter Model: GDM-8034) and an auto ranging microvolt (Model: 197A DMM-Keithley) was used to measure the current passing through and voltage across the sample. The breakdown field (E_B) was measured at a current density of 1.0 mA/cm^2, and the leakage current density (J_L) was measured at $0.80 E_B$. In addition, the non-ohmic coefficient (α) was determined from the current–voltage curve plotted on a log–log scale, from which the slope of the curve gives the value α [8].

Also, for doing microstructure investigations, both surfaces of the samples were lapped and ground with SiC paper and polished with 0.3 μm Al$_2$O$_3$ powder to mirror-like surfaces. The polished samples were chemically etched using 1HCl:1000H$_2$O for 40 s and then coated with gold. The surface microstructure was then examined by a scanning electron microscope (SEM, JEOL, JSM5410, Japan). The average grain size (D) was determined by the linear intercept method using the expression $D = 1.56L/(MN)$ [33], where L is the random line length on the micrograph; M is the magnification of the micrograph; and N is the number of the grain boundaries intercepted by lines. The compositional elemental analysis of the selected areas on micrographs was determined by an attached energy dispersive X-ray analysis (EDX) system. The crystalline phases were identified by powder X-ray diffraction (XRD, X'pert-PRO MPD, Analytical,

Almelo, the Netherlands) with Cu Kα radiation.

3 Results and discussion

3.1 Microstructure investigations

3.1.1 Density and microhardness

Figure 1 shows the compositional dependences of both the sintered density ρ and the Vicker's microhardness H_V of xMn$_3$O$_4$–0.5V$_2$O$_5$–(99.5–x)ZnO ceramics sintered at 900 ℃ for 5 h, where x changes from 0 to 1.00 mol%. This figure reveals that ρ and H_V are rapidly decreased with increasing the Mn$_3$O$_4$ content up to 0.15 mol%. Then, the density continues its decreasing up to 0.50 mol% of Mn$_3$O$_4$, while beyond this content it decreases slowly with the gradual increase of Mn$_3$O$_4$ up to 1.00 mol%. H_V has also slowly decreased as x increases from 0.15 mol% to 1.00 mol%, the end of tested range. Typical average results for both ρ and H_V are listed in Table 1. The obtained values for ρ in Table 1 are comparable with those reported for the same materials [17]. The above dependences of both ρ and H_V on the mole percentage of Mn$_3$O$_4$ content can be interpreted as follows. Although gradual replacement of ZnO (molecular weight 81.369) by the same amount of Mn$_3$O$_4$ (molecular weight 228.812) in the ZnO based varistor matrix is supposed to increase the molecular weight of prepared samples, the bulk density is decreased all over the tested range as seen in Fig. 1. Also, it is

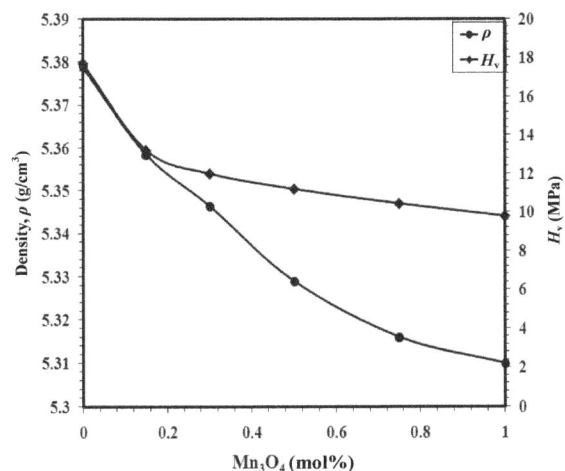

Fig. 1 Compositional dependences of the sintered density ρ and the Vicker's microhardness H_V of xMn$_3$O$_4$–0.5V$_2$O$_5$–(99.5–x)ZnO ceramics sintered at 900 ℃ for 5 h, where x changes from 0 to 1.00 mol%.

Table 1 Average calculated values of density ρ, Vicker's microhardness H_V, molar volume V and grain size D as found from both XRD and SEM results

x (mol%)	ρ (g/cm^3)	H_V (MPa)	Molar volume V (cm^3/mol)	Average grain size (XRD) D (nm)	Average grain size (SEM) D (μm)
0.00	5.379	17.726	15.224	34.502	27.513
0.15	5.359	13.209	15.324	31.110	21.303
0.30	5.347	12.456	15.399	29.002	20.458
0.50	5.329	11.212	15.506	27.211	19.545
0.75	5.316	10.440	15.613	34.129	24.982
1.00	5.310	9.799	15.700	30.510	23.670

known that the density equals the molecular weight divided by the molar volume. This volume which increases due to the weak connectivity of the ceramic structure results in a decrease in microhardness of the varistor ceramics with increasing the Mn$_3$O$_4$ content (Table 1) in addition to the formation of some pores in the ZnO polycrystalline grains of varistor matrix by sintering at high temperature 900 ℃ (as will be seen in the SEM micrographs in Fig. 3). Thus, the observed decrease in density could be attributed analogously to an increase in the molar volume and the obtained decrease in microhardness due to the weakness of the structural connectivity and coupling between the ZnO grains [13]. On the other hand, it has been assumed that the decrease of sintered density is attributed to the volatility of V species for V$_2$O$_5$ with low melting point 690 ℃ [28].

3.1.2 XRD analysis

Figure 2 shows the dependence of XRD patterns of tested varistors on the mole percentage of Mn$_3$O$_4$ content. As indicated in Fig. 2, the observed diffraction graphs for all tested samples confirm the presence of the wurtzite hexagonal structure of the main ZnO primary phase. All diffraction peaks corresponding to this phase are indexed according to their reflection planes of polycrystalline Miler orientations (100), (002) and (101) [7,13,34] as indicated in Fig. 2 with (101) index having the high-intensity major peak. Besides, there are some secondary phases, such as Zn$_3$(VO$_4$)$_2$ appeared as a result of the presence of ZnO–V$_2$O$_5$ which acts as a liquid phase sintering aid for the formation of it [17,27–30] and Mn-rich Mn$_3$O$_4$ (for samples with $x = 0.75$ mol% and $x = 1.00$ mol%) [25,30]. The average crystallite size D is calculated in terms of X-ray line broadening at full width half maximum (FWHM) by using Scherer's equation [7,13,24] for all tested samples and the average values are summarized in Table 1.

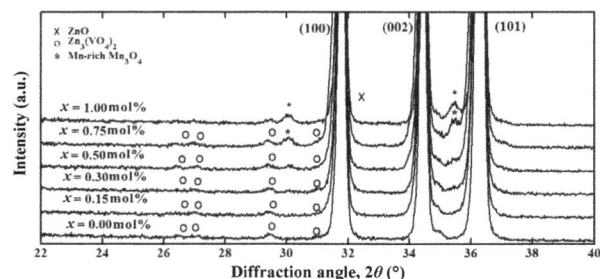

Fig. 2 Dependence of XRD patterns of tested ceramic varistors on the mole percentage of Mn$_3$O$_4$ content.

3.1.3 SEM and EDX analyses

Figure 3 shows the SEM micrographs of the samples with different amounts of Mn$_3$O$_4$ (x from 0 to 1.00 mol%). Inspection of this figure reveals the occurrence of ZnO major phase of large grains, and other different minor phases with small grains disperse randomly in the former major one such as Zn$_3$(VO$_4$)$_2$ and Mn-rich Mn$_3$O$_4$ in addition to some pores (black regions), in agreement with Refs. [17,20] and the above XRD results (Fig. 2). Observed grains are of different sizes and are randomly distributed throughout the samples. The average value of grain size D of observed grains was calculated for all specimens according to the equation [33], and the typical average values are listed in Table1. This table indicates that the values of D, even found from either XRD or SEM, have similarly changed with the gradual doping of Mn$_3$O$_4$, in which D values decrease from 34.502 nm to 27.211 nm (XRD) and from 27.513 μm to 19.545 μm (SEM) with increasing the content of Mn$_3$O$_4$ from 0 up to 0.50 mol%. Then they increase up to 34.129 nm (XRD) and 24.982 μm (SEM) upon increasing the Mn$_3$O$_4$ content to 0.75 mol%, and finally a slight decrease is observed in both D values to 30.510 nm (XRD) and 23.670 μm (SEM) with increasing the content of Mn$_3$O$_4$ to 1.00 mol%, the end of tested range.

Figure 4 shows the EDX analyses for the ceramic varistors with $x = 0.15$, 0.50 and 1.00 mol% of Mn$_3$O$_4$ at the interior of ZnO grain as well as on the grain boundary. Within EDX detection limit, all figures show the existence of elements V, Mn and Zn both at the interior of ZnO grain and on the grain boundary, which indicates that both V and Mn species are dissolved in ZnO grains and reside on their boundaries as well. This will prove their effects on the electrical parameters as will be seen in the Section 3.2.2.

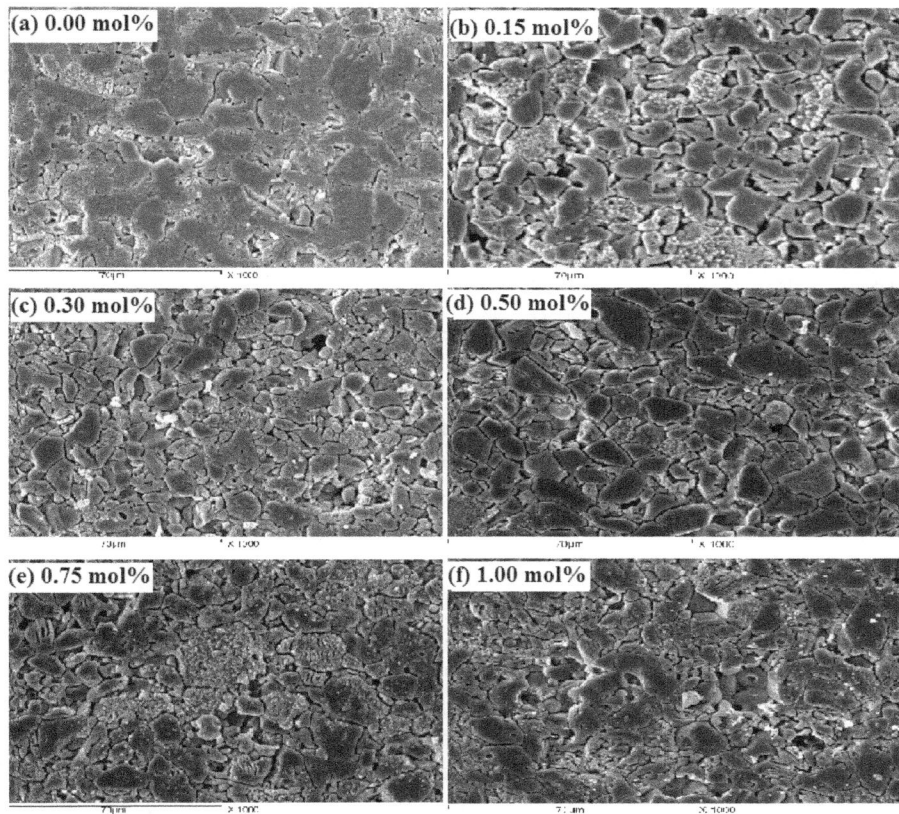

Fig. 3   SEM micrographs of xMn$_3$O$_4$–0.5V$_2$O$_5$–(99.5–x)ZnO samples: (a) x = 0.00 mol%, (b) x = 0.15 mol%, (c) x = 0.30 mol%, (d) x = 0.50 mol%, (e) x = 0.75 mol% and (f) x = 1.00 mol%.

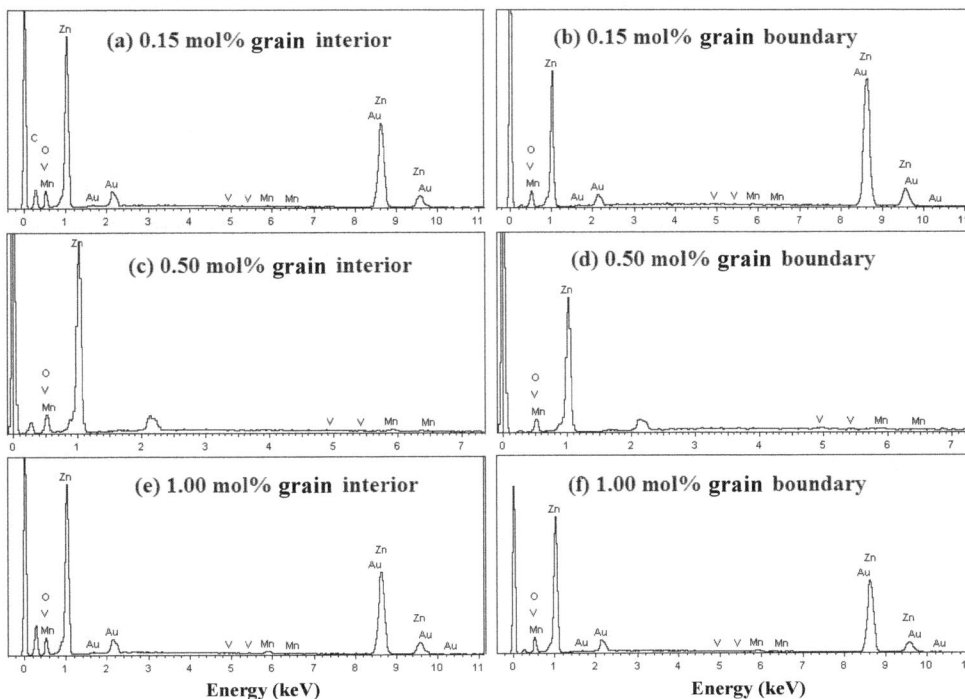

Fig. 4   EDX analyses for the ceramic varistors with x = 0.15, 0.50 and 1.00 mol% of Mn$_3$O$_4$ at the interior of ZnO grain ((a), (c) and (e)) and on the grain boundary ((b), (d) and (f)).

It is worth mentioning that the comparison of the compositional dependences of the grain size between EDX results observed here and those reported [17] for the same composition samples, indicates that the grain size is only consistent with those reported results in the x range from 0 to 0.50 mol% of Mn_3O_4 and differs beyond this content, where the previous reported results show that D is nearly constant in the range from 0.5 mol% to 2 mol% of Mn_3O_4 [17]. On the other hand, EDX results of this work show the existence of both V and Mn elements on the boundaries as well as in the ZnO grain (Fig. 4), while it was reported that no V spiece was found in the ZnO grains [17]. The above differences in both results may be because the preparation of specimens (grinding methods) and sintering times in the two works are different (5 h here and 3 h there), and only one specimen with $x = 2$ mol% of Mn_3O_4 was presented there [17] for EDX analysis while specimens with lower contents $x = 0.15$, 0.50 and 1.00 mol% of Mn_3O_4 are investigated in the present work.

3.2   Electrical characterization

3.2.1   Current density–electric field (J–E) characteristics

Figure 5 shows the compositional dependence of the current density–electric field (J–E) characteristics of xMn_3O_4–$0.5V_2O_5$–$(99.5-x)ZnO$ ceramics prepared by solid state reaction technique and sintered at 900 ℃ for 5 h, where $x = 0$, 0.15, 0.30, 0.50, 0.75 and 1.00 mol%. Inspection of Fig. 5 reveals that the varistor properties are basically characterized by non-ohmicity in all obtained J–E plots and each plot is divided into two regions, an ohmic region before the breakdown and a non-ohmic one after the breakdown with some overlap between them. However, the sharper the knee of the curves between the two regions is, the better the nonlinear properties are. Also, it can be noted from Fig. 5 that the plot corresponding to the un-doped sample

($x = 0$ mol% of Mn_3O_4) shows very poor non-ohmic properties, but with increasing the Mn_3O_4 content, in other tested samples up to 0.50 mol%, the knee gradually becomes more pronounced and the non-ohmic properties are enhanced where the breakdown electric field E_B increases, and after the above content of Mn_3O_4, the J–E characteristics show an overlap with more doping up to 1.00 mol%, which again results in a decrease in the breakdown field as seen in Fig. 5.

Using the J–E characteristics of Fig. 5, the breakdown electric field E_B, nonlinear coefficient α, leakage current density J_L, barrier height φ_B, donor concentration N_d, density of states N_s, barrier width ω and non-ohmic conductivity σ are calculated and their typical average values are summarized in Table 2.

3.2.2   Breakdown field and grain size

The compositional dependences of both the breakdown field E_B and average grain size D are shown in Fig. 6. From this figure, converse behaviors of changes are observed for the two parameters, as functions of the mole percentage of Mn_3O_4 content, i.e., E_B has increased and shows a peak value (1409 V/cm) at 0.50 mol% of Mn_3O_4, whereas D decreases and has a

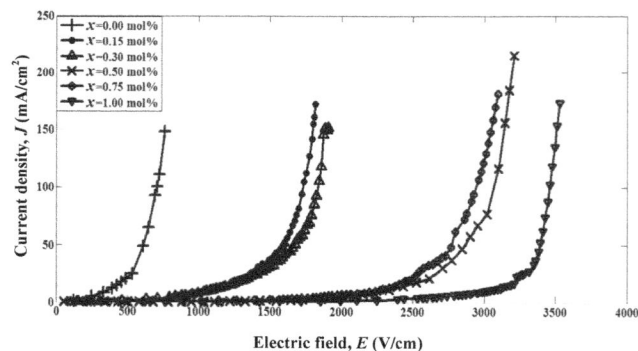

Fig. 5   Compositional dependence of J–E characteristics of tested varistor ceramics xMn_3O_4–$0.5V_2O_5$–$(99.5-x)ZnO$, where $x = 0$, 0.15, 0.30, 0.50, 0.75 and 1.00 mol%.

Table 2   Average calculated values of the breakdown field E_B, nonlinear coefficient α, leakage current density J_L, barrier height φ_B, donor concentration N_d, density of states N_s, barrier width ω and non-ohmic conductivity σ

x (mol%)	E_B (V/cm)	α	J_L (mA/cm^2)	φ_B (eV)	N_d (10^{18}cm^{-3})	N_s (10^{12}cm^{-2})	Barrier width ω (nm)	Non-ohmic conductivity σ (10^{-4}(Ω·m)$^{-1}$)
0.00	118.30	5.33	0.6611	0.837	1.13	2.98	26.40	89.30
0.15	229.10	5.63	0.7713	0.856	0.88	2.71	30.66	5.33
0.30	493	11.36	0.6026	0.879	0.83	2.62	31.52	3.28
0.50	1409	16.29	0.4419	0.916	0.77	2.57	33.47	0.18
0.75	1138	7.62	0.5844	0.894	0.20	1.30	64.74	0.38
1.00	1907	14.66	0.5028	0.909	0.16	1.18	72.51	0.09

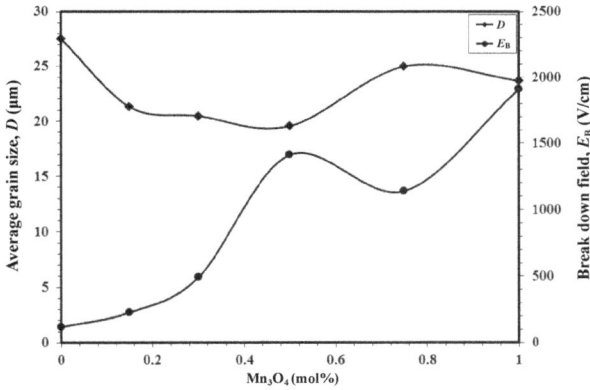

Fig. 6 Compositional dependences of both the breakdown field E_B and average grain size D of xMn$_3$O$_4$–0.5V$_2$O$_5$–(99.5–x)ZnO ceramics.

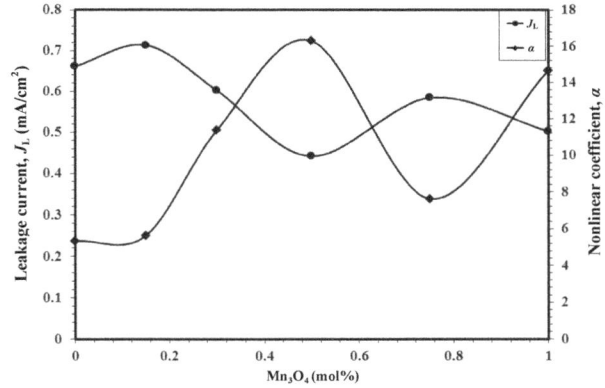

Fig. 7 Variations of average values of the nonlinear coefficient α and the leakage current density J_L as functions of mole percentage of Mn$_3$O$_4$ content.

trough value (19.545 μm), and vice versa is occurred at 0.75 mol% of Mn$_3$O$_4$. On the other hand, on increasing the mole percentage of Mn$_3$O$_4$ beyond 0.75 mol% up to 1.00 mol%, E_B increases and D decreases (Tables 1 and 2). This obtained behavior of change of E_B could be interpreted based on the changes observed in the average value of grain size D, i.e., the initial increase of E_B with the increase of Mn$_3$O$_4$ content up to 0.50 mol% is owing to the increase in the number of grain boundaries [17,26–30] which is resulted from the corresponding decrease in the average grain size and vice versa in the final part, the rest of tested range up to 1.00 mol%.

3.2.3 Nonlinear coefficient and leakage current density

Variations of the nonlinear coefficient α and the leakage current density J_L as functions of the mole percentage of Mn$_3$O$_4$ content of tested ceramics are shown in Fig. 7. Inspection of this figure reveals that α increases from 5.33 for the un-doped sample to 16.29 reaching this maximum for that which doped with 0.50 mol% of Mn$_3$O$_4$, whereas further additions of Mn$_3$O$_4$ have led α to decrease and reach the value of 7.62 at 0.75 mol% of Mn$_3$O$_4$ and then increase again to 14.66 at 1.00 mol% of Mn$_3$O$_4$ content. On the other hand, J_L value decreases from 0.6611 mA/cm^2 with the increase of Mn$_3$O$_4$ amount from $x = 0$ up to 0.50 mol% reaching a minimum value, 0.4419 mA/cm^2 at $x =$ 0.50 mol%, whereas further additions of Mn$_3$O$_4$ have led J_L to increase to 0.5844 mA/cm^2 at 0.75 mol% and decrease to 0.5028 mA/cm^2 at $x = 1.00$ mol% of Mn$_3$O$_4$ content to show an opposite behavior to that observed for α. The behaviors of both the nonlinear coefficient α and the leakage current density J_L mentioned above are related to the variation of the density of interface states

N_s at the grain boundaries, and subsequently they are dependent on the Schottky barrier height φ_B according to the variation of the concentration of electronic donor states N_d (Table 2). It is worth mentioning that the observed value of $\alpha = 16.29$ for the ceramic doped with 0.50 mol% of Mn$_3$O$_4$ in this work is less than that reported for the same ceramic which was 22.4 [17]. This is perhaps due to different conditions of preparation and soaking time (5 h here and 3 h there) in spite of the two works were done at the same sintering temperature 900 ℃.

3.2.4 Barrier height and donor concentration

The donor concentration N_d and barrier height φ_B were determined by using the grain boundary defect model [5,6] defined previously in the introduction part of this work as

$$J = AT^2 \exp[(\beta E^{1/2} - \varphi_B)/(kT)]$$

where by considering the measured current density–electric field values in the ohmic region for all samples (Fig. 5), keeping the temperature constant (at room temperature 27 ℃) and plotting ln$J$ versus $E^{1/2}$, the values of β and φ_B can be derived from the obtained slopes and intersections of the extrapolated regression lines with the ln$J$ axis, respectively. Also, since [32,34]

$$\beta = [(1/(r^*\omega))(2e^3/(4\varepsilon_0\varepsilon_r))]^{1/2}$$

where r^* is the number of grains per unit length (it could be obtained from SEM micrographs (Fig. 3); ω is the barrier width; e is the electron charge; ε_0 is the vacuum dielectric constant (8.85×10^{-14} F/cm); and ε_r is the relative dielectric constant (8.5 for ZnO), then N_d can be deduced from the equation [35]:

$$\omega^2 = (2\varphi_B\varepsilon_0\varepsilon_r)/(e^2N_d)$$

and thus the density of states $N_s = N_d \cdot \omega$ [28] can also be determined and typical values for these parameters are listed in Table 2. The variations of both the barrier height and donor density with the mole percentage of Mn_3O_4 content in tested varistors are shown in Fig. 8.

The illustrated figure (Fig. 8) indicates that φ_B increases from 0.837 eV to 0.916 eV by Mn_3O_4 addition up to 0.50 mol%, then decreases to 0.894 eV with the gradual doping of Mn_3O_4 up to 0.75 mol% and then increases to 0.909 eV at 1.00 mol% of Mn_3O_4 content. This behavior agrees with that one where it has been reported that the incorporation of Mn_3O_4 above 0.50 mol% causes the electronic inactivity of grain boundaries [17]. Anyhow, the nonlinear electrical conductivity σ shows an opposite behavior to that observed for the barrier height φ_B (Table 2). On the other hand, the donor density N_d shows a decrease from 1.13×10^{18} cm^{-3} to 0.12×10^{18} cm^{-3} with increasing of Mn_3O_4 content from 0 to 1.00 mol%.

3.2.5　General discussion

In general, the above characterizations of physical properties, density and microhardness in addition to microstructural and electrical properties of the tested xMn_3O_4–$0.5V_2O_5$–$(99.5-x)ZnO$ ceramics, x changing from 0 to 1.00 mol%, reveal that all investigated properties are dependent on the ceramic composition rather than the secondary phases, i.e., on the mole percentage of Mn_3O_4 doping content, the different measurements obtained from employed techniques in this work show no monotonic behaviors of changes for investigated properties. Anyhow, the effect on the properties depends on the manner in which the Mn ions enter the ZnO major matrix of the hexagonal structure wurtzite phase. When Mn is dissolved in these ZnO grains, it can behave as a deep donor [13]

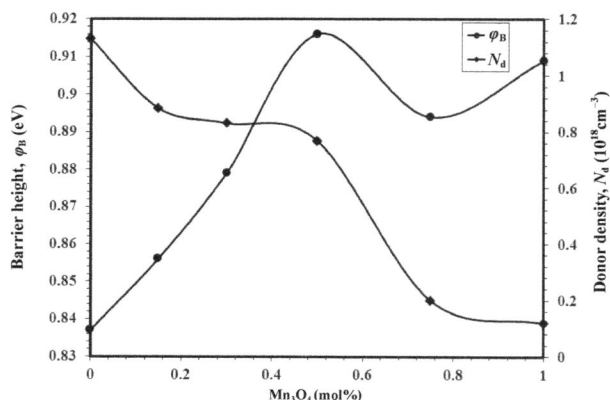

Fig. 8　Variations of the barrier height φ_B and the donor density N_d with the mole percentage of Mn_3O_4 content.

and depress the concentrations of the intrinsic donors (oxygen vacancies). In this way, the conductivity, mainly arising from ionization of intrinsic donors, is lowered on doping up to 0.50 mol% of Mn_3O_4 content. This helps to increase the nonlinear coefficient α and the barrier height φ_B, and decrease the donor density N_d and the conductivity σ (Figs. 7 and 8 and Table 2). On the other hand, with increasing Mn content through the interstitial positions, it can behave as an acceptor and has its effect to decrease the nonlinear coefficient and the height barrier φ_B as observed in the above mentioned figures. Also, the behavior of nonlinear coefficient α, in accordance with the incorporation of mole percentage of Mn_3O_4, can be related to the variation of the Schottky barrier height φ_B according to the variation of the density of electronic states N_s, the donor concentration N_d and the barrier width ω at the grain boundaries (Table 2). The doping of Mn_3O_4 into the ceramic matrix has decreased the density of interface states N_s, donor concentration N_d and increased the barrier width ω as seen in Table 2. It is therefore believed that the decrease of donor concentration N_d with the increase of Mn_3O_4 content is associated with a decrease of partial pressure of oxygen in the materials [17]. The variation in α with further increase in Mn_3O_4 is attributed to the variation of barrier height φ_B as observed in Table 2.

The ceramic microstructure has also been affected by the incorporation of Mn_3O_4 into the ceramic matrix, where obtained results show that the gradual substitution of ZnO by the same amount of Mn_3O_4 up to 0.50 mol% has led to:

(1) Decreasing the ceramic density and the Vicker's microhardness due to the produced increase in the molar volume and the decrease in the connectivity of the microstructure as well as weak coupling between the ZnO grains, respectively (Fig. 1 and Table 1). These results are consistent with SEM photographs (Fig. 3) where the grains are not close together through the weak link regions, i.e., the dark region (pores) between them.

(2) Decreasing the average grain size D due to the resulted increase in the number of grain boundaries and hence the breakdown electric field E_B is increased. However, further increases in Mn_3O_4 content from 0.50 mol% to 0.75 mol% and from 0.75 mol% to 1.00 mol% has increased and decreased the average grain size, respectively, to show their reverse behaviors for the obtained breakdown electric field.

4 Conclusions

Microstructure, sintered density, Vicker's micro-hardness and electrical properties of varistor ceramics prepared by solid state reaction having compositional formula xMn$_3$O$_4$–0.5V$_2$O$_5$–(99.5–x) ZnO, $x = 0$, 0.15, 0.30, 0.50, 0.75 and 1.00 mol%, were investigated. Both the microhardness and the sintered density of varistor samples have decreased with increasing the Mn$_3$O$_4$ content. XRD and SEM investigations show the presence of the major ZnO wurtzite phase of hexagonal structure and some secondary phases, such as Zn$_3$(VO$_4$)$_2$ and Mn-rich Mn$_3$O$_4$.

Also, the results obtained from J–E show that, all electrical properties: breakdown field, nonlinear coefficient, the Schottky barrier height, barrier width, the conductivity of nonlinear region, donor concentration and density of surface states at ZnO grain boundaries of the samples tested are dependent on the microstructure, i.e., on the grain size of existed phases, mentioned above, which affect the breakdown field. It is worth mentioning that the 99ZnO–0.5V$_2$O$_5$ ceramic which doped with 0.50 mol% Mn$_3$O$_4$ and sintered at 900 ℃ for 5 h has exhibited optimum value for the nonlinear coefficient (16.29), the lowest leakage current density (0.4419 mA/cm^2) and the highest potential barrier height (0.916 eV). This shows that the above sample can be used in the manufacture of low-voltage varistors and multilayer components.

References

[1] Gupta TK. Application of zinc oxide varistors. *J Am Ceram Soc* 1990, **73**: 1817–1840.

[2] Peiteado M, Fernánde JF, Caballero AC. Processing strategies to control grain growth in ZnO based varistors. *J Eur Ceram Soc* 2005, **25**: 2999–3003.

[3] Levinson LM, Philipp HR. Zinc oxide varistors—A review. *Am Ceram Soc Bull* 1986, **65**: 639–646.

[4] Deshpande VV, Patil MM, Ravi V. Low voltage varistors based on CeO$_2$. *Ceram Int* 2006, **32**: 85–87.

[5] Pianaro SA, Bueno PR, Longo E, *et al*. A new SnO$_2$-based varistor system. *J Mater Sci Lett* 1995, **14**: 692–694.

[6] Dhage SR, Ravi V, Yang OB. Low voltage varistor ceramics based on SnO$_2$. *Bull Mater Sci* 2007, **30**: 583–586.

[7] Sedky A, Al-Sawalha A, Yassin AM. Enhancement of electrical conductivity of ZnO ceramic varistor by Al doping. *Egyptian Journal of Solids* 2008, **31**: 205–215.

[8] Matsuoka M. Nonohmic properties of zinc oxide ceramics. *Jpn J Appl Phys* 1971, **10**: 736.

[9] Glot AB. A model of non-ohmic conduction in ZnO varistors. *J Mater Sci: Mater El* 2006, **17**: 755–765.

[10] Inada M. Formation mechanism of nonohmic zinc oxide ceramics. *Jpn J Appl Phys* 1980, **19**: 409.

[11] Houabes M, Bernik S, Talhi C, *et al*. The effects of aluminium oxide on the residual voltage of ZnO varistors. *Ceram Int* 2005, **31**: 783–789.

[12] Bernik S, Daneu N. Characteristics of ZnO-based varistor ceramics doped with Al$_2$O$_3$. *J Eur Ceram Soc* 2007, **27**: 3161–3170.

[13] Sedky A, El-Suheel E. A comparative study between the effects of magnetic and nonmagnetic dopants on the properties of ZnO varistors. *Physics Research International* 2010, **2010**: 120672.

[14] Sedky A, Abu-Abdeen M, Almulhem AA. Nonlinear I–V characteristics in doped ZnO based-ceramic varistor. *Physica B* 2007, **388**: 266–273.

[15] Sawalha A, Abu-Abdeen M, Sedky A. Electrical conductivity study in pure and doped ZnO ceramic system. *Physica B* 2009, **404**: 1316–1320.

[16] Gupta TK. Microstructural engineering through donor and acceptor doping in the grain and grain boundary of a polycrystalline semiconducting ceramics. *J Mater Res* 1992, **7**: 3280–3295.

[17] Nahm CW. Improvement of electrical properties of V$_2$O$_5$ modified ZnO ceramics by Mn-doping for varistor applications. *J Mater Sci: Mater El* 2008, **19**: 1023–1029.

[18] Nahm C-W. The nonlinear properties and stability of ZnO–Pr$_6$O$_{11}$–CoO–Cr$_2$O$_3$–Er$_2$O$_3$ ceramic varistors. *Mater Lett* 2001, **47**: 182–187.

[19] Nahm C-W. ZnO–Pr$_6$O$_{11}$–CoO–Cr$_2$O$_3$–Er$_2$O$_3$-based ceramic varistors with high stability of nonlinear properties. *J Mater Sci Lett* 2002, **21**: 201–204.

[20] Tsai JK, Wu TB. Non-ohmic characteristics of ZnO–V$_2$O$_5$ ceramics. *J Appl Phys* 1994, **76**: 4817.

[21] Tsai J-K, Wu T-B. Microstructure and nonohmic properties of binary ZnO–V$_2$O$_5$ ceramics sintered at 900 ℃. *Mater Lett* 1996, **26**: 199–203.

[22] Kuo C-T, Chen C-S, Lin I-N. Microstructure and nonlinear properties of microwave-sintered ZnO–V$_2$O$_5$ varistors: I, Effect of V$_2$O$_5$ doping. *J Am Ceram Soc* 1998, **81**: 2942–2948.

[23] Hng HH, Knowles KM. Characterisation of $Zn_3(VO_4)_2$ phases in V_2O_5-doped ZnO varistors. *J Eur Ceram Soc* 1999, **19**: 721–726.

[24] Hng HH, Halim L. Grain growth in sintered ZnO–1 mol% V_2O_5 ceramics. *Mater Lett* 2003, **57**: 1411–1416.

[25] Nahm C-W. Effect of Mn doping on electrical properties and accelerated ageing behaviours of ternary ZVM varistors. *Bull Mater Sci* 2011, **34**: 1385–1391.

[26] Nahm C-W. Nb_2O_5 doping effect on electrical properties of $ZnO–V_2O_5–Mn_3O_4$. *Ceram Int* 2012, **38**: 5281–5285.

[27] Nahm C-W. Non-omic properties and impulse aging behavior of quaternary $ZnO–V_2O_5–Mn_3O_4–Er_2O_3$ semiconducting varistors with sintering processing. *Mat Sci Semicon Proc* 2013, **16**: 1308–1315.

[28] Nahm C-W. Characteristics of $ZnO–V_2O_5–MnO–Nb_2O_5–Er_2O_3$ semiconducting varistors with sintering processing. *Mat Sci Semicon Proc* 2013, **16**: 778–785.

[29] Nahm C-W. Effect of erbium on varistor characteristics of vanadium oxide-doped zinc oxide ceramics. *J Mater Sci: Mater El* 2013, **24**: 27–35.

[30] Nahm C-W. Varistor characteristics of vanadmium oxide-doped zinc oxide ceramics modified with bismuth oxide. *J Mater Sci: Mater El* 2013, **24**: 70–78.

[31] Hng HH, Chan PL. Microstructure and current–voltage characteristics of $ZnO–V_2O_5–MnO_2$ varistor system. *Ceram Int* 2004, **30**: 1647–1653.

[32] Liu H-Y, Kong H, Ma X-M, *et al.* Microstructure and electrical properties of ZnO-based varistors prepared by high-energy ball milling. *J Mater Sci* 2007, **42**: 2637–2642.

[33] Wurst JC, Nelson JA. Lineal intercept technique for measuring grain size in two-phase polycrystalline ceramics. *J Am Ceram Soc* 1972, **55**: 109–111.

[34] Senthilkumaar S, Rajendran K, Banerjee S, *et al.* Influence of Mn doping on the microstructure and optical property of ZnO. *Mat Sci Semicon Proc* 2008, **11**: 6–12.

[35] Morris WG. Physical properties of the electrical barriers in varistors. *J Vac Sci Technol* 1976, **13**: 926–931.

16

Dielectric behavior of $CaCu_3Ti_4O_{12}$ electro-ceramic doped with La, Mn and Ni synthesized by modified citrate-gel route

Laxman SINGH[a], U. S. RAI[a], Alok Kumar RAI[b], K. D. MANDAL[c],*

[a]Department of Chemistry, Centre of Advanced Study, Faculty of Science, Banaras Hindu University, Varanasi 221005, U.P., India
[b]Department of Materials Science and Engineering, Chonnam National University, 300 Yongbong-dong, Bukgu, Gwangju 500-757, Republic of Korea
[c]Department of Applied Chemistry, Indian Institute of Technology, Banaras Hindu University, Varanasi 221005, U.P., India

Abstract: The effect of La^{3+}, Mn^{2+} and Ni^{2+} doped calcium copper titanate, $CaCu_3Ti_4O_{12}$ (CCTO), at higher concentrations (CR1 and CR2 with 5 mol% and 10 mol%, respectively), has been examined by semi-wet route at relatively lower temperature. This semi-wet route employs citrate–nitrate gel chemical method using TiO_2 solid powders. X-ray diffraction (XRD) analysis confirms the formation of single phase in the doped samples sintered at 900 ℃ for 8 h. Scanning electron micrographs (SEM) show that the average grain size for CR2 is larger than that of CR1 composition. The energy dispersive X-ray spectroscopy (EDX) is used to study the percentage compositions of different ions present in both ceramics. Dielectric constant (ε_r) and dielectric loss ($\tan\delta$) values of CR1 are comparatively higher than those of CR2 ceramic at all measured frequencies and temperatures. The nature of temperature-dependent relaxation behavior of the ceramics is also studied by impedance, modulus spectroscopic analysis and confirms Maxwell–Wagner relaxation.

Keywords: oxide materials; chemical synthesis; dielectric properties; grain boundaries; X-ray diffraction (XRD)

1 Introduction

Recently, much attention has been paid to an unusual cubic perovskite oxide material $CaCu_3Ti_4O_{12}$ (CCTO), due to its high dielectric constant (10^4–10^5) and good stability over a wide temperature range from 100 K to 600 K [1–4]. The large dielectric constant is very unusual because it does not show ferroelectric behavior. Giant dielectric constant of CCTO has been attributed to Maxwell–Wagner effect [5] at the interface of grains and grain boundaries in earlier studies, while Lunkenheimer et al. [6] believed that it is due to the electrode polarization effect. The materials with high ε_r can store more charge and the smaller electronic circuits can be designed. It is considered as a very potential material for applications in microelectronics, especially in capacitive components and varistors. Partial substitution of appropriate cations in CCTO is considered as an effective method to improve the physical properties of the materials. The radius, valancy and co-ordination number of an element are important parameters to determine the site it occupies

* Corresponding author.
E-mail: kdmandal.apc@itbhu.ac.in

in the parent compound and the composition range of solid solution formation [7]. It is well known that the dielectric properties of CCTO are strongly dependent upon the processing condition as well as doping effects [8,9]. A lot of research work has been done on partial substitution of Cu and Ti ions in CCTO ceramic in order to improve the dielectric properties for understanding the origin of dielectric response in these materials [10,11]. In the previous work, various substitutions were made on different sites of CCTO to investigate the effect of substitution on dielectric properties, e.g., Ca by La [12], Cu by Mn [13] and Ti by Ni [14]. In the present communication, La^{3+}, Mn^{2+} and Ni^{2+} cations doped calcium copper titanate, CCTO, at higher concentration level (5 mol% and 10 mol%), were synthesized by semi-wet route at relatively lower temperature and abbreviated as CR1 and CR2 for 5 mol% and 10 mol%, respectively. The surface morphology, dielectric properties, relaxation behavior of the doped ceramics have been examined.

2 Experimental procedures

Analytical reagent grade chemicals $Ca(NO_3)_2 \cdot 4H_2O$, $La(NO_3)_3 \cdot 6H_2O$, $Cu(NO_3)_2 \cdot 3H_2O$, $Ni(NO_3)_2 \cdot 6H_2O$, $Mn(CH_3COO)_2 \cdot 4H_2O$, titanium dioxide ($TiO_2$) and citric acid, all having purity of 99.95% obtained from Merck, India, were used as starting materials. Standard solutions of metal nitrates were prepared using distilled water. Solutions of metal nitrates in stoichiometric amount of these metal ions were mixed in a beaker. Calculated amount of TiO_2 and citric acid equivalent to metal ions were added to the solution. The solution was heated on a hot plate using a magnetic stirrer at 70–90 ℃ to evaporate water and dried at 100–120 ℃ in hot air oven for 12 h to yield a blue gel. The gel was calcined in air at 800 ℃ for 6 h in a muffle furnace. The resultant mixtures were ground into fine powders using a pestle and a mortar. Cylindrical pellets were made using a hydraulic press. The pellets were sintered at 900 ℃ for 8 h in air. In order to perform the electrical measurements, silver paste was applied to both sides of the circular face of the ceramic pellets, then dried at 600 ℃ for 20 min, and cooled naturally to room temperature. The crystalline phases of the sintered samples were identified using an X-ray diffractometer (Rich-Siefert, ID-3000) employing Cu Kα radiation.

The microstructures of the fractured surfaces were examined using a scanning electron microscope (SEM, JEOL JSM5410). The energy dispersive X-ray analyzer (EDX, Kevex, Sigma KS3) was used for elemental analysis of the sintered samples. The dielectric and impedance data of the doped CCTO ceramics were collected using the LCR (inductance, capacitance and resistance) meter (PSM 1735, Newton 4th Ltd, U.K.) with variation in frequency (100 Hz–5 MHz) and temperature (300–500 K).

3 Results and discussion

The X-ray diffraction (XRD) patterns of CR1 and CR2 ceramics sintered at 900 ℃ for 8 h are shown in Fig. 1. The doped CCTO precursor dry powders were calcined at 800 ℃ for 6 h and then sintered at 900 ℃ for 8 h. XRD data were indexed on the basis of cubic unit cell similar to the undoped CCTO. The similarity of the doped X-ray data to the undoped CCTO (JCPDS 75-2188), confirms the formation of a single-phase material. There is no evidence of the presence of a secondary phase in the ceramic. The lattice parameter and unit cell volume were determined using the least-square refinement method. The value of lattice parameter and its unit cell volume for both compositions are given in Table 1. The lattice parameter increases with the dopant concentration

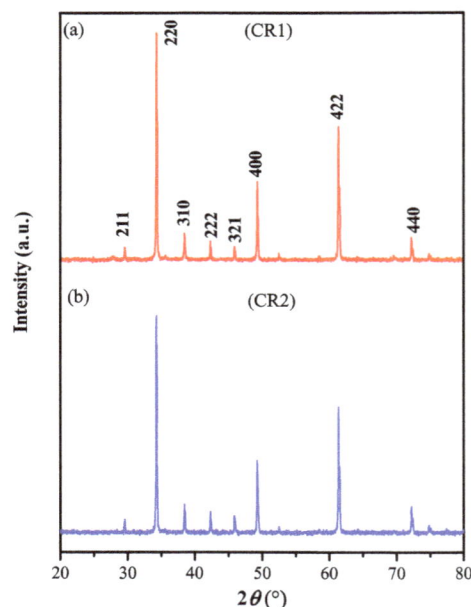

Fig. 1 XRD patterns of (a) CR1 and (b) CR2 sintered at 900 ℃ for 8 h.

Table 1   Crystal system, lattice parameter and unit cell volume for CR1 and CR2 ceramics

System	Composition x	Crystal system	Lattice parameter a (Å)	Unit cell volume a^3 (Å^3)
$Ca_{1-x}La_xCu_{3-x}Mn_xTi_{4-x}Ni_xO_{12}$	0.05	Cubic	7.3877	403.2051
	0.10	Cubic	7.6591	449.2992

increasing. The slight increase in lattice parameter and unit cell volume is due to the small difference in ionic radii of the substituent ions and the host ions [15].

To investigate the role of microstructure and composition on the dielectric properties of these ceramics, SEM and EDX studies were performed for both materials. Figure 2 shows the SEM images of the fractured surface of both compositions sintered at 900 ℃ for 8 h. The SEM image for CR1 shows faceted-packed morphology having grain size in the range of 1–2 μm. The SEM image for CR2 shows faceted grains with cubical appearance morphology. It is clearly seen that the microstructure of the doped CCTO ceramic changes significantly with the value of x increasing. The grain size for CR2 composition is found to be in the range of 1–3 μm and there is an indication of agglomeration of a secondary phase along the grain boundaries. This is due to the transition of

copper oxide into the liquid phase during sintering treatment [16,17]. This suggests that the specimens prepared in the present study have a polycrystalline structure. From Figs. 2(a) and 2(b), it seems that the grain size increases remarkably on the increasing of the dopant concentration x. The modification of grain size affects the dielectric response in CCTO as reported earlier, but it should not be greater than 20 μm [18]. The grain size of these ceramics synthesized by this method is relatively smaller than those synthesized by the conventional ceramic methods [19,20].

The EDX spectra obtained for CR1 and CR2 (Fig. 3), clearly show the presence of Ca^{2+}, La^{3+}, Cu^{2+}, Mn^{2+}, Ti^{4+} and Ni^{2+} in both ceramics. The atomic percentages of the major ions, i.e., Ca^{2+}, Cu^{2+}, Ti^{4+} and O^{2-} present in CR1 ceramic are 7.37%, 27.75%, 36.32% and 28.51%, respectively, whereas in CR2 their respective compositions are 7.25%, 26.05%, 34.27% and 29.92%. In both compositions, the major ions are present as per stoichiometric ratio and confirm the purity of CR1 and CR2 ceramics prepared by the semi-wet route.

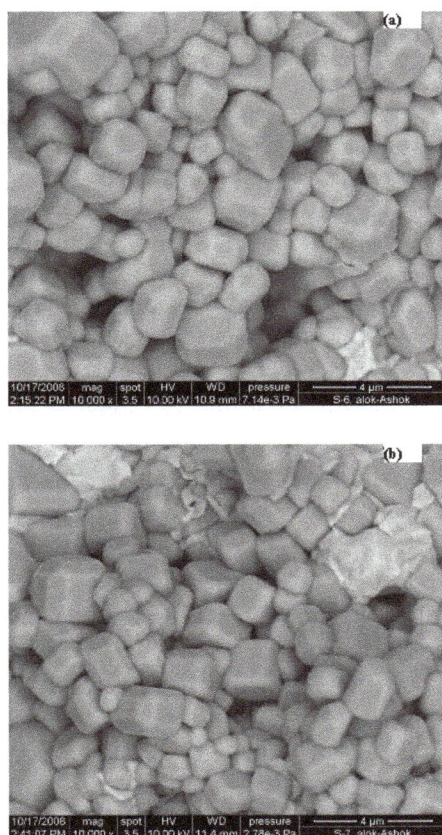

Fig. 2   SEM micrographs of (a) CR1 and (b) CR2 sintered at 900 ℃ for 8 h.

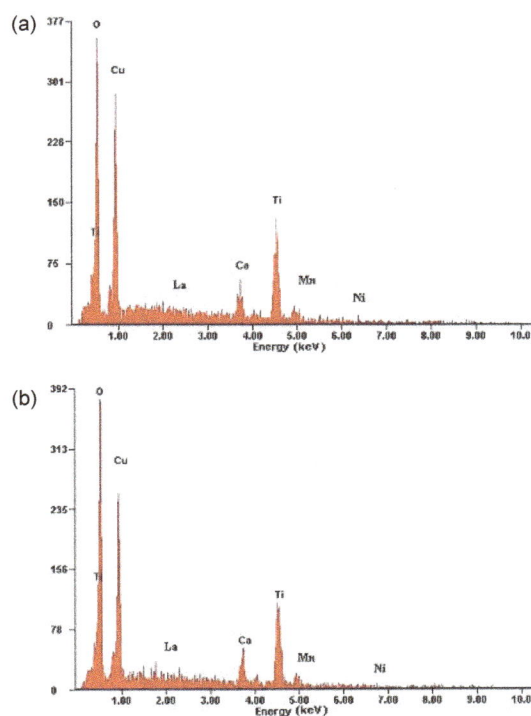

Fig. 3   EDX spectra of (a) CR1 and (b) CR2 sintered at 900 ℃ for 8 h.

The temperature dependence of dielectric constant (ε_r) and dielectric loss (tanδ) of the doped CCTO samples under different frequencies between 100 Hz and 100 kHz are shown in Figs. 4 and 5. It is noted from Fig. 4 that the dielectric constant of CR1 is ~ 1164 at room temperature (310 K) and 100 Hz,

which increases continuously with temperature increasing and becomes more rapid beyond 425 K. Dielectric constant of CR2 is ~ 510 at room temperature and 100 Hz, and also increases with temperature increasing as shown in Fig. 5.

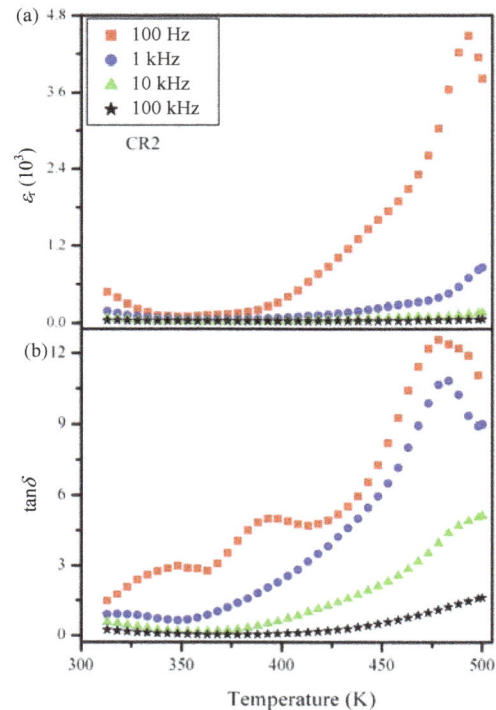

Fig. 4   The variation of dielectric constant (ε_r) and dielectric loss (tanδ) with temperature at a few selected frequencies for CR1 ceramic.

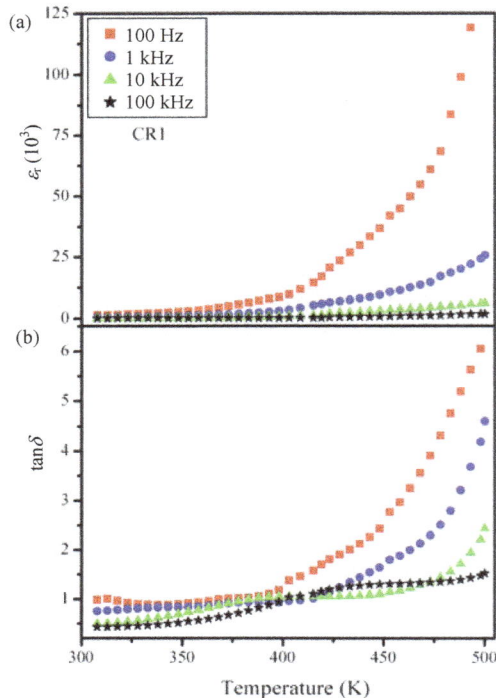

Fig. 5   The variation of dielectric constant (ε_r) and dielectric loss (tanδ) with temperature at a few selected frequencies for CR2 ceramic.

It is found that dielectric constant (ε_r) at low frequency is quite higher than that at high frequency. The result is similar to that of dielectric properties of CCTO as a function of temperature in the lower-frequency region [21]. The dielectric constants of CR1 and CR2 ceramics at 10 kHz and 100 kHz are almost independent in the temperature range of 300–425 K, and then increase steadily. It is also observed that the dielectric constants of CR1 are always higher than those of CR2 ceramic for all measured frequencies. This is due to the larger grain size of CR2 than CR1. It is noted that the tanδ values of CR1 and CR2 ceramics at room temperature and 100 kHz are 0.46 and 0.02, respectively. The tanδ values for CR2 remain constant in the temperature range of 310–425 K, and then increase steadily with the rise in temperature at high frequency (100 kHz). The plots of tanδ at low frequency show anomaly at 480 K.

The plots of variation of dielectric constant (ε_r) and dielectric loss (tanδ) of these samples with frequency at a few selected temperatures are shown in Figs. 6 and 7. As seen from the figures, dielectric constants of both ceramics decrease with the increase in frequency.

The dielectric constant of CR1 is always higher than that of CR2 ceramic in the measured frequency range of 100 Hz–5 MHz. The rate of decrease of ε_r with frequency increases with temperature increasing. It is observed from Fig. 6 that the dielectric loss also decreases with frequency increasing. As we increase the temperature, the dielectric loss also increases in low-frequency regions. It is observed from Fig. 7 that the broadening of the peaks shows typical relaxation behavior found in CR2 ceramic at low frequency with temperature increasing. These peaks shift to lower-frequency region with temperature decreasing [22]. This relaxation behavior arises due to space charge polarization present in these ceramics. The grain boundary barrier layer capacitor consists of conducting or semi-conducting grains and insulating grain boundary layers. Therefore, relaxation behavior arises whenever two phases with different electrical conductivities are in contact.

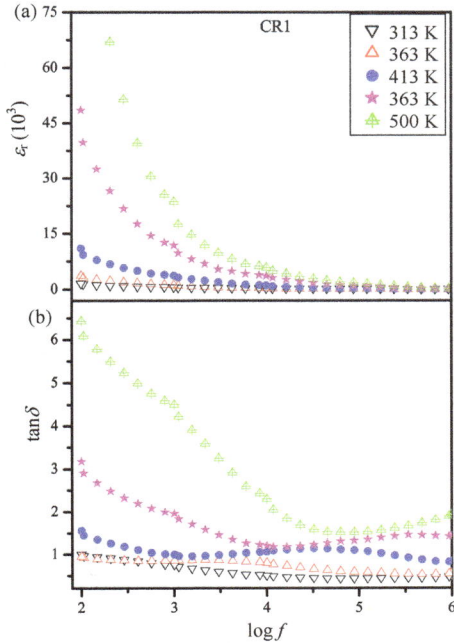

Fig. 6   The variation of dielectric constant (ε_r) and dielectric loss (tanδ) vs. log f for CR1 ceramic at a few selected temperatures.

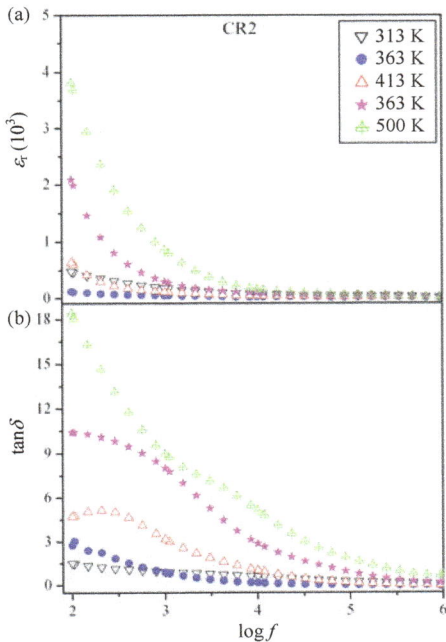

Fig. 7   The variation of dielectric constant (ε_r) and dielectric loss (tanδ) vs. log$f$ for CR2 ceramic at a few selected temperatures.

The investigation of dielectric relaxation and formation of barrier layer in this ceramic is confirmed by complex plane impedance (Z^*) and complex plane modulus (M^*) analysis, which separates the contributions of grains, grain boundaries and electrode specimen interface to the total resistance and capacitance, respectively, and correlates the electrical behavior of the ceramic to its microstructure [23]. This structure can be modeled using an equivalent impedance circuit that consists of two parallel resistance–capacitance (RC) elements in series. One pair of the parallel RC elements represents the grain boundary layers, while the other pair of parallel RC elements represents the grains. The capacitance C_{gb} and resistance R_{gb} of the grain boundary layers are usually much larger than the capacitance C_g and resistance R_g of the grains. Therefore, the relaxation time $\tau_{gb} = R_{gb}C_{gb}$ of the grain boundary layers is much larger than that of the grains $\tau_g = R_gC_g$. In this case, two parallel RC elements in series produce two semicircular arcs in a complex impedance plane Z^* plot [24] and the impedance can be calculated from the following equations:

$$Z^* = \frac{1}{R_g^{-1} + i\omega C_g} + \frac{1}{R_{gb}^{-1} + i\omega C_{gb}} = Z' - iZ'' \quad (1)$$

where

$$Z' = \frac{R_g}{1 + (\omega R_g C_g)^2} + \frac{R_{gb}}{1 + (\omega R_{gb} C_{gb})^2} \quad (2)$$

and

$$Z'' = R_g\left[\frac{\omega R_g C_g}{1 + (\omega R_g C_g)^2}\right] + R_{gb}\left[\frac{\omega R_{gb} C_{gb}}{1 + (\omega R_{gb} C_{gb})^2}\right] \quad (3)$$

Z' is the real part of the complex impedance and Z'' is the imaginary part. The large arc is due to the grain boundary responses at the low frequencies, and the small arc is due to the grain responses at the high frequencies. At the maximum of an arc $\omega\tau = 2\pi f_{max}\tau = 1$, where $\omega = 2\pi f$, f is the frequency of the applied field. In actual measurements, it is difficult to obtain both semicircular arcs in one Z^* plot, especially for the small arc associated with the grains. Its f_{max} at room temperature is well beyond the highest available frequency of impedance analyzers used. Figure 8 shows the complex impedance plane Z^* plots of CR1 and CR2 ceramics obtained at temperatures 313 K and 413 K, respectively. The grain boundary response of CR1 dominates at 313 K as shown in the inset of Fig. 8(a). The grain boundary resistivity of CR1 and CR2 are 122.44×10^5 $\Omega \cdot$cm and 151.48×10^5 $\Omega \cdot$cm, respectively, at room temperature (313 K). The resistivity of grain boundary (R_{gb}) for CR1 and CR2 are 3.25×10^5 Ω·cm and 12.36×10^5 Ω·cm at 413 K, respectively. It is observed from Fig. 8(b) and its inset that the grain resistivity of CR2 are 70.64×10^5 Ω·cm and 10.68×10^5 Ω·cm at 313 K and 413 K, respectively. It is observed that on increasing the dopants concentration, the resistivity of CR2 is higher than that of CR1 at both temperatures. In the case of CR1, the existence of single arc can be taken to interpret both the grain and grain boundary conduction processes have identical time constant at both temperatures, while CR2 shows the semi-circle arcs at both temperatures but overlapped. The overlapped semicircle arcs between grain and grain boundary may be due to the molten grain and the formation of undefined grain boundaries [25].

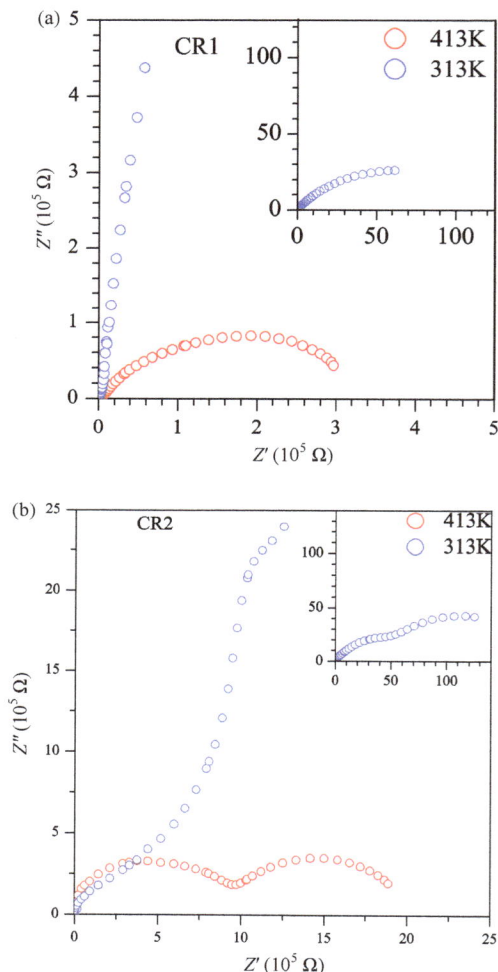

Fig. 8 Complex impedance plane plots vs. measuring temperature for (a) CR1 and (b) CR2 ceramics. The inset in (a) and (b) show the expended view of CR1 and CR2 at 313K, respectively.

The variation of the imaginary part of the impedance Z'' with frequency at a few selected temperatures is shown in Figs. 9 and 10 for CR1 and CR2, respectively. The inset figures are the expanded view of Figs. 9 and 10 at high temperature. It clearly shows that the relaxation peaks are observed at all measured temperatures for both ceramics. In the case of CR2 ceramic, they show relaxation peaks at low frequency and high frequency. The peaks are suppressed and slightly shifted to high-frequency region on increasing temperature, which confirms the existence of a temperature-dependent dielectric relaxation. The dispersion of curves in the low-frequency region at different temperatures is very clear and appears to be

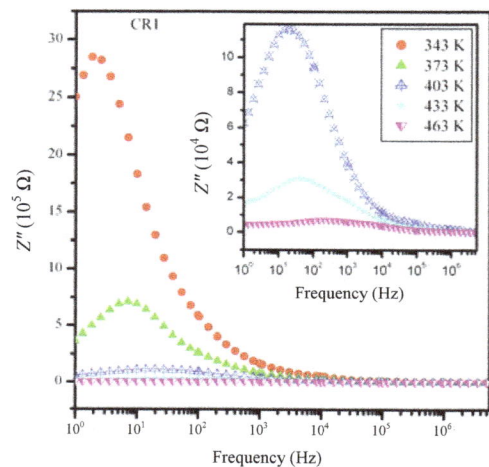

Fig. 9 Impedance plane plots Z'' vs. frequency at a few selected temperatures for CR1 sintered at 900 ℃ for 8 h. The inset shows an expended view at high temperature.

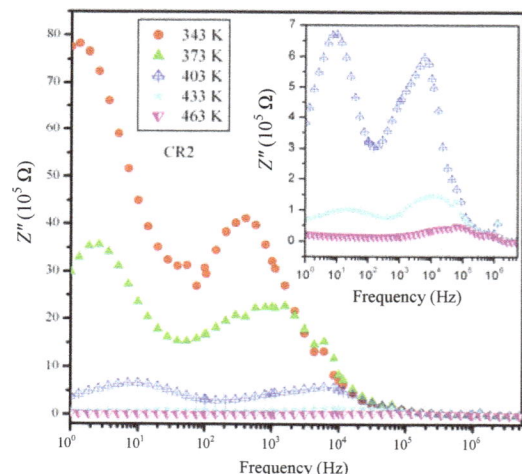

Fig. 10 Impedance plane plots Z'' vs. frequency at a few selected temperatures for CR2 sintered at 900 ℃ for 8 h. The inset shows an expended view at high temperature.

merging at higher frequency with temperature. This behavior is apparently due to the presence of space charge polarization effect at lower frequency which is definitely eliminated at higher frequency.

In order to investigate the temperature dependence of relaxation processes in these ceramics, the electric modulus studies were also carried out. The complex electric modulus (M^*) is defined in terms of the complex dielectric constant (ε^*) and is represented as

$$M^* = (\varepsilon^*)^{-1} \tag{4}$$

$$M' + iM'' = \varepsilon' / \varepsilon'^2 + \varepsilon''^2 + i\varepsilon'' / \varepsilon'^2 + \varepsilon''^2 \tag{5}$$

where M', M'' and ε', ε'' are the real and imaginary parts of the electric modulus and dielectric constants, respectively. The imaginary parts of the modulus at various temperatures are calculated using Eq. (5) for these ceramics. Figs. 11 and 12 show the plots of

imaginary part of electric modulus (M'') with variation of frequency at few selected temperatures for CR1 and CR2. The relaxation M'' peaks are observed which shift to higher-frequency region with temperature increasing and give a direct evidence of temperature-dependent relaxation.

The impedance data were also used to evaluate the relaxation time of the electrical phenomena in CR1 and CR2 ceramics using the relation $\tau = 1/\omega = 1/2\pi f$, where f is the relaxation frequency. The nature of variation of the relaxation time (τ) with reciprocal of temperature ($1/T$) is shown in Fig. 13. The activation energy in the relaxation process is determined by the temperature-dependent relaxation time constant τ_{gb} which obey Arrhenius law given by the following equation:

$$\tau = \tau_0 \exp(E_a / k_B T) \tag{6}$$

where E_a is the activation energy involved in the dielectric relaxation process; τ_0 is the pre-exponential factor; k_B is Boltzmann constant; and T is the absolute temperature. The grain boundary activation energy evaluated from the slope of $\ln\tau_{gb}$ against $1/T$ curve is found to be 0.42 eV and 0.38 eV for CR1 and CR2, respectively. This value agrees with the value reported earlier for Maxwell–Wagner relaxation arising from the interfacial polarization for CCTO ceramic which confirms the temperature dependence of dielectric properties [26,27].

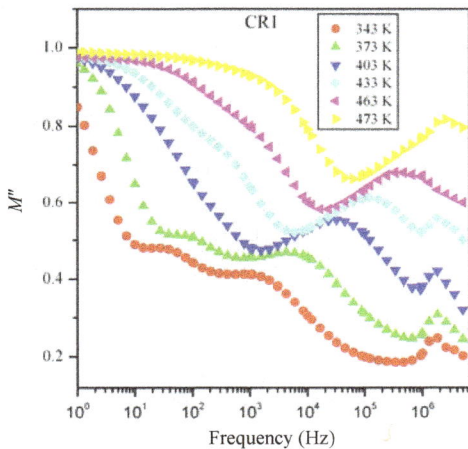

Fig. 11 Electric modulus plots of imaginary M'' vs. frequency at a few selected temperatures for CR1 sintered at 900 ℃ for 8 h.

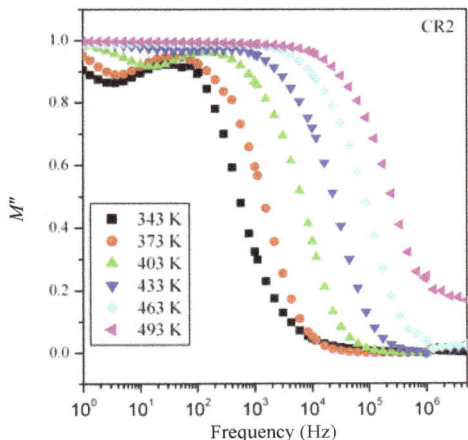

Fig. 12 Electric modulus plots of imaginary M'' vs. frequency at a few selected temperatures for CR2 sintered at 900 ℃ for 8 h.

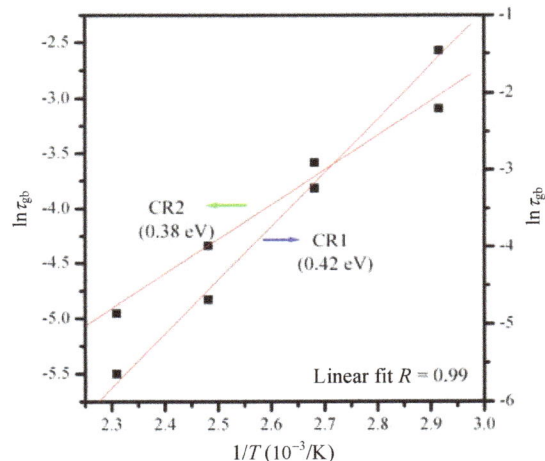

Fig. 13 Variation of relaxation time (τ_{gb}) with the inverse of temperature ($1/T$) for CR1 and CR2 ceramics.

4 Conclusions

Calcium copper titanate ceramics, CaCu$_3$Ti$_4$O$_{12}$ (CCTO), doped with La^{3+}, Mn^{2+} and Ni^{2+} at higher concentration (5 mol% and 10 mol%) were synthesized by semi-wet route at relatively lower temperature. XRD confirms the single-phase formation of the ceramics and the scanning electron micrographs show their surface morphology with grain size in the range of 1–3 μm. The stoichiometry of the ceramics is confirmed by EDX studies. With the increase in concentration of the dopant ions in CCTO ceramics, the values of dielectric constant as well as the dielectric loss decrease. Impedance analyses of both the ceramics show that grain boundary resistance dominates the total resistance of the ceramics. The dielectric constant responses of both the ceramics arise due to the space charge polarisation. The value of resistance of the grain boundary is found to be several orders of magnitude larger than that of the grain. The nature of temperature-dependent relaxation behavior of the ceramics studied by impedance and modulus spectroscopic analyses confirms Maxwell–Wagner relaxation.

Acknowledgements

One of the authors would like to thank Prof. Om Parkash, the head of Department of Ceramic Engineering, Indian Institute of Technology, B.H.U., for extending XRD facility.

References

[1] Subramanian MA, Li D, Duan N, et al. High dielectric constant in $ACu_3Ti_4O_{12}$ and $ACu_3Ti_3FeO_{12}$ phases. *J Solid State Chem* 2000, **151**: 323–325.

[2] Ramirez AP, Subramanian MA, Gardel M, et al. Giant dielectric constant response in a copper-titanate. *Solid State Commun* 2000, **115**: 217–220.

[3] Sinclair DC, Adams TB, Morrison FD, et al. $CaCu_3Ti_4O_{12}$: One-step internal barrier layer capacitor. *Appl Phys Lett* 2002, **80**: 2153.

[4] Homes CC, Vogt T, Shapiro SM, et al. Optical response of high-dielectric-constant perovskite-related oxide. *Science* 2001, **293**: 673–676.

[5] Lunkenheimer P, Bobnar V, Pronin AV, et al. Origin of apparent colossal dielectric constants. *Phys Rev B* 2002, **66**: 052105.

[6] Lunkenheimer P, Fichtl R, Ebbinghaus SG, et al. Nonintrinsic origin of the colossal dielectric constants in $CaCu_3Ti_4O_{12}$. *Phys Rev B* 2004, **70**: 172102.

[7] Buscaglia MT, Viviani M, Buscaglia V, et al. Incorporation of Er^{3+} into $BaTiO_3$. *J Am Ceram Soc* 2002, **85**: 1569–1575.

[8] Bender BA, Pan MJ. The effect of processing on the giant dielectric properties of $CaCu_3Ti_4O_{12}$. *Mat Sci Eng B* 2005, **117**: 339–347.

[9] Smith AE, Calvarese TG, Sleight AW, et al. An anion substitution route to low loss colossal dielectric $CaCu_3Ti_4O_{12}$. *J Solid State Chem* 2009, **182**: 409–411.

[10] Singh L, Rai US, Mandal KD. Preparation and characterization of nanostructured $CaCu_{2.90}Zn_{0.10}Ti_4O_{12}$ ceramic. *Nanomater Nanotechnol* 2011, **1**: 59–66.

[11] Leret P, Fernandez J, de Frutos J, et al. Nonlinear *I–V* electrical behaviour of doped $CaCu_3Ti_4O_{12}$ ceramics. *J Eur Ceram Soc* 2007, **27**: 3901–3905.

[12] Mandal KD, Rai AK, Kumar D, et al. Dielectric properties of the $Ca_{1-x}La_xCu_3Ti_{4-x}Co_xO_{12}$ system ($x = 0.10$, 0.20 and 0.30) synthesized by semi-wet route. *J Alloys Compd* 2009, **478**: 771–776.

[13] Li M, Feteira A, Sinclair DC, et al. Influence of Mn doping on the semiconducting properties of $CaCu_3Ti_4O_{12}$ ceramics. *Appl Phys Lett* 2006, **88**: 232903.

[14] Rai AK, Mandal KD, Kumar D, et al. Dielectric properties of lanthanum-doped $CaCu_3Ti_4O_{12}$ synthesized by semi-wet route. *J Phys Chem Solids* 2009, **70**: 834–839.

[15] Shannon RD, Prewitt CT. Revised values of effective ionic radii. *Acta Cryst* 1970, **B26**: 1046–1048.

[16] Kim DW, Kim TG, Hong KS, et al. Low-firing of CuO-doped anatase. *Mater Res Bull* 1999, **34**: 771–781.

[17] Huang CL, Chen YC. Low temperature sintering and microwave dielectric properties of $SmAlO_3$ ceramics. *Mater Res Bull* 2002, **37**: 563–574.

[18] Ni L, Chen XM, Liu XQ. Structure and modified giant dielectric response in $CaCu_3(Ti_{1-x}Sn_x)_4O_{12}$ ceramics. *Mater Chem Phys* 2010, **124**: 982–986.

[19] Jha P, Arora P, Ganguli AK. Polymeric citrate precursor route to the synthesis of the high dielectric constant oxide, $CaCu_3Ti_4O_{12}$. *Mater Lett* 2003, **57**: 2443–2446.

[20] Singh L, Rai US, Mandal KD. Influence of Zn

doping on microstructures and dielectric properties in $CaCu_3Ti_4O_{12}$ ceramic synthesised by semiwet route. *Adv Appl Ceram* 2012, **111**: 374–380.

[21] Shao SF, Zhang JL, Zheng P, *et al*. Microstructure and electrical properties of $CaCu_3Ti_4O_{12}$ ceramics. *J Appl Phys* 2006, **99**: 084106.

[22] Yu HT, Liu HX, Hao H, *et al*. Grain size dependence of relaxor behavior in $CaCu_3Ti_4O_{12}$. *Appl Phys Lett* 2007, **91**: 222911.

[23] Hodge IM, Ingram MD, West AR. Impedance and modulus spectroscopy of polycrystalline solid electrolytes. *J Electroanal Chem Interfacial Electroch* 1976, **74**: 125–143.

[24] Fang TT, Lin WJ, Lin CY. Evidence of the ultrahigh dielectric constant of $CaSiO_3$-doped $CaCu_3Ti_4O_{12}$ from its dielectric response, impedance spectroscopy, and microstructure. *Phys Rev B* 2007, **76**: 045115.

[25] Mazni M, Daud WM, Talib SA, *et al*. AC conductivity of $Ca_{1-x}A_xCu_3Ti_4O_{12}$ (A = Sr or Ba) with $x = 0.0$ and 0.2 ceramics. *Solid State Sci Tech* 2009, **17**: 222–228.

[26] Chanmal CV, Jog JP. Dielectric relaxations in PVDF/BaTiO3 nanocomposites. *Express Polym Lett* 2008, **2**: 294–301.

[27] Zhang L, Shan XB, Wu PX, *et al*. Dielectric characteristics of $CaCu_3Ti_4O_{12}$/P(VDF-TrFE) nanocomposites. *Appl Phys A* 2012, **107**: 597–602.

17

Thermal properties of AlN polycrystals obtained by pulse plasma sintering method

Paweł J. RUTKOWSKI[*], Dariusz KATA

Department of Ceramics and Refractories, Faculty of Materials Science and Ceramics, AGH University of Science and Technology, al. Mickiewicza 30, 30-059 Krakow, Poland

Abstract: Aluminum nitride (AlN) polycrystals were prepared by pulse plasma sintering (PPS) technique. The starting AlN powder mixtures were composed with 3.0 wt%, 5.0 wt% and 10 wt% of yttrium oxide (Y_2O_3), respectively. Relative density of each polycrystal was measured by hydrostatic method and evaluated higher than 97%. X-ray diffraction (XRD) method was used for phase examination of the samples after heat treatment. Microstructure examination supported by computer-aided analysis was performed by scanning electron microscopy (SEM) and energy dispersive spectrometry (EDS). The results were correlated with thermal conductivity of the samples carried out by laser pulse method (LFA). The influence of the rapid sintering technique and yttrium oxide additive content on the thermal conductivity and microstructure appearance of AlN polycrystals was clearly shown.

Keywords: aluminum nitride (AlN); thermal conductivity; microstructure; thermal diffusivity; specific heat

1 Introduction

Aluminum nitride (AlN) is considered as a challenging material for structural and functional applications, because it has a very high thermal conductivity among other ceramic compounds. In the form of monocrystal or polycrystal, AlN is used in electronic devices, thus squeezing out conventional materials like SiO_2, Al_2O_3 or BeO, which are harmful for human body [1,2]. AlN is applied for sensors and heat exchangers to improve their sensitivity and efficiency [3]. Due to its thermal properties, it is also used in arms industry as a part of high-power and high-resolution radars [4]. The thermal conductivity (λ) of polycrystalline AlN mainly depends on the level of oxygen content dissolved in the crystallographic structure, microstructure appearance and nature of points and linear defects. These factors can radically reduce λ from 320 W/(m·K) to 70 W/(m·K) [5] unless they are controlled well at every stage of manufacturing procedure [6]. Preparation of dense AlN polycrystals requires the addition of liquid phases for sintering. Commercially, MgO, CaO or rare-earth metal oxides are used, but the most promising addition is yttrium oxide (Y_2O_3) resulting in the highest value of thermal conductivity. AlN grains are usually covered by thin layers of alumina (Al_2O_3); therefore, during sintering, yttrium oxide forms several compounds with alumina, e.g., YAP ($YAlO_3$), YAG ($Y_3Al_5O_{12}$) and YAM ($Y_4Al_2O_9$). The garnet formation

---

* Corresponding author.
E-mail: pawelr@agh.edu.pl

promotes densification process but additionally leads to the removal of oxygen from AlN structure at temperatures above 1750 ℃ [7]. This is related to improving thermal conductivity and density of the polycrystals. The introduction of liquid phases is associated with grain growth and decreasing in concentration of grain boundaries to improve thermal properties [8]. An effective method of preparation AlN polycrystals having higher λ is pressureless sintering; however, it needs a long time of heat treatment, usually 2 h at 1800 ℃ [9,10].

The aim of this study is to obtain AlN samples by a rapid densification process—pulse plasma sintering (PPS, modification of SPS (spark plasma sintering)). The literature about such preparation of AlN polycrystals is rather limited. Therefore, the results given below fulfill this gap and make good starting points for further scientific discussion.

2 Materials and methods

To produce AlN sinters, commercially available AlN powders (H.C. Starck GmbH, Germany) were used. The powder morphology observed by scanning electron microscopy (SEM, Nova NANOSEM 200 from FEI), and grain size distribution measured with laser diffraction method in alcohol environment (Mastersizer 2000 of Melvern Intruments) are shown in Figs. 1 and 2, respectively. It is detected that the average size of grains is 1.7 μm, which confirms D_{50} in the range of 0.8–1.8 μm taken from the company's analysis. Yttrium oxide produced by the same company having medium-sized grains of 0.8 μm was added to prepare mixtures for sintering.

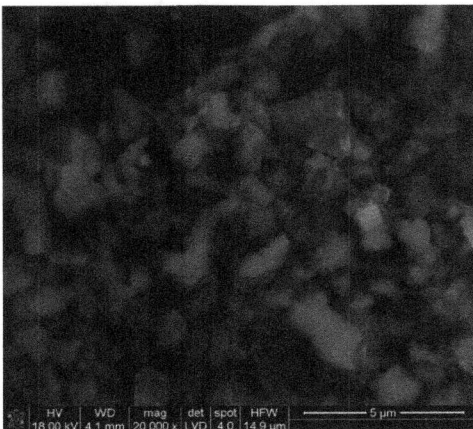

Fig. 1 Morphology of commercial AlN powders.

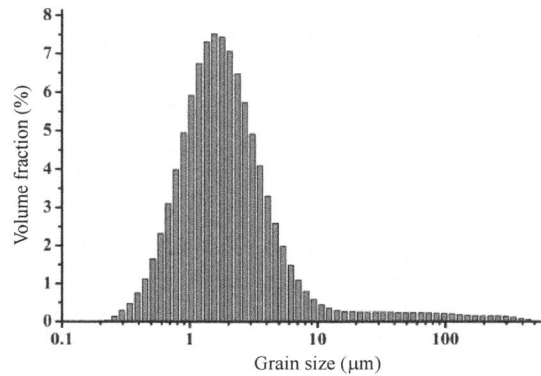

Fig. 2 Grain size distribution of commercial AlN powders.

AlN powders were mixed with 3.0 wt%, 5.0 wt% and 10 wt% of Y_2O_3. The mixtures were homogenized in isopropyl for 24 h by silicon nitride grinding media. The as-prepared powders were granulated and compacted in carbon mould, and then sintered by PPS in nitrogen flow [11]. The heating rate of 170 ℃/min was associated with the pressure of 81 MPa. The maximum temperature of 1500 ℃ was dwelled for 600 s. The apparent density of the sintered AlN was measured using hydrostatic method, and compared to the theoretical density calculated on the base of X-ray diffraction (XRD) analysis. The phase composition of AlN polycrystals was qualitatively and quantitatively examined by XRD method. The thermal measurements were performed by Netzsch LFA 427 apparatus. The thermal diffusivity and specific heat were determined by laser pulse method (LFA) at temperatures from 25 ℃ to 600 ℃ in step of 100 ℃ in argon flow with "Cape–Lehmann + pulse correction" computational model. The thermal expansion coefficient was measured by a Netzsch DIL 402C dilatometer at temperature range from 20 ℃ to 600 ℃. The thermal conductivity was calculated on the base of the above measured thermal properties. All the samples were polished and then mechano-chemically etched to reveal microstructure details. Microstructure observation was performed by SEM. The grain size of the sintered materials was estimated on SEM images with computer-aided analysis—Aphelion Image Processing Programme.

3 Results and discussion

The relative density and phase composition of AlN

polycrystals are summarized in Table 1. All the samples' relative densities exceed 97%, and two phases are detected in the XRD patterns, i.e., AlN and Y_2O_3 garnets. Increasing quantity of Y_2O_3 in starting mixtures leads to higher concentration of YAP and YAM phases. Reaction between Al_2O_3 originated from grain surface and Y_2O_3 leads to the formation of garnets during sintering. It is associated with the decreasing concentration of oxygen in AlN structure resulting in higher thermal conductivity of the obtained polycrystals. It is shown that a short time of heat treatment, e.g., 600 s, and low temperature are enough to prepare full dense AlN polycrystals by PPS technique.

LFA-measured values of specific heat, thermal diffusivity and thermal conductivity of the polycrystals are collected in Table 2. The highest λ is achieved for the samples with 5 wt% of yttrium oxide. All the polycrystals have rather good thermal conductivity ranging from 78 W/(m·K) to 100 W/(m·K) at 25 ℃. However, these results are not as high as those of pressureless sintered AlN [10,12]. This is due to a very fine microstructure of the obtained samples and high concentration of intergranular boundaries. Other concentrations of yttrium oxide dopant (3 wt% and 10 wt%) cause lower thermal conductivity. In the case for 3 wt% of yttrium oxide addition, AlN structure contaminated by oxygen is a plausible cause of lower thermal conductivity. On the other hand, the increased porosity of AlN polycrystals sintered with 10 wt% of yttrium oxide is a reason of lower λ compared to 5 wt% of yttrium oxide content.

Table 1   Relative density and phase composition of polycrystalline AlN

Y_2O_3 addition (wt%)	Apparent density (%)	Phase composition (wt%)				
		AlN	YAG	YAP	YAM	Gamma AlON
3.0	99.9	94.5	5.0	—	—	0.5
5.0	97.9	91.2	5.8	3.0	—	—
10	97.2	87.1	—	6.3	6.6	—

Table 2   Thermal properties of AlN sinters with various additions of yttrium oxide

Temperature (℃)	Thermal diffusivity (mm²/s)			Specific heat (J/(g·K))			Thermal conductivity (W/(m·K))		
	3 wt%	5 wt%	10 wt%	3 wt%	5 wt%	10 wt%	3 wt%	5 wt%	10 wt%
25	27.9	31.7	30.7	0.84	0.94	0.92	78.8	100.1	93.8
100	22.3	25.0	24.2	0.99	1.14	1.10	71.1	93.5	87.1
200	17.8	19.0	18.7	1.10	1.24	1.19	65.3	81.5	76.5
300	14.8	15.8	15.3	1.32	1.49	1.43	61.6	72.1	67.5
400	12.8	13.5	13.0	1.28	1.45	1.39	59.6	64.8	60.1
600	10.0	10.5	10.1	1.33	1.52	1.46	56.6	53.3	49.4

For the obtained sinters, thermal expansion coefficients (CTE) are measured. CTE values for the temperature range of 40–600 ℃, are estimated to 5.72×10^{-6} (℃)$^{-1}$, 5.41×10^{-6} (℃)$^{-1}$ and 5.54×10^{-6} (℃)$^{-1}$ with 3 wt%, 5 wt% and 10 wt% Y_2O_3, respectively. The various additions of sintering aid do not influence significantly on thermal expansion of AlN. These results are similar to the literature data [13].

The microstructure appearance of AlN samples are shown in Figs. 3–5. They are very similar in shape of grains for each content of yttrium oxide dopant. Two phases are clearly visible. The grey fields are corresponded to pure AlN grains and light spots are attributed to yttrium-containing phases. EDS examination shown in Fig. 3 and XRD analysis indicate the formation of YAG at grain boundary for the sample containing 3 wt% of yttrium oxide dopant. It can be concluded that a short time of rapid densification by PPS gives good crystallized garnet phase located at grain boundary. This phase is distributed uniformly at all images shown in Figs. 3–5.

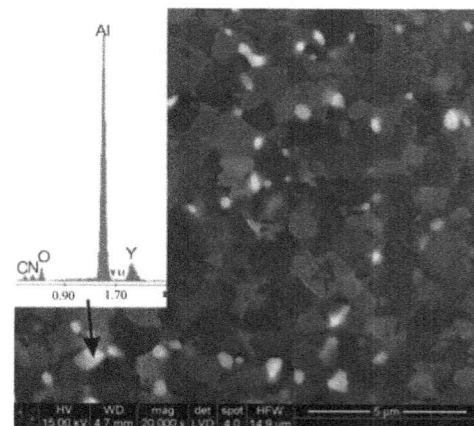

Fig. 3   Microstructure of sintered AlN polycrystals (3 wt% Y_2O_3, sintered at 1500 ℃ for 600 s).

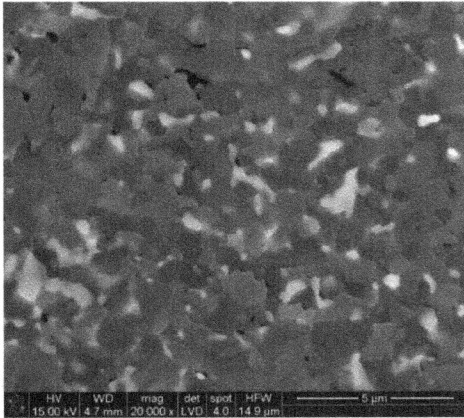

Fig. 4   Microstructure of sintered AlN polycrystals (5 wt% Y_2O_3, sintered at 1500 ℃ for 600 s).

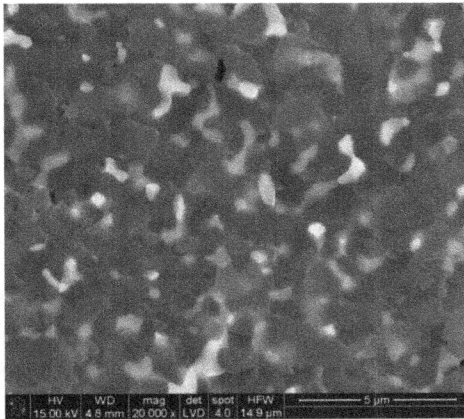

Fig. 5   Microstructure of sintered AlN polycrystals (10 wt% Y_2O_3, sintered at 1500 ℃ for 600 s).

To calculate grain size distribution, the image shown in Fig. 4 is binarized (Fig. 6) and then examined by computer-aided analysis. 40% of the grains have an average grain size estimated to 1.0 μm as shown in Fig. 7. This value is similar to that of the initial powders, and thus it is concluded that grain growth does not occur during PPS process significantly. For other microstructures, the results are very similar; thus we believe that the amount of yttrium oxide dopant does not control AlN microstructure appearance sintered by PPS. The difference in thermal conductivity of the samples can be explained by porosity changes or specific YAG, YAM or YAP formation. This reaction can reduce the oxygen content inside AlN grains, and therefore each sample has different thermal conductivity. The obtained thermal results for lower temperature sintered polycrystals (PPS) are similar to SPS-sintered materials [14], whose authors used maximum 3 wt% various additive mixtures (Sn_2O_3–Li_2O–Y_2O_3). However, AlN

polycrystals described in the present work were prepared at lower temperature of about 250 ℃.

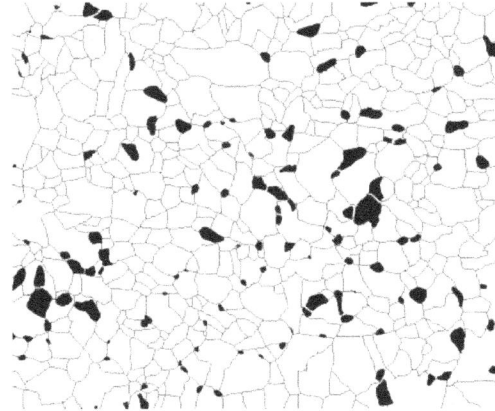

Fig. 6   Binary microstructure image of sintered AlN polycrystals (5 wt% Y_2O_3, sintered at 1500 ℃ for 600 s).

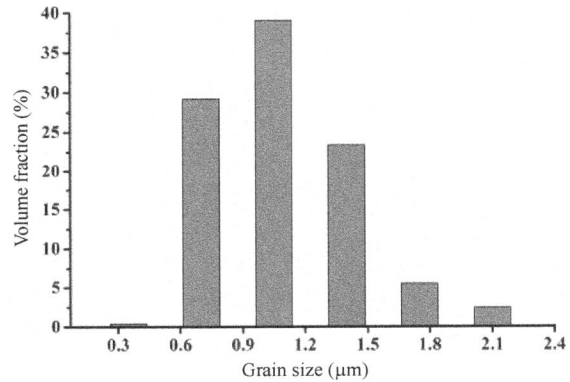

Fig. 7   Grain size distribution of sintered AlN polycrystals (5 wt% Y_2O_3, sintered at 1500 ℃ for 600 s).

4   Conclusions

It is possible to obtain dense polycrystals of AlN by PPS method at temperature of 1500 ℃ in a very short time. The obtained AlN polycrystals are characterized with high density (above 97%) and good thermal properties. All the samples have fine microstructure with grain size estimated to 1.0 μm, which can have an influence on the mechanical property improvement. The thermal conductivity of AlN sinters is comparable to polycrystals prepared by SPS method.

Acknowledgements

The research work was carried out within the project "New Construction Materials with High Thermal

Conductivity" (No. UDA-POIG.01.01.02-00-97/09-01). Thermal conductivity measurements were performed in collaboration with Faculty Laboratory of Thermophysical Measurements of Materials Science and Ceramics Faculty, AGH University of Science and Technology in Krakow. Materials were obtained in collaboration with Laboratory of Warsaw University of Technology. Thanks to Dr. Lukasz Ciupinski from Faculty of Materials Science and Engineering, Warsaw University of Technology, for sample sintering by PPS process, and to Kamil Jankowski from Faculty of Material Science and Ceramics, AGH University of Science and Technology, for powder mixture preparation.

References

[1] Baik K, Drew RAL. Aluminum nitride: Processing and applications. *Key Eng Mat* 1996, **122–124**: 553–570.

[2] Axelbaum RL, Lottes CR, Huertas JI, *et al.* Gas-phase combustion synthesis of aluminum nitride powder. Twenty-Sixth Symposium (International) on Combustion/The Combustion Institute, 1996: 1891–1897.

[3] Campbell CK. Applications of surface acoustic and shallow bulk acoustic wave devices. *Proc IEEE* 1989, **77**: 1453–1484.

[4] Tiegs TN, Kiggans Jr. JO. High thermal conductivity lossy dielectric using codensified multilayer configuration. U.S. Patent 6579393, June 2003.

[5] Rutkowski P, Kata D. The influence of aluminum nitride polycrystal annealing time on its thermal conductivity. *Ceram Mater* 2013, **65** (in Polish, in press).

[6] Komeya K, Tatami J. Liquid phase sintering of aluminum nitride. *Mater Sci Forum* 2007, **554**: 181–188.

[7] Jankowski K, Kata D, Lis J. Preparation of polycrystalline aluminum nitride with yttria additive. *Ceram Mater* 2012, **64**: 214–218 (in Polish).

[8] Jackson TB, Virkar AV, More KL, *et al.* High-thermal-conductivity aluminum nitride ceramics: The effect of thermodynamics, kinetics and microstructural factors. *J Am Ceram Soc* 1997, **80**: 1421–1435.

[9] Xu GF, Olorunyolemi T, Wilson OC, *et al.* Microwave sintering of high-density, high thermal conductivity AlN. *J Mater Res* 2002, **17**: 2837–2845.

[10] Rutkowski P, Kata D, Lis J. Thermal conductivity of polycrystalline AlN. *Ceram Mater* 2012, **64**: 572–576 (in Polish).

[11] Michalski A. Pulse plasma sintering of ceramics materials. *Ceramics* 2005, **91**: 379–385.

[12] Bellosi A, Esposito L, Scafè E, *et al.* The influence of microstructure on the thermal conductivity of aluminum nitride. *J Mater Sci* 1994, **29**: 5014–5022.

[13] Craft S, Moody B, Dalmau R, *et al.* Thermal expansion engineering for polycrystalline aluminum nitride sintered bodies. U.S. Patent 2012/0146023 A1, June 2012.

[14] Li MJ, Zhang LM, Shen Q, *et al.* Microstructure and properties of spark plasma sintered AlN ceramics. *J Mater Sci* 2006, **41**: 7934–7938.

18

Transport and physical properties of V_2O_5–P_2O_5–B_2O_3 glasses doped with Dy_2O_3

R. V. BARDE[a], S. A. WAGHULEY[b,*]

[a]*Department of Engineering Physics, Shri Hanuman Vyayam Prasarak Mandal's College of Engineering and Technology, Amravati 444 605, India*
[b]*Department of Physics, Sant Gadge Baba Amravati University, Amravati 444 602, India*

Abstract: We investigate the DC transport properties of $60V_2O_5$–$5P_2O_5$–$(35-x)B_2O_3$–xDy_2O_3 (x = 0.4, 0.6, 0.8, 1.0 and 1.2 mol%) glasses as function of temperature which were prepared using the conventional melt-quenching method. These glasses are characterised by thermo gravimetric-differential thermal analysis (TG-DTA). Activation energy (E_{DC}) is obtained from Arrhenius plots of temperature-dependent DC conductivity, and it is found to be 0.30 eV for high conducting glass. In order to understand the role of Dy_2O_3 in these glasses, the density and molar volume are investigated. The results show that molar volume of the glass increases with the increasing of Dy_2O_3 concentration. The ionic conductivity is found to be dominant over the electronic conductivity and varies between 82% and 96%.

Keywords: transport properties; melt quenching; Arrhenius plot; glasses

1 Introduction

Transition metal oxide glasses have been studied because of their interesting semiconducting properties, which are due to the hopping of "polarons" from the higher to the lower valence states of the transition metal ions [1]. In these glasses, strong electron–phonon interaction is responsible for the formation of small polarons [2,3]. Vanadate glasses contain V^{4+} and V^{5+} ions, where the electrical conduction is endorsed to the hopping of $3d^1$ unpaired electron from V^{4+} to V^{5+} sites. These glasses have been considered as a new branch in semiconducting glasses because of their wider glass-forming region and possible technological applications [4–8]. In B_2O_3 glasses, two groups of bands such as trigonal BO_3 and tetrahedral BO_4 are obtained [9]. When transition metal ions are added to the borate glasses, they exhibit specific physical properties. When these glasses are grafted with alkaline earth ions, the resultant glasses have several applications [10–13]. Phosphate glasses hold abundant advantages such as high thermal conductivity, low melting and softening temperature and high thermal expansion coefficient over silicate and borate glasses [14].

In the last three decades, several attempts were made to develop fast ion conducting glasses because of their prospective applications as high energy density batteries [15]. The physical properties of the phosphate glasses can be improved by the addition of different heavy metal oxides [16,17].

Glasses containing rare earth and transition metal

* Corresponding author.
E-mail: sandeepwaghuley@sgbau.ac.in

have been widely studied using structural and optical spectroscopy due to their many potential applications, like optical amplifiers in telecommunication [18], phosphorescence materials and electrochemical batteries [19].

The aim of this work is to prepare the glass systems $60V_2O_5–5P_2O_5–(35–x)B_2O_3–xDy_2O_3$ ($x = 0.4$, 0.6, 0.8, 1.0 and 1.2 mol%) to analyze the DC electrical, thermal and physical properties. The characterization technique, thermo gravimetric-differential thermal analysis (TG-DTA) is employed to study the structural properties of the glass samples.

2 Experiment

The preparation of $60V_2O_5–5P_2O_5–(35–x)B_2O_3–xDy_2O_3$ ($x = 0.4$, 0.6, 0.8, 1.0 and 1.2 mol%) glasses has been described and its characterization details were reported in Ref. [20]. For the electrical measurements, the samples were polished and conducting silver paste was deposited on both sides. The sample area was taken to be the area exposed to the electrode surface. The area under electrode ranged between 0.22–0.33 cm^2, and the thickness of samples varied between 0.29–0.36 cm. Measurements of DC conductivity as a function of temperature in all the samples were made by two-probe technique in the temperature range of 303–473 K, as the glass systems $60V_2O_5–5P_2O_5–(35–x)B_2O_3–xDy_2O_3$ have low Johnson noise due to high conductance at high temperature [21]. For ionic transference number measurement, the samples were polished and graphite electrode (blocking electrode) was deposited on both sides of the samples. The samples were held between the sample holders. The sample holder was provided with spring-mechanical pressure to ensure good electrical contact. A constant voltage (DC) of 6 V was applied to the samples. The measurement was performed at room temperature. The total ionic (t_{ion}) and electronic (t_{ele}) transference numbers were calculated from Eqs. (1) and (2) [22–25]:

$$t_{ion} = \frac{I_i - I_f}{I_i} \qquad (1)$$

$$t_{ele} = 100 - t_{ion} \qquad (2)$$

where I_i is the initial value of the current at the start; and I_f is the current on reaching saturation.

The density of the glass bits free of air bubbles and cracks was determined at room temperature through Archimedes principle, by using xylene ($\rho = 0.863$ g/ml). The density was estimated by using Eq. (3):

$$\rho = \left(\frac{W_a}{W_a - W_l} \right) \times \rho_l \qquad (3)$$

where ρ is the density of the sample; W_a is the weight of the sample in air; W_l is the weight of the sample fully immersed in xylene; and ρ_l is the density of xylene.

The molar volume V_m was calculated from Eq. (4) [26–30]:

$$V_m = \frac{M_T}{\rho} \qquad (4)$$

where M_T is the molecular weight of the glass calculated by multiplying x times the molecular weights of the various constituents.

The oxygen packing density (OPD) was determined from Eq. (5) [31]:

$$D_0 = \left(\frac{\rho}{M_T} \right) \times \text{number of oxygen atoms per formula unit} \qquad (5)$$

3 Results and discussion

3.1 Characterization of materials

From the previous results by X-ray diffraction analysis, it is found that all the samples possess the amorphous nature; the thermal stability was studied by thermo gravimetric differential analysis, and the thermal stability of the glass is found to be excellent for the composition with $x = 0.6$ [20].

3.2 DC conductivity

The DC conductivity (σ_{DC}) for all samples was calculated using the sample dimensions. The reciprocal temperature dependence of the DC conductivity is shown in Fig. 1. The plots show DC conductivity exhibits an Arrhenius-type temperature dependence given by the relation:

$$\sigma_{DC} = \sigma_0 \exp\left(-\frac{E_{DC}}{KT} \right) \qquad (6)$$

where E_{DC} is the activation energy which is calculated from the least square straight line fitting of plots; and σ_0 is the pre-exponential factor.

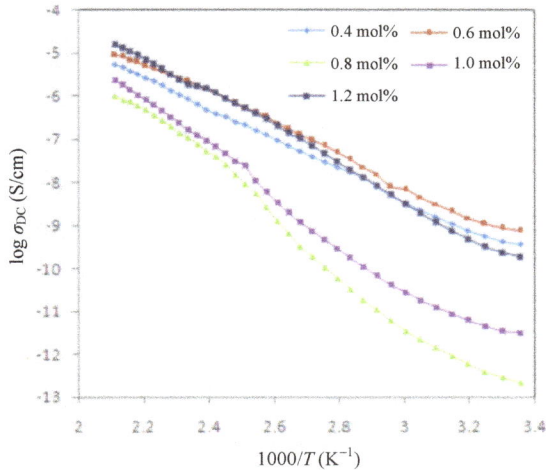

Fig. 1 Temparature dependence of DC conductivity.

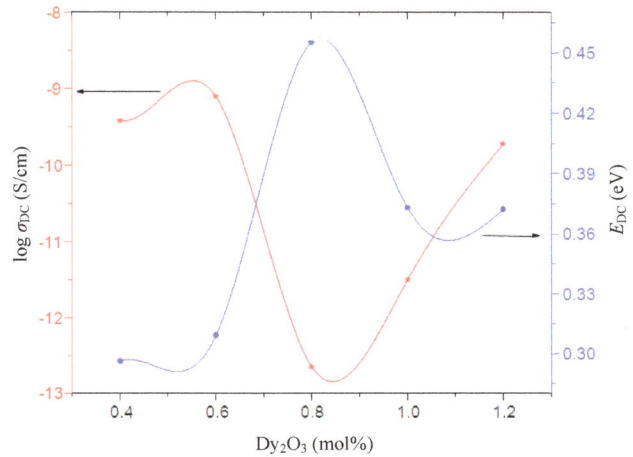

Fig. 2 Variation of σ_{DC} and E_{DC} with molar fraction of Dy_2O_3.

The variation of σ_{DC} and E_{DC} as functions of molar fraction of Dy_2O_3 are depicted in Fig. 2. It is observed that the conductivity shows a random nature. At room temperature, it is maximum for 0.6 mol% of Dy_2O_3 and decreases for 0.8 mol% of Dy_2O_3. If we further increase the molar fraction of Dy_2O_3, the conductivity increases. The highest conductivity is found to be 8.22×10^{-3} S/cm at 473 K. The value of activation energy ranges between 0.30–0.52 eV. The maximum value of conductivity corresponds with the minimum value of activation energy. The explanation for the enhancement in the conductivity is given on the basis of Anderson and Stuart model. According to this model, as one of the network ions is substituted by another glass modifier ion, the average interionic bond distance becomes larger or smaller according to whether the substituting ion is larger or smaller. In the present case, Dy being slightly larger in size than boron, the substitution of boron by Dy will increase the interionic bond distance. Thus with the addition of 0.4 mol% Dy_2O_3, the structure becomes loose and hence the conductivity increases [32]. The decrease in conductivity beyond 0.8 mol% Dy_2O_3 is similar to that reported in alkali alumino silicate glass system [33]. In the present study, the addition of Dy_2O_3 eliminates the number of non-bridging oxygens (NBOs) and simultaneously creates bridging oxygens (BOs). This may decrease the open structure, through which charge carrier can move with lower mobility [34]. The further enhancement in conductivity may be due to the creation of NBOs.

3.3 Transference number measurement

The transference number gives the information of the extent of ionic and electronic contribution to the total conductivity. The total ionic and electronic transference numbers (t_{ion} and t_{ele}) were measured using DC polarization technique by sandwiching the sample between graphite (blocking) electrodes. The polarization current has been monitored as a function of time by applying fixed DC potential across the sample. The total ionic and electronic transference numbers have been calculated from the plot using Eqs. (1) and (2) [22–25].

The current versus time plots of all samples are obtained which exhibit typical behavior of ionic charge transport. Figure 3 shows the plots of current versus time for all the compositions of $60V_2O_5–5P_2O_5–$

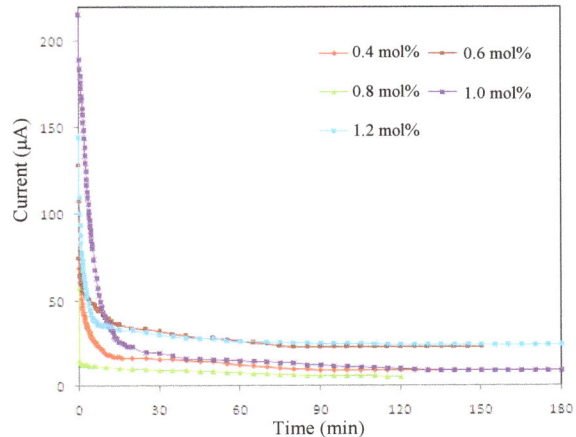

Fig. 3 Plots of DC current versus time for $60V_2O_5–5P_2O_5–(35-x)B_2O_3–xDy_2O_3$.

$(35-x)B_2O_3-xDy_2O_3$. The total current becomes nearly constant at some non-zero value after some time. The final residual current is mainly due to electrons/holes. The values of transference numbers are found to be in the range of 0.82 to 0.96. This suggests that the charge transport in all the samples are predominantly due to ions.

3.4 Density, molar volume and oxygen packing density

The density is a powerful tool capable of exploring the changes in the structure of glasses. The density is affected by the structural softening/compactness, change in geometrical configuration, coordination number, cross-link density, and dimension of interstitial spaces of the glass. In the studied glasses, it is noted that the density decreases from 2.88 g/cm^3 to 2.74 g/cm^3, congruent with an increase in the molar volume from 49.32 cm^3/mol to 52.36 cm^3/mol as the Dy_2O_3 content increases on the expense of B_2O_3 content as shown in Fig. 4 and listed in Table 1. It is expected that the density and molar volume should show opposite behavior to each other, and so in the studied glasses the molar volume increase with the decrease in the density as the Dy_2O_3 content increases [35]. The increase in the molar volume can be attributed to the larger packing factor of Dy_2O_3 than that of B_2O_3. Accordingly, the structure of the studied glasses will be expanded, but with compactness and high number of covalent bonds. Figure 5 shows the variation of molar volume as well as OPD with Dy_2O_3 content. It is noted that the molar volume increases and OPD decreases with Dy_2O_3 content [36–38].

4 Conclusions

The melt-quenching technique is a very simple method for the preparation of the conducting glasses. The DC conductivity shows Arrhenius-type temperature dependence. The maximum value of conductivity and

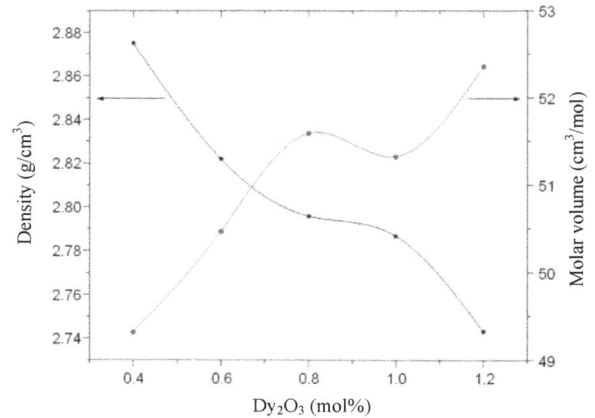

Fig. 4   Variation of the density and molar volume with molar fraction of Dy_2O_3.

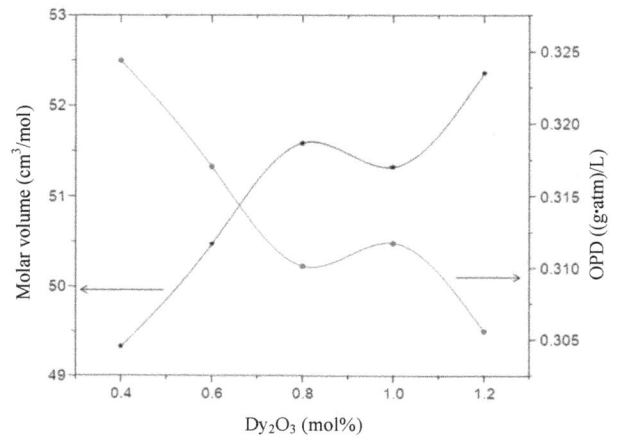

Fig. 5   Variation of molar volume and OPD with molar fraction of Dy_2O_3.

minimum value of activation energy are found to be in the range of 8.22×10^{-3} S/cm at 473 K and 0.30–0.52 eV, respectively. The ionic conductivity is found to be dominant over the electronic conductivity and varies between 82% and 96%. It is realised that the structural modification depends on the density and molar volume. The density is found to be decreased from 2.88 g/cm^3 to 2.74 g/cm^3, congruent with the increase in molar volume from 49.32 cm^3/mol to 52.36 cm^3/mol as the Dy_2O_3 content increases.

Table 1   Physical properties of $60V_2O_5-5P_2O_5-(35-x)B_2O_3-xDy_2O_3$

Dy_2O_3 (mol%)	Density, ρ (g/cm^3)	Molecular weight, M_T (g/mol)	Molar volume, V_m (cm^3/mol)	OPD, D_O ((g·atm)/L)
0.4	2.88	141.80	49.32	0.320
0.6	2.82	142.41	50.47	0.317
0.8	2.79	144.23	51.59	0.310
1.0	2.78	143.02	51.32	0.311
1.2	2.74	143.63	52.36	0.305

Acknowledgements

Authors are thankful to the head of Department of Physics, Sant Gadge Baba Amravati University, Amravati, and the principal of Shri Hanuman Vyayam Prasarak Mandal's College of Engineering and Technology, Amravati, for providing necessary facilities.

References

[1] Al-Hajry A, Soliman AA, El-Desoky MM. Electrical and thermal properties of semiconducting Fe_2O_3–Bi_2O_3–$Na_2B_4O_7$ glasses. *Thermochim Acta* 2005, **427**: 181–186.

[2] El-Desoky MM, Al-Shahrani A. Variable—Range hopping in Fe_2O_3–Bi_2O_3–$K_2B_4O_7$ glasses. *J Mater Sci: Mater El* 2005, **16**: 221–224.

[3] Al-Hajry A, Tashtoush N, El-Desoky MM. Characterization and transport properties of semiconducting Fe_2O_3–Bi_2O_3–$Na_2B_4O_7$ glasses. *Physica B* 2005, **368**: 51–57.

[4] Saddeek YB, Shaaban ER, Aly KA, *et al.* Characterization of some lead vanadate glasses. *J Alloys Compd* 2009, **478**: 447–452.

[5] El-Desoky MM. Small polaron transport in V_2O_5–NiO–TeO_2 glasses. *J Mater Sci: Mater El* 2003, **14**: 215–221.

[6] Al-Shahrani A, Al-Hajry A, El-Desoky MM. Non-adiabatic small polaron hopping conduction in sodium borate tungstate glasses. *Phys Status Solidi a* 2003, **200**: 378–387.

[7] El-Desoky MM, Tashtoush NM, Habib MH. Characterization and electrical properties of semiconducting Fe_2O_3–Bi_2O_3–$K_2B_4O_7$ glasses. *J Mater Sci: Mater El* 2005, **16**: 533–539.

[8] El-Desoky MM, Al-Hajry A, Tokunaga M, *et al.* Effect of sulfur addition on the redox state of iron in iron phosphate glasses. *Hyperfine Interact* 2004, **156–157**: 547–553.

[9] Mahmoud KH, Abdel-Rahim FM, Atef K, *et al.* Dielectric dispersion in lithim–bismuth–borate glasses. *Curr Appl Phys* 2011, **11**: 55–60.

[10] Terczynska-Madej A, Cholewa-Kowalska K, Laczka M. The effect of silicate network modifiers on colour and electron spectra of transition metal ions. *Opt Mater* 2010, **32**: 1456–1462.

[11] Venkat Reddy P, Laxmi Kanth C, Prasanth Kumar V, *et al.* Optical and thermoluminescence properties of R_2O–RF–B_2O_3 glass systems doped with MnO. *J Non-Cryst Solids* 2005, **351**: 3752–3759.

[12] Ramesh Babu A, RajyaSree C, Srinivasa Rao P, *et al.* Vanadyl ions influence on spectroscopic and dielectric properties of glass network. *J Mol Struct* 2011, **1005**: 83–90.

[13] Ramesh Babu A, RajyaSree C, Vinaya Teja PM, *et al.* Influence of manganese ions on spectroscopic and dielectric properties of LiF–SrO–B_2O_3 glasses. *J Non-Cryst Solids* 2012, **358**: 1391–1398.

[14] Srinivasa Rao P, Bala Murali Krishna S, Yusub S, *et al.* Spectroscopic and dielectric investigations of tungsten ions doped zinc bismuth phosphate glass-ceramics. *J Mol Struct* 2013, **1036**: 452–463.

[15] Takahashi H, Karasawa T, Sakuma T, *et al.* Electrical conduction in the vitreous and crystallized Li_2O–V_2O_5–P_2O_5 system. *Solid State Ionics* 2010, **181**: 27–32.

[16] Kiran N, Kesavulu CR, Suresh Kumar A, *et al.* Spectral studies on Cr^{3+} ions doped in sodium–lead borophosphate glasses. *Physica B* 2011, **406**: 1897–1901.

[17] Kim CE, Hwang HC, Yoon MY, *et al.* Fabrication of a high lithium ion conducting lithium borosilicate glass. *J Non-Cryst Solids* 2011, **357**: 2863–2867.

[18] Pisarski WA, Goryczka T, Wodecka-Dus B, *et al.* Structure and properties of rare earth-doped lead borate glasses. *Mat Sci Eng B* 2005, **122**: 94–99.

[19] Qiu J, Igarashi H, Makishima A. Long-lasting phosphorescence in Mn^{2+}:Zn_2GeO_4 crystallites containing transparent GeO_2–B_2O_3–ZnO glass-ceramics. *Sci Technol Adv Mater* 2005, **6**: 431–434.

[20] Barde RV, Waghuley SA. Study of AC electrical properties of V_2O_5–P_2O_5–B_2O_3–Dy_2O_3 glasses. *Ceram Int* 2013, **39**: 6303–6311.

[21] Yoon HJ, Jun DH, Yang JH, *et al.* Carbon dioxide gas sensor using a graphene sheet. *Sensor Actuat B: Chem* 2011, **157**: 310–313.

[22] Sekhon SS, Chandra S. Mixed cation effect in silver borate ion conducting glass. *J Mater Sci* 1999, **34**: 2899–2902.

[23] Das SS, Srivastava V, Singh P. Ionic transport sodium phosphate glasses doped with chloride of Co, Cd and Ag. *Indian J Eng Mater Sci* 2006, **13**: 455–461.

[24] Sheha E. Ionic conductivity and dielectric properties

of plasticized $PVA_{0.7}(LiBr)_{0.3}(H_2SO_4)_{2.7M}$ solid acid membrane and its performance in a magnesium battery. *Solid State Ionics* 2009, **180**: 1575–1579.

[25] Reddy J, Ramesh Ch, Kumar S, *et al.* Conductivity study of PEO complex with Mg^{2+} borate glass polymer electrolyte—Its application as electrochemical cell. *Int J Appl Engng Res Dindigul* 2011, **2**: 147–156.

[26] Govindaraj G, Mariappan CR. Synthesis, characterization and ion dynamic studies of NASICON type glasses. *Solid State Ionics* 2002, **147**: 49–59.

[27] Abid M, Et-tabirou M, Taibi M. Structure and DC conductivity of lead sodium ultraphosphate glasses. *Mat Sci Eng B* 2003, **97**: 20–24.

[28] Mansour E, El-Damrawi GM, Moustafa YM, *et al.* Polaronic conduction in barium borate glasses containing iron oxide. *Physica B* 2001, **293**: 268–275.

[29] Ramadevudu G, Laxmi Srinivasa Rao S, Hameed A, *et al.* FTIR and some physical properties of alkaline earth borate glasses containing heavy metal oxides. *Int J Eng Sci Tech* 2011, **3**: 6998–7005.

[30] Insiripong S, Chimalawong P, Kaewkhao J, *et al.* Optical and physical properties of bismuth borate glasses doped with Dy^{3+}. *Am J Applied Sci* 2011, **8**: 574–578.

[31] Bale S, Rahman S. Spectroscopic and physical properties of Bi_2O_3–B_2O_3–ZnO–Li_2O glasses. *Int Sch Res Netw Spectrosc* 2012, DOI: 10.5402/2012/634571.

[32] Gedam RS, Deshpande VK. An anomalous enhancement in the electrical conductivity of Li_2O:B_2O_3:Al_2O_3 glasses. *Solid State Ionics* 2006, **177**: 2589–2592.

[33] Alexander MN, Onorato PIK, Struck CW, *et al.* Structure of alkali (alumino) silicate glasses: I. Tl^+ luminescence and the nonbridging oxygen issue. *J Non-Cryst Solids* 1986, **79**: 137–154.

[34] El-Desoky MM, Ibrahim FA, Mostafa AG, *et al.* Effect of nanocrystallization on the electrical conductivity enhancement and Mössbauer hyperfine parameters of iron based glasses. *Mater Res Bull* 2010, **45**: 1122–1126.

[35] Abid M, Et-tabirou M, Taibi M. Structure and DC conductivity of lead sodium ultraphosphate glasses. *Mat Sci Eng B* 2003, **97**: 20–24.

[36] Abdel-Baki M, El-Diasty F. Role of oxygen on the optical properties of borate glass doped with ZnO. *J Solid State Chem* 2011, **184**: 2762–2769.

[37] Veeranna Gowda VC, Anavekar RV. Elastic properties and spectroscopic studies of lithium lead borate glasses. *Ionics* 2004, **10**: 103–108.

[38] Silim HA. Composition effect on some physical properties and FTIR spectra of alumino–borate glasses containing lithium, sodium, potassium and barium oxides. *Egypt J Solids* 2006, **29**: 293–301.

19

Effect of sintering temperature on microstructure and electrical properties of $Sr_{1-x}(Na_{0.5}Bi_{0.5})_xBi_2Nb_2O_9$ solid solutions

Hana NACEUR[*], Adel MEGRICHE, Mohamed EL MAAOUI

Laboratory of Applied Mineral Chemistry, Department of Chemistry, Faculty of Sciences, University Tunis ElManar, Campus 2092, Tunis, Tunisia

Abstract: In this study, we investigated the effect of sintering temperature on densification, grain size, conductivity and dielectric properties of $Sr_{1-x}(Na_{0.5}Bi_{0.5})_xBi_2Nb_2O_9$ ceramics, prepared by hydrothermal method. Pellets were sintered at different temperatures. Density increased with sintering temperature, reaching up to 96% at 800 ℃. A grain growth was observed with increasing sintering temperature. Impedance spectroscopy analyses of the sintered samples at various temperatures were performed. Increase in dielectric constant and in Curie temperature with sintering was discussed. Electrical conductivity and activation energy were calculated and attributed to the microstructural factors.

Keywords: ceramics; sintering; light scattering; dielectric properties

1 Introduction

Dielectric and ferroelectric ceramic materials are mature and ubiquitous materials for advanced technology. These ceramics are active elements in a range of dielectric devices and performing functions such as sensing and actuation. Recently, ferroelectric material of strontium bismuth niobate $SrBi_2Nb_2O_9$ (SBN) and its solid solutions have attracted much attention for potential applications in the field of nonvolatile random access memory (NVRAM) due to their excellent fatigue endurance and low switching voltage. The performance of these materials is closely related to their microstructures and, for this reason, to

the ways they have been processed. Preparation condition, especially sintering temperature is an important factor in the fabrication process of ceramic bodies, which can significantly affect the mechanical and electrical properties of materials [1,2]. For improvement of these properties, parameters affected by sintering such as density, porosity, grain size and their distribution must be controlled [3,4]. In fact, sintering, as one of the most important processes for the production of ceramic materials, usually goes through a sequence of essential phenomena such as neck formation, pore, shrinkage, and grain growth. It results in a solid compact with a suitable microstructure. During the sintering of a powder compact, both densification and grain growth occur simultaneously [5]. It is necessary to control the total densification process of the material as well as the pore presence and to avoid grain growth, which require a full understanding of the material constitutive laws

---

* Corresponding author.
E-mail: Naceur.hana@live.fr

governing the sintering process.

There are numerous reports on the synthesis of bismuth niobate and its solid solutions by processes such as solid-state reaction [6–8], aqueous solution method [9,10], organic precursor decomposition [11,12], co-precipitation [13–15], sol–gel [16,17], and combustion synthesis [18] which often results in high agglomeration of powders. For these several chemical synthesis routes, a high sintering temperature is usually necessary for the densification of ceramics [19]. It often causes a variation in the stoichiometry ceramics and in the production of a liquid phase, caused by volatilization of Na_2O, Bi_2O_3, etc.

Density can be improved through the use of sintering aids [20]. In fact, further studies on sintering behavior reveal that adding excess amounts of Bi_2O_3 in the processing of many compounds could facilitate the densification process [21]. Increasing the amount of Bi_2O_3 addition and sintering temperature enhancs the development of the preferred orientation and coarsening of grains [21]. In a recent work for $BaTiO_3$ compound, the addition of small amounts of Li_2O to various ceramics has been reported to lower the sintering temperature by various mechanisms [22]. In another recent work, an addition of H_3BO_3 effectively lowers the sintering temperature of $Li_2ZnTi_3O_8$ ceramics from 1075 ℃ to 880 ℃ [23]. Many reports reviewed the hot pressing and the spark plasma (sintering) [24–26] for densification of ceramics; however, these methods are unsuitable for stress and impurity-free ceramics. One of the main challenges is the difficulty of producing dense ceramics using ordinary firing methods. A solid state double sintering method is employed in many works involving two-step firing [27–29]. The first step, at relatively low temperature, is to promote the chemical reactions among the constituent compounds so as to form a single phase layered perovskite. The second step, firing at high temperature, is to achieve high densification. In order to use moderate sintering temperatures and attain low porosity, recently, a first study on microwave sintered $SrBi_2Ta_2O_9$ ceramic was reported [30].

Hydrothermal synthesis of ceramic powders has gained considerable popularity. This process progresses in a closed system at a high autogenous pressure and leads to high-quality powders, with well-controlled morphology and narrow particle size distribution [31]. The hydrothermal synthesis is sometimes called "soft chemical processing solution" because of the mild reacting conditions under which

the products are achieved, leading to controlled particle size and morphology. These characteristics decrease the sintering temperatures and, consequently, low-price products can be obtained [32]. It is thus regarded as a promising way to prepare bismuth niobate powders with a low sintering temperature, which is beneficial to obtain easily controlled-size grains, adjust the compound stoichiometry and reduce the volatilization of alkaline elements [33]. A novel hydrothermal apparatus with ball milling system is developed by Hotta et al. to produce nanosized $BaTiO_3$ powders with low-temperature sintering capability [34].

Therefore, there is a considerable interest in the search of new dielectric materials with low sintering temperature [35,36]. Our recent work stated a particular point concerning the sintering temperature for $Sr_{1-x}(Na_{0.5}Bi_{0.5})_xBi_2Nb_2O_9$ pellets at 800 ℃ [37]. Powders synthesized by hydrothermal method are lowly sinterable. This temperature appears to be sufficient to perform a good densification despite the fact that samples are usually sintered at temperature higher than 1000 ℃ [35,38].

The main objective of this work is to develop the effect of sintering temperature on densification, shrinkage and grain size. These parameters have pronounced influence on the device properties of materials. Density of $Sr_{1-x}(Na_{0.5}Bi_{0.5})_xBi_2Nb_2O_9$ pellets at different sintering temperatures is determined. The morphology of different sintered powders is discussed. Their electrical conductivity and activation energy are characterized and compared. The variation of Curie temperature and dielectric constant in function of sintering temperatures for all compositions is evaluated.

2 Experiment

The $Sr_{1-x}(Na_{0.5}Bi_{0.5})_xBi_2Nb_2O_9$ powders with different x values ($x = 0.0$, 0.2, 0.5, 0.8 and 1.0) were synthesized by the hydrothermal process. The preparation was detailed in our recent report [37].

Pellets were prepared by pressing the powder samples at 10 tons. The thickness and the diameter were about 1.6 mm and 13 mm, respectively. These disks were sintered at different temperatures in a programmable furnace (Vulcan, Model 3-550) and the heating rate was 5 ℃/min.

The pellet densities were measured before sintering (called green density ρ_g) and after sintering (called sintered density ρ_s), in water using the Archimedes principle as follows:

$$\rho_{g/s} = \frac{M_a}{(M_b - M_w) \times \rho_w} \qquad (1)$$

where M_a is the mass of the sample in air; M_b is the mass of the sample in air after submerged from water; M_w is the mass of the sample in water; ρ_w is the density of water; and $\rho_{g/s}$ is the green or the sintered density.

The theoretical density (ρ_t) was calculated from the diffraction patterns applying the formula:

$$\rho_t = \frac{\text{Cell mass}}{\text{Cell volume}} = \frac{n \times M_s}{N \times a \times b \times c}(\text{g/cm}^3) \qquad (2)$$

where n is the number of atoms per unit cell; M_s is the molecular weight of the compound; N is the Avogadro number; and a, b and c are the lattice parameters of orthorhombic sample refined by the least square method from the powder data.

The density percentage ($\rho_s(\%)$) after sintering was carried out using the following formula:

$$\rho_s(\%) = \left(\frac{\rho_s}{\rho_t}\right) \times 100 \qquad (3)$$

A densification factor (DF) for all sintered pellets can be related with sintered density, green density and theoretical density, with the following expression:

$$\text{DF} = \frac{\rho_s - \rho_g}{\rho_t - \rho_g} \qquad (4)$$

It is worthwhile giving thought on the changes that occur in a ceramic body during sintering. The shrinkage measurements were determined for all samples by the diameter and thickness reduction. The percentages of the diameter and thickness reduction were calculated using these two equations:

$$\text{Diameter shrinkage} = \left(\frac{d_0 - d_f}{d_0}\right) \times 100\% \qquad (5)$$

$$\text{Thickness shrinkage} = \left(\frac{h_0 - h_f}{h_0}\right) \times 100\% \qquad (6)$$

where d_0 is the diameter before sintering; d_f is the diameter after sintering; h_0 is the height before sintering; and h_f is the height after sintering.

In order to estimate average grain size, a suspension of the obtained product was measured by a dynamic light scattering (DLS) technique, calculated by the Stocks–Einstein relation [39]:

$$G_i = \frac{k_B \times T}{3\pi \times \eta_s \times D} \qquad (7)$$

where D is the diffusion coefficient; k_B is the Boltzmann constant; T is the absolute temperature; η_s is the viscosity coefficient of solvent; and G_i is the mean grain size.

Measurements were carried out at 90° over a temperature range of 10–50 ℃ using a variable angle light scattering instrument equipped with an Excel 3000 argon ion laser (1200 mW at 514.5 nm) as the source and a Brookhaven Digital BI 2030AT correlator.

The dielectric constant and the Curie temperature were deducted for the sintered ceramic pellets, using an HP 4192A impedance gain phase analyzer. The electrical conductivities of the samples were measured. The activation energies were calculated using traditional Arrhenius equation:

$$\sigma = (\sigma_0 / T)\exp[-E_a / (k_B \times T)] \qquad (8)$$

where E_a is the activation energy for conduction; T is the absolute temperature; k_B is the Boltzmann constant; and σ_0 is the pre-exponential factor.

3 Results and discussion

3.1 Densification and microstructure

3.1.1 Densification

In order to improve the properties of ceramics for either structural or functional applications, reducing the grain size and limiting the formation of defects are important. During the sintering period, a considerable microstructural development takes place. Garcia et al. [40] reported a detailed study for the relation between the microstructure and densification of ceramics and the control of the process responsible for the rates of densification and grain growth, by controlling the time/temperature conditions employed during sintering of the compact powders.

Firstly, theoretical densities were calculated for all solid solutions by means of their lattice parameters (detected in our previous study) [37]. It is noted that these densities increase from 7.251 g/cm³ for SBN to 7.375 g/cm³, 7.502 g/cm³, 7.608 g/cm³ and 7.678 g/cm³ for $x = 0.2$, 0.5, 0.8 and 1.0, respectively.

All samples with different dopings were pressed into pellets and sintered for 2 h at different temperatures from 700 ℃ to 900 ℃. Green and sintered density values were performed by the Archimedes method. Evolution of sintered density percentage for different compositions as function of sintering temperature is plotted in Fig. 1. It could be observed that the sintering curves of all solid solutions are similar; their densities increase initially with increasing sintering temperature until 800 ℃ reaching their maximum values and remain constant up to 820 ℃, then decrease slightly for all solid solutions. Thus, the sintering of $Sr_{1-x}(Na_{0.5}Bi_{0.5})_xBi_2Nb_2O_9$ ceramics is completed at the low temperature of 800 ℃. It is well known that the process of sintering has three stages [41–43]: the initial, the intermediate, and the final one. In the initial stage, the green body has a low density and is generally lacking in physical integrity. There is a small degree of adhesion between adjacent particles. Then, the necks begin to form at the contact points. The final stage of sintering begins when most of the pores are closed.

However, beyond 820 ℃, a decrease on the density occurs on the pellets for all compositions. This phenomenon might be associated with the loss of bismuth oxide during the sintering stage. In other words, for SBN and its solid solutions, the increase of the sintering temperature is not favorable due to possible Bi loss; the melting point of Bi_2O_3 is about 825 ℃ [44] which significantly changes the density of the final ceramic. Looking away, the more the sintering temperature increases, the more the density decreases. The deviation from the stoichiometry is possible because of the vapor pressure of elements like Na^+. Na_2O loss starts to appear only above 850 ℃ and at a much higher rate at higher temperatures (>1000 ℃) [45]. For $x = 1.0$ at 840 ℃, a variation of density at

different sintering durations for 2 h, 3 h, 4 h and 5 h is plotted in Fig. 2. The increase of the sintering duration induces a decrease in density, which confirms the increase in loss of bismuth. As a result, sintering above 800 ℃ corresponds to saturation density value; all compositions are fully densified by sintering at this temperature. Thus, the hydrothermal synthesis is promising to obtain a high density at low sintering temperature. Therefore, sintered samples prepared from SBN powders, produced by hydrothermal synthesis, show a higher density than that prepared from solid-state synthesis [7]. In recent data [46], it has been found that co-precipitated $La_{0.8}Ca_{0.2}Cr_{0.9}Al_{0.1}O_3$ powders achieve a relative density of 92% after sintering at 1400 ℃ for 5 h in air. This relative density is lower than that obtained by the hydrothermally derived fine $La_{0.8}Ca_{0.2}Cr_{0.9}Al_{0.1}O_3$ powders (97%). It is obvious that generally densification is affected by the used method of preparation of the starting powders. This implies that hydrothermal method offers the potential of enhanced densification and decreased sintering temperature [46,47]. Furthermore, porosity of the crude pellet resulting from hydrothermal powder is much smaller than that attained by pressing larger grains (from other synthesis methods).

From Fig. 1, it can be also observed that sintered materials show densities significantly dependent on the composition. The saturated density values at 800 ℃ for $x = 0.0$, 0.2, 0.5, 0.8 and 1.0 are 6.933 g/cm^3, 7.057 g/cm^3, 7.181 g/cm^3, 7.295 g/cm^3 and 7.371 g/cm^3, respectively, which are equivalent to 95.62%, 95.7%, 95.73%, 95.89% and 96% of the theoretical density.

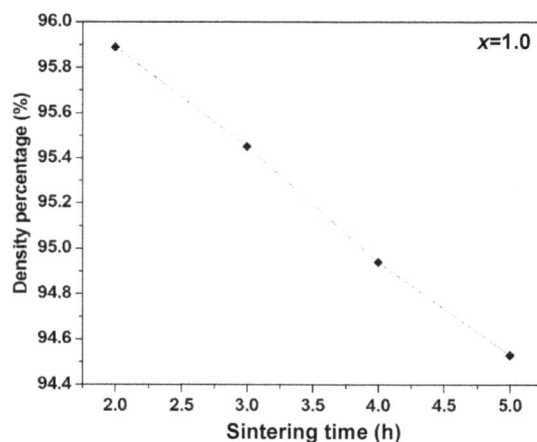

Fig. 1 Sintered density percentage as function of sintering temperature for $Sr_{1-x}(Na_{0.5}Bi_{0.5})_xBi_2Nb_2O_9$ solid solutions.

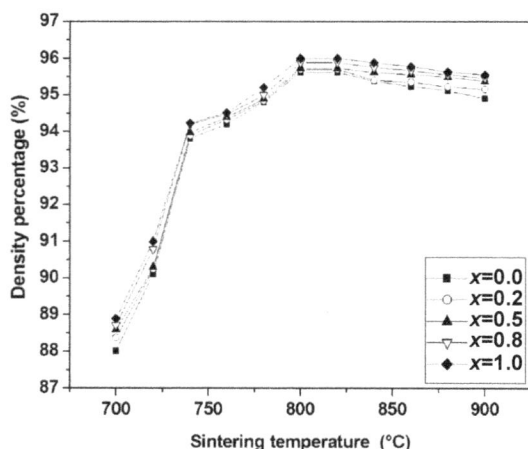

Fig. 2 Sintered density percentage as function of sintering duration at 840 ℃ for $Na_{0.5}Bi_{2.5}Nb_2O_9$ ceramic.

One can consider that the variation from 95.62% to 96% is not significant being into the error interval. In fact, three series of measurements confirm the obtained values. It appears that the addition of $Na_{0.5}Bi_{0.5}$ improves the densification process, and there is higher value of about 96% of the theoretical value for $x = 1$ than the un-doped SBN. The introduction of $Na_{0.5}Bi_{0.5}$ is thus beneficial to the densification of pellets. The addition of $Na_{0.5}Bi_{0.5}$ into the system suggests that this doping can suppress the occurrence of defects [37].

Figure 3 presents the densification factor (DF) as function of sintering temperature for all solid solutions. Two comments are revealed: the first is that DF increases with $Na_{0.5}Bi_{0.5}$ increasing; the second is DF increases with increasing temperature until 800 ℃, where it remains constant then decreases for sintering temperatures higher than 820 ℃ for all solid solutions. A positive DF indicates shrinkage. The shrinkage during sintering is a result of densification mechanisms, and during cooling it is a result of thermal contraction. Yahya et al. [48] proved that shrinkage is the phenomenon of sintering. The shrinkage is thus only dependent on thermal dilatation. It is determined by the diameter and thickness reduction. Figure 4 shows the thickness shrinkage percentage of the pellets as function of sintering temperature. An increasing progress from 4.14% to 5.2% as the sintering temperature is increased from 700 ℃ to 800 ℃ for SBN , 4.58% to 5.48%, 4.61% to 5.55%, 4.64% to 5.58% and 4.73% to 5.64% for $x = 0.2, 0.5, 0.8$ and 1.0, respectively. The diameter shrinkage percentage is plotted in Fig. 5. As shown, at the same range of temperature, it increases from 5.28% to 5.72% for SBN, 5.44% to 5.93%, 5.49% to 6.22%, 5.73% to

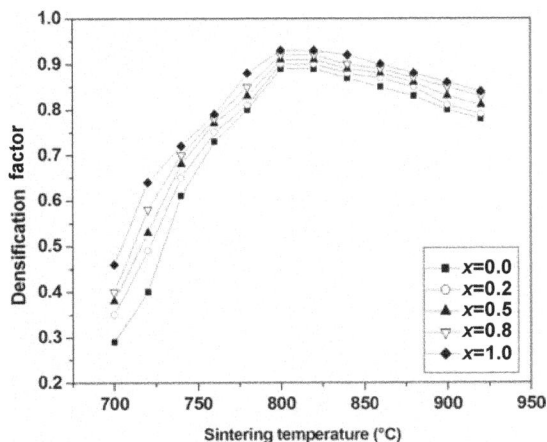

Fig. 4   Thickness shrinkage percentage as function of sintering temperature for $Sr_{1-x}(Na_{0.5}Bi_{0.5})_xBi_2Nb_2O_9$ solid solutions.

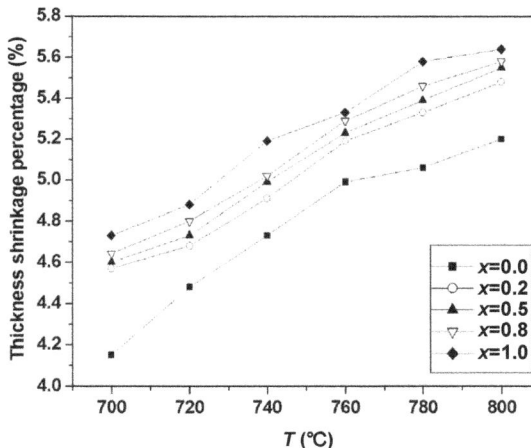

Fig. 5   Diameter shrinkage percentage as function of sintering temperature for $Sr_{1-x}(Na_{0.5}Bi_{0.5})_xBi_2Nb_2O_9$ solid solutions.

6.29%, 5.83% to 6.54% for $x = 0.2, 0.5, 0.8$ and 1.0, respectively. That means the higher the temperature and the $Na_{0.5}Bi_{0.5}$ content are, the denser the products are; in fact, as the mass is supported to be constant, shrinkage is observed when the pellet volume decreases. This is obviously due to the pore collapse. The increased shrinkage of $Na_{0.5}Bi_{2.5}Nb_2O_9$ sample at 800 ℃ implies that both the $Na_{0.5}Bi_{0.5}$ dopant and the increase in sintering temperature promote the densification.

3. 1. 2   Grain size measurements

The grain sizes of $Sr_{1-x}(Na_{0.5}Bi_{0.5})_xBi_2Nb_2O_9$ solid solutions, determined from DLS measurements before sintering are 13 nm for SBN, 25 nm, 34 nm, 48 nm and

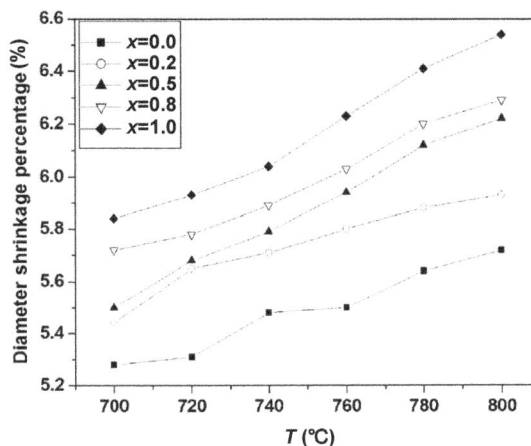

Fig. 3   Densification factor as function of sintering temperature for $Sr_{1-x}(Na_{0.5}Bi_{0.5})_xBi_2Nb_2O_9$ solid solutions.

65 nm for $x = 0.2$, 0.5, 0.8 and 1.0, respectively.

Mean grain sizes after sintering for all compounds as function of sintering temperature are plotted in Fig. 6. Two main observations can be inferred from this graph. First, the grain size increases with increasing $Na_{0.5}Bi_{0.5}$. This anisotropic behavior of grain growth was noted in our recent work for the sintering temperature equal to 800 ℃ [37]. This grain growth behavior is general whatever the sintering temperature is. This result suggests that $Na_{0.5}Bi_{0.5}$ dopant speeds up the grain growth [37]. It is due to a flux effect resulting from the introduced Na^+ and Bi^{3+}. These cations are well known for their flux behavior. It is also possible to involve the effect of solute drag on the grain boundaries [49]. Thus, migration of cation impurities to the grain boundaries contributes to lower the melting point of these boundaries. In this case Sr^{2+} ($r = 1.44$ Å) is replaced by Na^+ ($r > 1.44$ Å) and Bi^{3+} ($r < 1.44$ Å) [50], so the rate of diffusion to grain boundaries of Sr^{2+} is lower than that of Bi^{3+} and faster than that of Na^+. So, Bi^{3+} will diffuse better to grain boundaries than Sr^{2+}. As Bi^{3+} is a flux, it enhances grain growth.

Second, the grain size for all compositions increases when sintering temperature increases up to 800 ℃. Grain growth is a temperature dependent phenomenon. While the rate of the ion diffusion to the grain boundaries increases, the sintering temperature increases. The important driving forces for the mechanism in general are: excess free energy in a grain boundary which makes the grain to minimize its local surface area, and the volume free energy difference between the neighboring grains on either side of a grain boundary [51]. Shirsath et al. [52] correlated the increasing particle size with increasing sintering

temperature to the coalescence that increases as sintering temperature increases. According to the phenomenological kinetic grain growth equation, the increase in sintering temperature increases the grain size [53]. This behavior of grain growth is in accord with the shrinkage increase, to explain that, shrinkage necessarily involves movement of the grain boundaries, and this movement must be in such a direction that the total area of boundaries decreases, so the shrinkage must thus be accompanied by grain growth [54]. In addition, the growth grain obeys the Arrhenius law [55]:

$$G_s = G_0 \exp[-A / (R \times T)] \qquad (9)$$

where G_0 is the grain size before sintering; G_s is the grain size after sintering; A is the activation energy of grain growth; R denotes the ideal gas constant; and T is the sintering temperature.

This Arrhenius model describes the interplay between densification and grain size; accordingly, with increasing sintering temperature at the same holding time, crystal boundaries migrate more quickly. Thus, the microstructure is denser and the average grain size increases with the increase of sintering temperature. The logarithms of average grain versus reciprocal of sintering temperature are plotted in Fig. 7 by linear fitting for all solid solutions. Therefore, the effective activation energies of grain growth can be obtained by calculating the slope of $\ln(G_s / G_0)$ versus $1000/T$ and the results are 135.64 kJ/mol, 122.46 kJ/mol, 99.78 kJ/mol, 93.25 kJ/mol and 63.21 kJ/mol for $x = 0.0$, 0.2, 0.5, 0.8 and 1.0, respectively. Compared to SBN, $Na_{0.5}Bi_{0.5}$ doping indeed reduces the activation energy as shown in Fig. 8. The more Sr is replaced by $Na_{0.5}Bi_{0.5}$, the lower the activation energy is and the easier the reaction of densification is. According to the sintering theory [56], the activation energy of sintering should be consistent with that of the diffusion of the rate-limiting species of sintering. In sintering polycrystalline ceramics, it is common that both grain boundary and lattice diffusion of the rate-limiting species can simultaneously contribute to densification [57]. In other words, generally the kinetics is limited either by the rate of chemical reaction or by the rate of diffusion phenomenon. In the former case, activation energy is very high. In our case, only the diffusion phenomenon is involved, which would explain why obtained activation energies for the grain growth of all solid solutions are relatively low (between 135 kJ/mol and 63 kJ/mol). The densification takes place through a faster diffusion process and results in an early lower

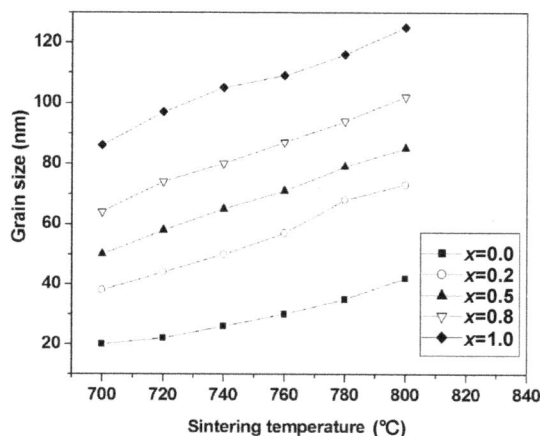

Fig. 6 Mean grain size as function of sintering temperature for $Sr_{1-x}(Na_{0.5}Bi_{0.5})_xBi_2Nb_2O_9$ solid solutions.

temperature sintering (800 ℃). This attitude is similar to that observed by Jardiel *et al.* [58], who indicated that activation energies of sintered $Bi_4Ti_3O_{12}$ prepared by coprecipitation method are lower than those energies calculated for compositions prepared by the solid state synthesis.

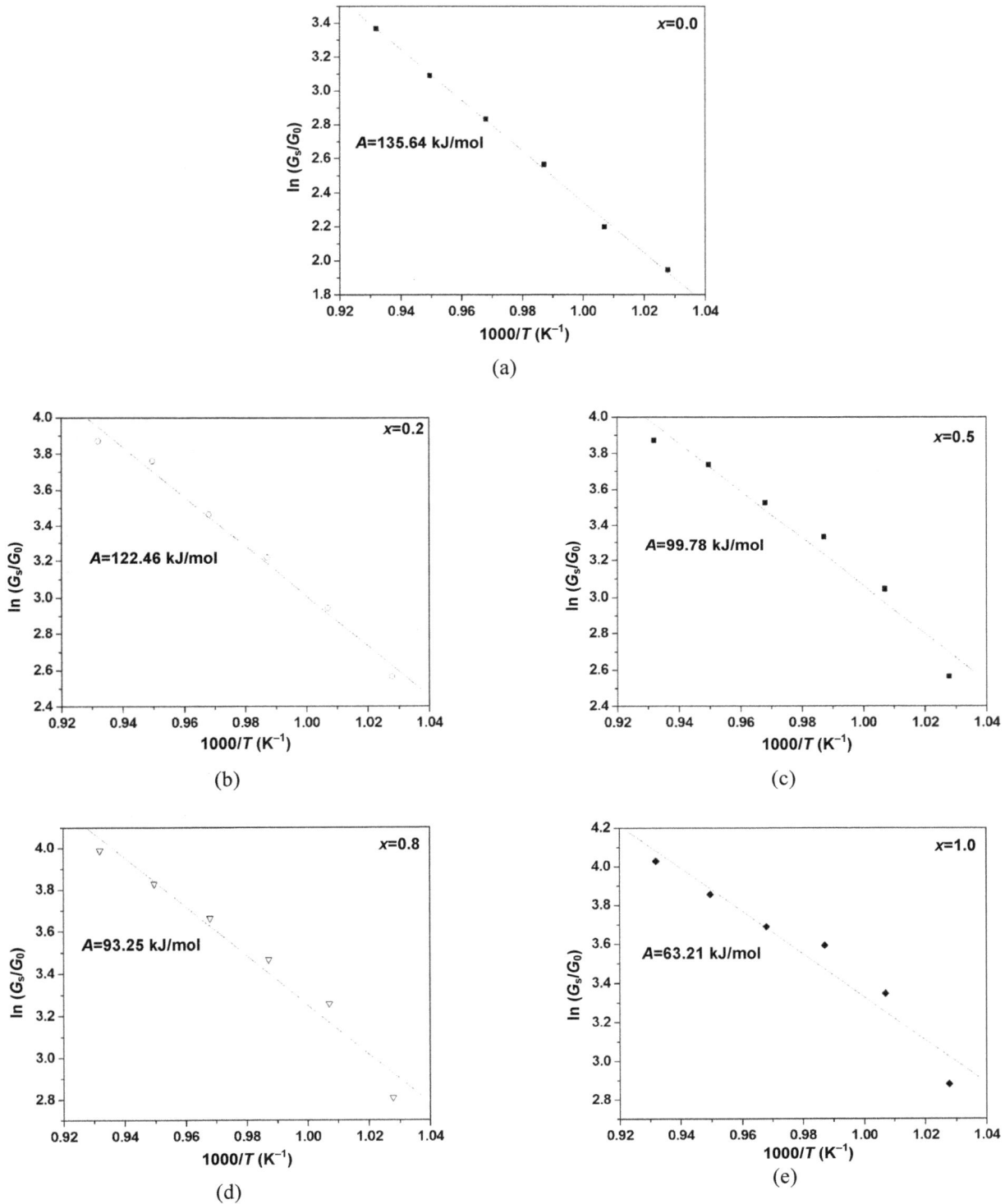

Fig. 7 Logarithm of average grain versus reciprocal of sintering temperature for $Sr_{1-x}(Na_{0.5}Bi_{0.5})_xBi_2Nb_2O_9$ solid solutions.

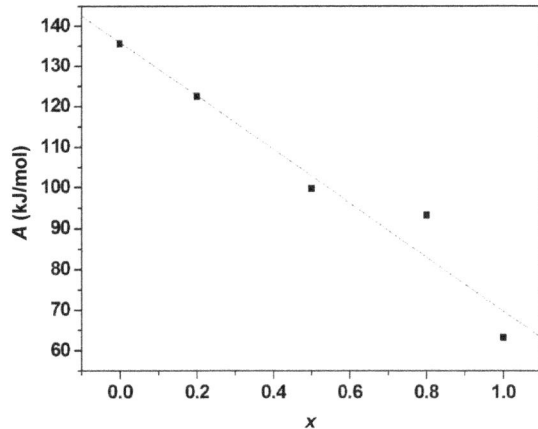

Fig. 8 Activation energy of grain growth as function of composition.

After 820 ℃, the sintering behavior begins to deteriorate due to exaggerated $Sr_{1-x}(Na_{0.5}Bi_{0.5})_xBi_2Nb_2O_9$ grain growth [59].

3. 2 Impedance spectroscopy

Figure 9 shows the sintering temperature dependence on dielectric constant (ε_r) at 1 kHz frequency for $Sr_{1-x}(Na_{0.5}Bi_{0.5})_xBi_2Nb_2O_9$ ceramics. As can be seen, the permittivity is affected significantly by sintering temperature and compositions. For each value of x, ε_r increases up to 800 ℃.

On one hand, the relationships between ε_r values and sintering temperatures exhibit the similar trend as that between sintered densities and sintering temperatures. This attitude can be explained by the increase in density and in grain size as the sintering

temperature is increased [60–62]. The raise in dielectric constant may be attributed to the mobile charge carriers related to oxygen vacancies [63]. In addition, agglomerations of powders become a resistant for ions to polarize between grains and grain boundaries [64]. According to our recent study, the dielectric constant of these compounds is only due to ionic polarization [37]. Theoretically, permittivity is related to ion polarizabilities by the Clausius–Mosotti equation [65]. It is believed that polarization of ions in grains may increase with grain necking growth.

On the other hand, the ε_r values increase with increasing the $Na_{0.5}Bi_{0.5}$ content. Maximum dielectric constant value of 384 is found for $Na_{0.5}Bi_{2.5}Nb_2O_9$ in ceramic sintered at 800 ℃. This behavior was explained in our recent work [37].

An increase in Curie temperature for $Sr_{1-x}(Na_{0.5}Bi_{0.5})_xBi_2Nb_2O_9$ ceramics with the increase of sintering temperature is observed (Fig. 10). Su et al. [66] determined that the behavior of Curie temperature can be related to microstructure, lattice defect, porosity, grain size, etc. Generally, the free energy of the ferroelectric phase increases and Curie temperature decreases with the increase of internal stress which can be relieved by pores [67]. Therefore, the ceramics with a lower density should have a higher T_C than the dense samples. In the dense ceramics with small amounts of pores, internal stress is mainly relieved by grain-boundary sliding [67]. In our case, the increase in Curie temperature with sintering suggests that the effect of internal stress relief by grain-boundary sliding

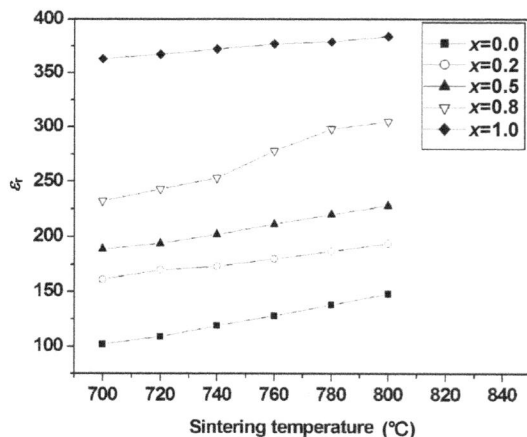

Fig. 9 Dielectric constant as function of sintering temperature for $Sr_{1-x}(Na_{0.5}Bi_{0.5})_xBi_2Nb_2O_9$ solid solutions.

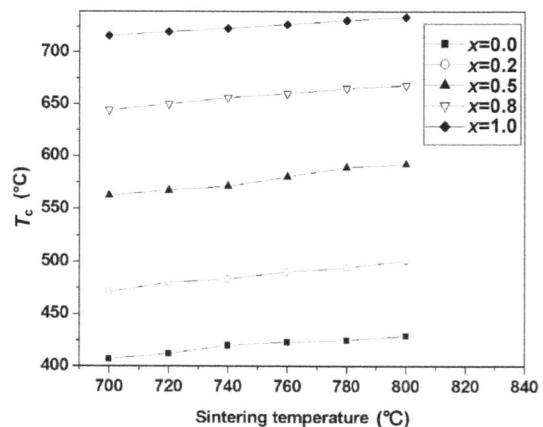

Fig. 10 Curie temperature as function of sintering temperature for $Sr_{1-x}(Na_{0.5}Bi_{0.5})_xBi_2Nb_2O_9$ solid solutions.

is more than the reduction of internal stress by pores. Mishra *et al.* [68] observed the similar behavior in BZT–BCT ceramics. The Curie temperature of the samples may be strongly affected by the grain size [69–71].

3. 3 Electrical studies

Firstly, it is worth knowing that a closer examination of the structure of SBN has revealed the existence of the vacancy; in order to compensate for the vacancy charge defects of opposite sign, oxygen vacancies are created to maintain the charge neutrality on the whole [72].

Electrical conductivities versus temperature at different sintering temperatures between 700 °C and 800 °C, for $Sr_{1-x}(Na_{0.5}Bi_{0.5})_xBi_2Nb_2O_9$ are plotted in

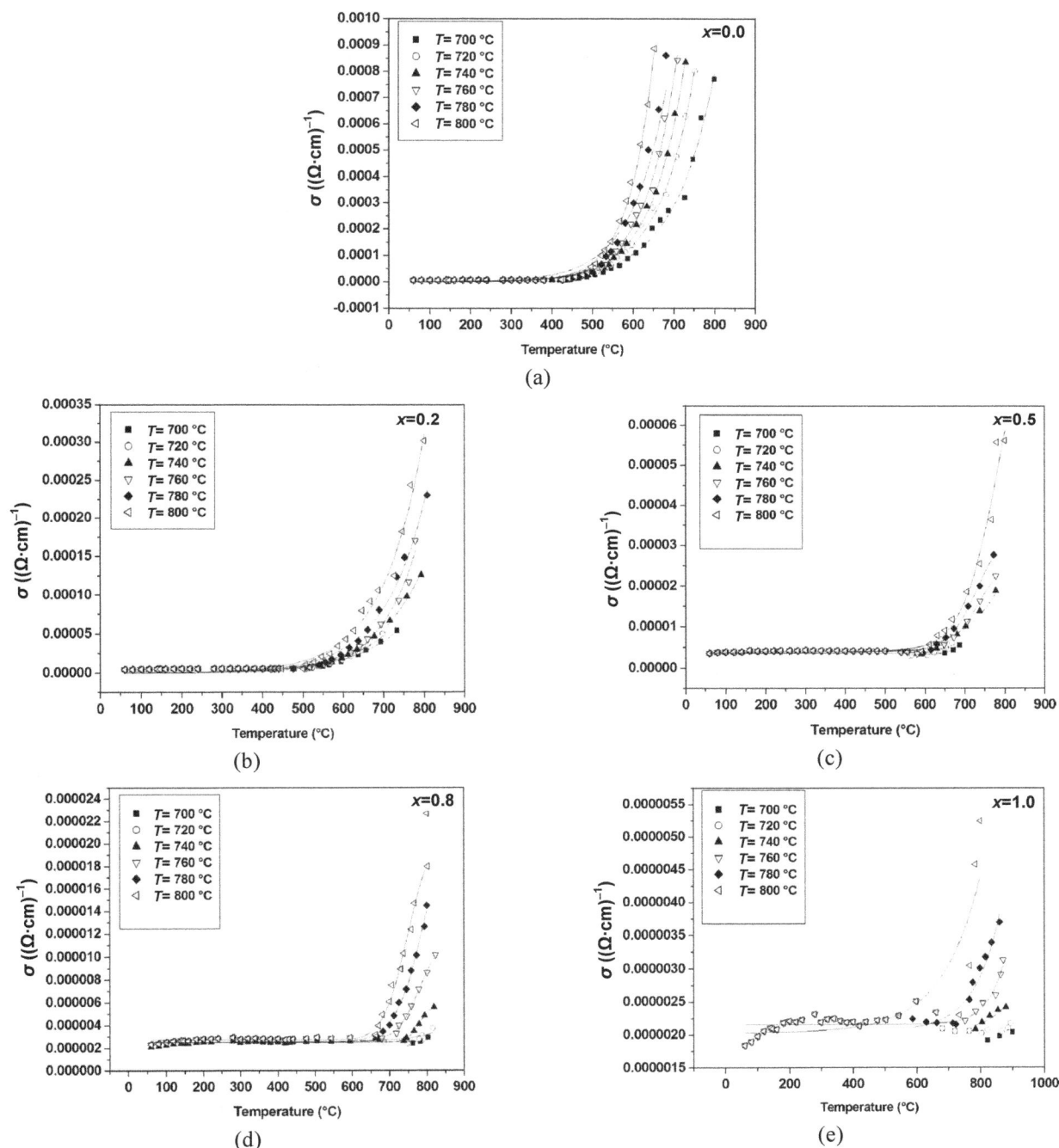

(a)

(b)   (c)

(d)   (e)

Fig. 11 Electrical conductivity as function of temperature at different sintering temperatures for $Sr_{1-x}(Na_{0.5}Bi_{0.5})_xBi_2Nb_2O_9$ solid solutions.

Fig. 11. The nature of curves shows that the conductivity increases with temperature. This suggests the presence of negative temperature coefficient of resistance (NTCR), a characteristic of dielectrics [73]. Activation energies (E_a) for conductivity are calculated from the Arrhenius relation for thermally activated conduction which is given as activated conduction (Eq. (8)). The linear fit is applied for the conductivity data using the least square fitting technique. It can be seen from Fig. 12 that for all

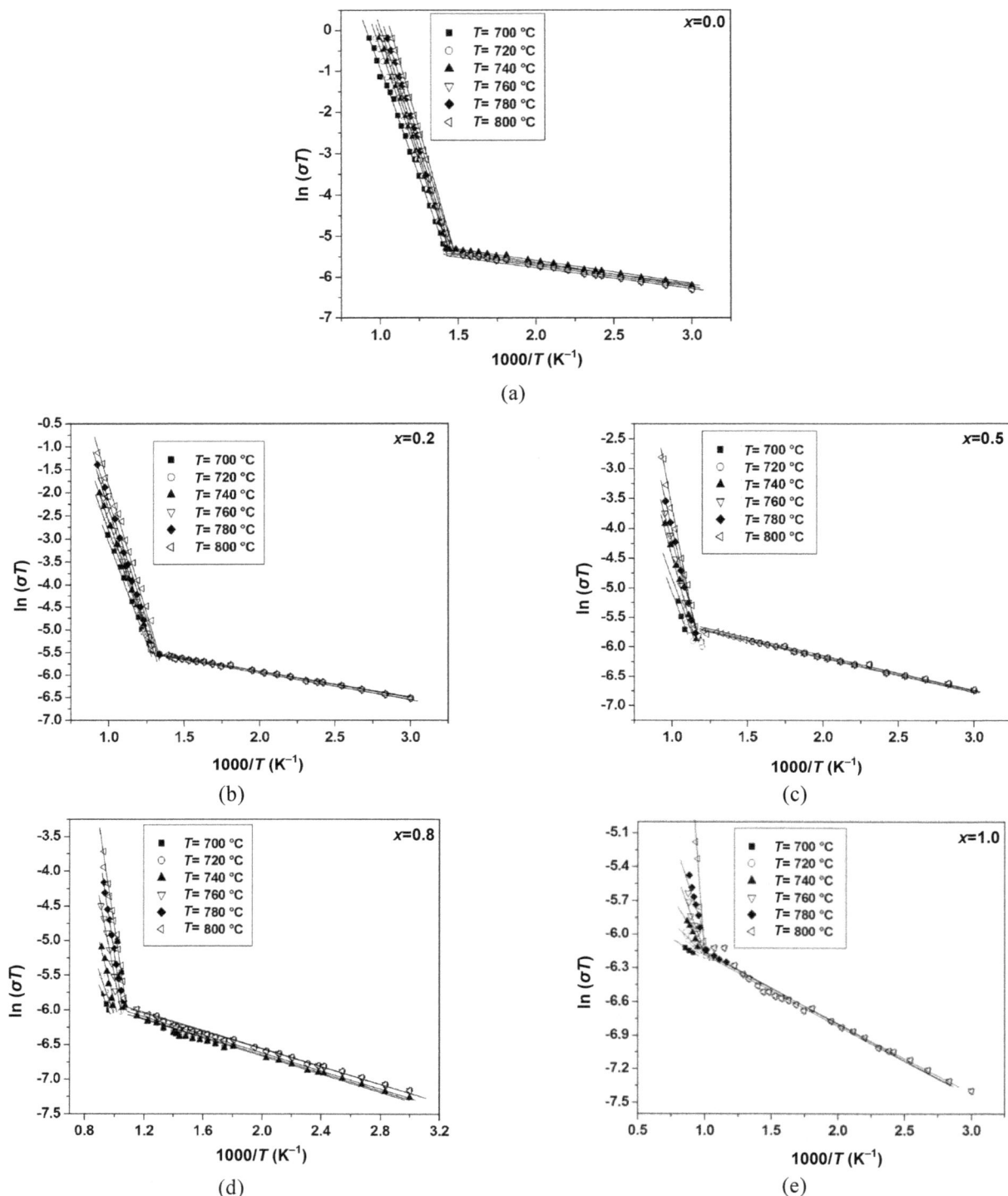

Fig. 12 Arrhenius plots of electrical conductivities as function of temperature for $Sr_{1-x}(Na_{0.5}Bi_{0.5})_xBi_2Nb_2O_9$ solid solutions.

samples the value of $\ln(\sigma T)$ increases almost linearly with increasing temperature up to a certain temperature at which a change of slope has occurred. This temperature corresponds to Curie point. It is observed that the transition temperatures are very close to those in impedance measurements. Such change in the slope is also due to the change of conduction mechanism from grain boundary to intragrain with the increase of temperature. There are two predominant conduction mechanisms. At low temperature region (ferroelectric phase), the conduction is dominant by extrinsic conduction, whereas the conduction at the high temperature (paraelectric phase) is dominated by the intrinsic conduction [74–76]. The activation energy is assumed to be the energy necessary to create and move vacancies [77,78]. At low temperatures, when the extrinsic conduction is predominant, the nominal activation energy equals diffusion activation energy (E_d). However, at high temperatures when intrinsic conduction predominates, the nominal activation is the sum of diffusion activation energy (E_d), and the formation energy of charge carrier (E_f) [10]:

$$E_a = E_d + E_f \tag{10}$$

The effect of sintering temperature in electrical conductivity can be also explained on the basis of microstructural changes, mainly coming from the enhancement of the densification [79].

It is also worth noting that, the activation energy of pure SBN of an earlier report by conventional sintered method [74,80] is higher than the current reported ceramic sintered at 800 ℃ after hydrothermal method.

4   Conclusions

The hydrothermal route has an interesting advantage for the synthesis of $Sr_{1-x}(Na_{0.5}Bi_{0.5})_xBi_2Nb_2O_9$ ($x = 0.0$, 0.2, 0.5, 0.8 and 1.0) powders consisting in low sintering temperature. An optimum lower sintering temperature equal to 800 ℃ is sufficient enough to give suitable properties. The effects of sintering at different temperatures on density, grain size and dielectric properties, for all solid solutions are studied. The results indicate that the values of sintered density, shrinkage and grain size are all in agreement. With the increase in sintering temperature, the density and the grain size increase. The activation energy for grain growth is determined. Values of ε_r increase with

the increase of sintering temperature. The Curie temperature is progressively higher with increasing grain size. The electrical data indicate that the conductivity with different sintering is essentially due to the oxygen vacancies. The low sintering temperature of $Sr_{1-x}(Na_{0.5}Bi_{0.5})_xBi_2Nb_2O_9$ ceramics seems to be attractive for industrial purpose. It will allow their application to multilayer devices and will play an important role in the fabrication process to meet the requirement of miniaturization and integration.

References

[1] Wang PE, Chaki TK. Sintering behaviour and mechanical properties of hydroxyapatite and dicalcium phosphate. *J Mater Sci: Mater M* 1993, **4**: 150–158.

[2] Gökçe A, Findik F. Mechanical and physical properties of sintered aluminum powders. *Journal of Achievements in Materials and Manufacturing Engineering* 2008, **30**: 157–164.

[3] Mohan CRK, Bajpai PK. Effect of sintering optimization on the electrical properties of bulk $Ba_xSr_{1-x}TiO_3$ ceramics. *Physica B* 2008, **403**: 2173–2188.

[4] Hallmann L, Ulmer P, Reusser E, *et al.* Effect of dopants and sintering temperature on microstructure and low temperature degradation of dental Y-TZP-zirconia. *J Eur Ceram Soc* 2012, **32**: 4091–4104.

[5] Shaw NJ. Densification and coarsening during solid state sintering of ceramics: A review of the models III. Coarsening. *Int J Powder Metall* 1989, **21**: 25–29.

[6] Coondoo I, Jha AK, Agarwal SK. Enhancement of dielectric characteristics in donor doped Aurivillius $SrBi_2Ta_2O_9$ ferroelectric ceramics. *J Eur Ceram Soc* 2007, **27**: 253–260.

[7] Kajewski D, Ujma Z, Szot K, *et al.* Dielectric properties and phase transition in $SrBi_2Nb_2O_9$–$SrBi_2Ta_2O_9$ solid solution. *Ceram Int* 2009, **35**: 2351–2355.

[8] Wu W, Liang S, Wang X, *et al.* Synthesis, structures and photocatalytic activities of microcrystalline

ABi$_2$Nb$_2$O$_9$ (A = Sr, Ba) powders. *J Solid State Chem* 2011, **184**: 81–88.

[9] Panda AB, Tarafdar A, Pramanik P. Synthesis, characterization and properties of nano-sized SrBi$_2$Ta$_2$O$_9$ ceramics prepared by chemical routes. *J Eur Ceram Soc* 2004, **24**: 3043–3048.

[10] Dhak D, Dhak P, Pramanik P. Influence of substitution on dielectric and impedance spectroscopy of Sr$_{1-x}$Bi$_{2+y}$Nb$_2$O$_9$ ferroelectric ceramics synthesized by chemical route. *Appl Surf Sci* 2008, **254**: 3078–3092.

[11] Prasanta D, Debasis D, Kausikisankar P, *et al.* Studies of structural and electrical properties of Ca$_{1-x}$Bi$_{2+y}$Nb$_2$O$_9$ [0.0 ⩽ x ⩽ 0.4; 0.000 ⩽ y ⩽ 0.266] ferroelectric ceramics prepared by organic precursor decomposition method. *Solid State Sci* 2008, **10**: 1936–1946.

[12] Júnior NLA, Simões AZ, Cavalheiro AA, *et al.* Structural and microstructural characterization of SrBi$_2$(Ta$_{0.5}$Nb$_{0.48}$W$_{0.02}$)$_2$O$_9$ powders. *J Alloys Compd* 2008, **454**: 61–65.

[13] Gaikwad SP, Dhage SR, Potdar HS, *et al.* Co-precipitation method for the preparation of nanocrystalline ferroelectric SrBi$_2$Nb$_2$O$_9$ ceramics. *J Electroceram* 2005, **14**: 83–87.

[14] Gaikwad SP, Potdar HS, Samuel V, *et al.* Co-precipitation method for the preparation of fine ferroelectric BaBi$_2$Nb$_2$O$_9$. *Ceram Int* 2005, **31**: 379–381.

[15] Radha R, Gupta UN, Samuel V, *et al.* A co-precipitation technique to prepare BiNbO$_4$ powders. *Ceram Int* 2008, **34**: 1565–1567.

[16] Kato K, Zheng C, Finder JM, *et al.* Sol–gel route to ferroelectric layer-structured perovskite SrBi$_2$Ta$_2$O$_9$ and SrBi$_2$Nb$_2$O$_9$ thin films. *J Am Ceram Soc* 1998, **81**: 1869–1875.

[17] Nelis D, Van Werde K, Mondelaers D, *et al.* Aqueous solution–gel synthesis of strontium bismuth niobate (SrBi$_2$Nb$_2$O$_9$). *J Sol–Gel Sci Technol* 2003, **26**: 1125–1129.

[18] Zanetti SM, Santiago EI, Bulhões LOS, *et al.* Preparation of characterization of nanosized SrBi$_2$Nb$_2$O$_9$ powder by the combustion synthesis. *Mater Lett* 2003, **57**: 2812–2816.

[19] Lu C-H, Chen Y-C. Sintering and decomposition of ferroelectric layered perovskites: Strontium bismuth tantalate ceramics. *J Eur Ceram Soc* 1999, **19**: 2909–2915.

[20] Chen D, Liu Y, Li Y, *et al.* Low-temperature sintering of M-type barium ferrite with BaCu(B$_2$O$_5$) additive. *J Magn Magn Mater* 2012, **324**: 449–452.

[21] Liu H, Li Q, Ma J, *et al.* Effects of Bi^{3+} content and

grain size on electrical properties of SrBi$_2$Ta$_2$O$_9$ ceramic. *Mater Lett* 2012, **76**: 21–24.

[22] Iqbal Y, Jamal A, Ullah R, *et al.* Effect of fluxing additive on sintering temperature, microstructure and properties of BaTiO$_3$. *Bull Mater Sci* 2012, **35**: 387–394.

[23] Zhang P, Hua Y, Xia W, *et al.* Effect of H$_3$BO$_3$ on the low temperature sintering and microwave dielectric properties of Li$_2$ZnTi$_3$O$_8$ ceramics. *J Alloys Compd* 2012, **534**: 9–12.

[24] Zhang G, Liu S, Yu Y, *et al.* Microstructure and electrical properties of (Pb$_{0.87}$Ba$_{0.1}$La$_{0.02}$)(Zr$_{0.68}$Sn$_{0.24}$Ti$_{0.08}$)O$_3$ anti-ferroelectric ceramics fabricated by the hot-press sintering method. *J Eur Ceram Soc* 2013, **33**: 113–121.

[25] Nygren M, Shen Z. On the preparation of bio-, nano- and structural ceramics and composites by spark plasma sintering. *Solid State Sci* 2003, **5**: 125–131.

[26] Li J-F, Wang K, Zhang B-P, *et al.* Ferroelectric and piezoelectric properties of fine-grained Na$_{0.5}$K$_{0.5}$NbO$_3$ lead-free piezoelectric ceramics prepared by spark plasma sintering. *J Am Ceram Soc* 2006, **89**: 706–709.

[27] Polotai A, Breece K, Dickey E, *et al.* A novel approach to sintering nanocrystalline barium titanate ceramics. *J Am Ceram Soc* 2005, **88**: 3008–3012.

[28] Wang X-H, Deng X-Y, Bai H-L, *et al.* Two-step sintering of ceramics with constant grain-size, II: BaTiO$_3$ and Ni–Cu–Zn ferrite. *J Am Ceram Soc* 2006, **89**: 438–443.

[29] Fang J, Wang X, Tian Z, *et al.* Two-step sintering: An approach to broaden the sintering temperature range of alkaline niobate-based lead-free piezoceramics. *J Am Ceram Soc* 2010, **93**: 3552–3555.

[30] Senthil V, Badapanda T, Bose AC, *et al.* Impedance and electrical modulus study of microwave-sintered SrBi$_2$Ta$_2$O$_9$ ceramic. *ISRN Ceramics* 2012, **2012**: 943734.

[31] Lu SW, Lee BI, Wang ZL, *et al.* Hydrothermal synthesis and structural characterization of BaTiO$_3$ nanocrystals. *J Cryst Growth* 2000, **219**: 269–276.

[32] Dias A, Buono VTL, Ciminelli VST, *et al.* Hydrothermal synthesis and sintering of electroceramics. *J Eur Ceram Soc* 1999, **19**: 1027–1030.

[33] Li L, Gong Y-Q, Gong L-J, *et al.* Low-temperature hydro/solvothermal synthesis of Ta-modified K$_{0.5}$Na$_{0.5}$NbO$_3$ powders and piezoelectric properties of corresponding ceramics. *Mater Design* 2012, **33**: 362–366.

[34] Hotta Y, Duran C, Sato K, *et al.* Densification and grain growth in BaTiO$_3$ ceramics fabricated from

nanopowders synthesized by ball-milling assisted hydrothermal reaction. *J Eur Ceram Soc* 2008, **28**: 599–604.

[35] Venkataraman BH, Varma KBR. Impedance and dielectric studies of ferroelectric $SrBi_2Nb_2O_9$ ceramics. *J Phys Chem Solids* 2003, **64**: 2105–2112.

[36] Pookmanee P, Rujijanagul G, Ananta S, et al. Effect of sintering temperature on microstructure of hydrothermally prepared bismuth sodium titanate ceramics. *J Eur Ceram Soc* 2004, **24**: 517–520.

[37] Naceur H, Megriche A, Maaoui ME. Structural distortion and dielectric properties of $Sr_{1-x}(Na_{0.5}Bi_{0.5})_xBi_2Nb_2O_9$ ($x = 0.0$, 0.2, 0.5, 0.8 and 1.0). *J Alloys Compd* 2013, **46**: 145–150.

[38] Aoyagi R, Takeda H, Okamura S, et al. Synthesis and electrical properties of sodium bismuth niobate $Na_{0.5}Bi_{2.5}Nb_2O_9$. *Mater Res Bull* 2003, **38**: 25–32.

[39] Einstein A. Eine neue Bestimmung der Moleküldimensionen. *Annalen der Physik* 1906, **324**: 289–306.

[40] Garcia DE, Klein AN, Hotza D. Advanced ceramics with dense and fine-grained microstructures through fast firing. *Rev Adv Mater Sci* 2012, **30**: 273–281.

[41] Wang Q, Wang Q, Zhang X, et al. The effect of sintering temperature on the structure and degradability of strontium-doped calcium polyphosphate bioceramic. *Ceram-Silikaty* 2010, **54**: 97–102.

[42] Krishnan K, Sahay SS, Singh S, et al. Modeling the accelerated cyclic annealing kinetics. *J Appl Phys* 2006, **100**: 093505.

[43] Hungría T, Galy J, Castro A. Spark plasma sintering as a useful technique to the nanostructuration of piezo-ferroelectric materials. *Adv Eng Mater* 2009, **11**: 615–631.

[44] Hu M, Luo C, Tian H, et al. Phase evolution, crystal structure and dielectric behavior of $(1-x)Nd(Zn_{0.5}Ti_{0.5})O_3+xBi(Zn_{0.5}Ti_{0.5})O_3$ compound ceramics. *J Alloys Compd* 2011, **509**: 2993–2999.

[45] Skidmore TA, Milne SJ. Phase development during mixed-oxide processing of a $[Na_{0.5}K_{0.5}NbO_3]_{1-x}-[LiTaO_3]_x$ powder. *J Mater Res* 2007, **22**: 2265–2272.

[46] Rivas-Vázquez LP, Rendón-Angeles JC, Rodríguez-Galicia JL, et al. Preparation of calcium doped $LaCrO_3$ fine powders by hydrothermal method and its sintering. *J Eur Ceram Soc* 2006, **26**: 81–88.

[47] Mandoki NT, Courtois C, Champagne P, et al. Hydrothermal synthesis of doped PZT powders: Sintering and ceramic properties. *Mater Lett* 2004, **58**: 2489–2493.

[48] Yahya N, Masoud RAH, Daud H, et al. Synthesis of $Al_3Fe_5O_{12}$ cubic structure by extremely low sintering temperature of sol gel technique. *Am J Engg & Applied Sci* 2009, **2**: 76–79.

[49] Cahn JW. The impurity-drag effect in grain boundary motion. *Acta Metall* 1962, **10**: 789–798.

[50] Shannon RD. Revised effective ionic radii and systematic studies of interatomic distances in halides and chalcogenides. *Acta Cryst* 1976, **A32**: 751–767.

[51] Floro JA, Thompson CV, Carel R, et al. Competition between strain and interface energy during epitaxial grain growth in Ag films on Ni(001). *J Mater Res* 1994, **9**: 2411–2424.

[52] Shirsath SE, Kadam RH, Gaikwad AS, et al. Effect of sintering temperature and the particle size on the structural and magnetic properties of nanocrystalline $Li_{0.5}Fe_{2.5}O_4$. *J Magn Magn Mater* 2011, **323**: 3104–3108.

[53] Cortés J, Valencia E. Phenomenological equations of the kinetics of heterogeneous adsorption with interaction between adsorbed molecules. *Phys Rev B* 1995, **51**: 2621–2623.

[54] Budworth DW. The selection of grain-growth control additives for the sintering of ceramic. *Mineral Mag* 1970, **37**: 833–838.

[55] Coble RL. Sintering crystalline solids. I. Intermediate and final state diffusion models. *J Appl Phys* 1961, **32**: 787–792.

[56] Dosdale T, Brook RJ. Comparison of diffusion data and of activation energies. *J Am Ceram Soc* 1983, **66**: 392–395.

[57] Fang T-T, Shiue J-T, Shiau F-S. On the evaluation of the activation energy of sintering. *Mater Chem Phys* 2003, **80**: 108–113.

[58] Jardiel T, Caballero AC, Villegas M. Sintering kinetic of $Bi_4Ti_3O_{12}$ based ceramics. *Bol Soc Esp Ceram* 2006, **45**: 202–206.

[59] Kan Y, Wang P, Li Y, et al. Low-temperature sintering of $Bi_4Ti_3O_{12}$ derived from co-precipitation method. *Mater Lett* 2002, **56**: 910–914.

[60] Wu A, Vilarinho PM, Salvado IMM, et al. Sol–gel preparation of lead zirconate titanate powders and ceramics: Effect of alkoxide stabilizers and lead precursors. *J Am Ceram Soc* 2000, **83**: 1379–1385.

[61] Shaikh PA, Kolekar YD. Study of microstructural, electrical and dielectric properties of perovskite (0.7) PMN–(0.3) PT ferroelectric at different sintering temperatures. *J Anal Appl Pyrol* 2012, **93**: 41–46.

[62] Prasad KVR, Raju AR, Varma KBR. Grain size effects on the dielectric properties of ferroelectric $Bi_2VO_{5.5}$ ceramics. *J Mater Sci* 1994, **29**: 2691–2696.

[63] Cheng ZX, Wang XL, Dou SX, et al. Ferroelectric properties of $Bi_{3.25}Sm_{0.75}V_{0.02}T_{2.98}O_{12}$ thin film at

elevated temperature. *Appl Phys Lett* 2007, **90**: 222902.

[64] Yan LC, Hassan J, Hashim M, *et al.* Effect of sintering temperatures on the microstructure and dielectric properties of $SrTiO_3$. *World Appl Sci J* 2011, **15**: 1614–1618.

[65] Shannon RD. Dielectric polarizabilities of ions in oxides and fluorides. *J Appl Phys* 1993, **73**: 348–366.

[66] Su H, Zhang H, Tang X, *et al.* High-permeability and high-Curie temperature NiCuZn ferrite. *J Magn Magn Mater* 2004, **283**: 157–163.

[67] Chen XM, Ma HY, Ding W, *et al.* Microstructure, dielectric, and piezoelectric properties of $Pb_{0.92}Ba_{0.08}Nb_2O_6$–0.25 wt% TiO_2 ceramics: Effect of sintering temperature. *J Am Ceram Soc* 2011, **94**: 3364–3372.

[68] Mishra P, Sonia, Kumar P. Effect of sintering temperature on dielectric, piezoelectric and ferroelectric properties of BZT–BCT 50/50 ceramics. *J Alloys Compd* 2012, **545**: 210–215.

[69] Bhuiyan MA, Hoque SM, Choudhury S. Effect of sintering temperature on microstructure and magnetic properties of $NiFe_2O_4$ prepared from nano size powder of NiO and Fe_2O_3. *Journal of Bangladesh Academy of Sciences* 2010, **34**: 189–195.

[70] Zhang Q, Zhang Y, Wang X, *et al.* Influence of sintering temperature on energy storage properties of $BaTiO_3$–$(Sr_{1-1.5x}Bi_x)TiO_3$ ceramics. *Ceram Int* 2012, **38**: 4765–4770.

[71] Wada N, Tanaka H, Hamaji Y, *et al.* Microstructures and dielectric properties of fine-grained $BaTiO_3$ ceramics. *Jpn J Appl Phys* 1996, **35**: 5141–5144.

[72] Chen T-C, Thio C-L, Desu SB. Impedance spectroscopy of $SrBi_2Ta_2O_9$ and $SrBi_2Nb_2O_9$

ceramics correlation with fatigue behavior. *J Mater Res* 1997, **12**: 2628–2637.

[73] Shrivastava V, Jha AK, Mendiratta RG. Structural and electrical studies in La-substituted $SrBi_2Nb_2O_9$ ferroelectric ceramics. *Physica B* 2006, **371**: 337–342.

[74] Forbess MJ, Seraji S, Wu Y, *et al.* Dielectric properties of layered perovskite $Sr_xA_{1-x}Bi_2Nb_2O_9$ ferroelectrics (with A = La, Ca and x = 0, 0.1). *Appl Phys Lett* 2000, **76**: 2934–2936.

[75] He LX, Yoo HI. Effects of B-site ion (M) substitution on the ionic conductivity of $(Li_{3x}La_{2/3-x})_{1+y/2}(M_yTi_{1-y})O_3$ (M = Al, Cr). *Electrochim Acta* 2003, **48**: 1357–1366.

[76] Kuang X, Allix MMB, Claridge JB, *et al.* Crystal structure, microwave dielectric properties and AC conductivity of B-cation deficient hexagonal perovskites $La_5M_xTi_{4-x}O_{15}$ (x = 0.5, 1; M = Zn, Mg, Ga, Al). *J Mater Chem* 2006, **16**: 1038–1045.

[77] Kröger FA. The chemistry of imperfect crystals. *J Appl Cryst* 1975, **8**: 497–498.

[78] Parkash OM, Mandal KD, Christopher CC, *et al.* Electrical behaviour of lanthanum- and cobalt-doped strontium stannate. *Bull Mater Sci* 1994, **17**: 253–257.

[79] Kant R, Singh K, Pandey OP. Microstructural and electrical behavior of $Bi_4V_{2-x}Cu_xO_{11-\delta}$ ($0 \leqslant x \leqslant 0.4$). *Ceram Int* 2009, **35**: 221–227.

[80] Wang XP, Fang QF. Mechanical and dielectric relaxation studies on the mechanism of oxygen ion diffusion in $La_2Mo_2O_9$. *Phys Rev B* 2002, **65**: 064304.

20

Tailoring of electrical properties of BiFeO$_3$ by praseodymium

Samita PATTANAYAK, Ashwasa PRIYADARSHAN, Ritesh SUBUDHI,
Ranjan Kumar NAYAK, Rajib PADHEE*

Department of Physics, Institute of Technical Education & Research, Siksha 'O' Anusandhan University, Bhubaneswar 751030, India

Abstract: The polycrystalline samples of $Bi_{1-x}Pr_xFeO_3$ ($x = 0$ and 0.1) were prepared by a solid-state reaction technique. Preliminary X-ray structural analysis has confirmed the formation of a single-phase compound. Studies of dielectric and impedance spectroscopy of the materials, carried out in wide frequency (1 kHz–1 MHz) and temperature (25–400 ℃) ranges, have provided many interesting results including significant decrement in tangent loss, structural stability, obeying Jonscher's universal power law, etc.

Keywords: ceramics; impedance spectroscopy; electrical conductivity

1 Introduction

Since the discovery of magnetism in some elements and alloys and ferroelectricity in organic/inorganic compounds, extensive works were carried out with many breakthroughs. Multiferroics is defined as materials which combine at least two "ferroic" properties, such as ferromagnetic order (spontaneous magnetic polarization that can be reversed by a magnetic field), ferroelectricity (e.g., spontaneous electric polarization that can be switched by an applied electric field), and ferroelasticity (change in electric polarization accompanied by a change in shape) [1,2]. Multiferroic compounds are promising candidates for designing emerging electronic devices like multiple-state memories, magnetic data-storage media, actuators, transducers, sensors, and spintronic devices for different technological applications [3,4]. Bismuth ferrite, BiFeO$_3$ (BFO) is the most widely studied compound of the family, since it exhibits multiferroicity at room temperature, with a coexistence of ferroelectricity and anti-ferromagnetism up to its Neel temperature of $T_N \approx 643$ K [5]. At room temperature, BFO is a rhombohedrally distorted ferroelectric perovskite with crystallographic space group $R3c$ [6]. Its ferroelectric Curie temperature (T_C) is 1100 K. Unfortunately, BFO has high leakage current, low dielectric constant and high tangent loss [7] that limit the material to be used for fabrication of devices. Above problems can be reduced or eliminated by substituting suitable rare-earth elements or other element(s) at the Bi-site of BFO [8,9]. Even though lots of experimental works have been done on rare-earth ions including praseodymium (Pr)-substituted BFO [10], not much work on resistance and/or impedance properties of the Pr-modified BFO has been done so far. Therefore, in the present work, we report the detailed electrical characterizations of $Bi_{1-x}Pr_xFeO_3$ ($x = 0$ and 0.1, BFO and BPFO respectively).

---

* Corresponding author.
E-mail: padhee4u@gmail.com

2　Experiment

The polycrystalline samples of $Bi_{1-x}Pr_xFeO_3$ ($x = 0$ and 0.1) were synthesized by using a high-temperature solid-state reaction technique with high-purity (analytical reagent grade) ingredients: Bi_2O_3, Pr_2O_3 and Fe_2O_3 (99.9%, all from M/s LOBA Chemie. Pvt. Ltd., India). These oxides were mechanically mixed in dry (air) and wet (methanol) medium for several hours in an agate mortar. The mixed materials were calcined in the temperature range of 780–810 ℃ for 4 h in an alumina crucible. X-ray diffraction (XRD) patterns of the calcined powders were recorded at room temperature using X-ray powder diffractometer (XRD 6000). The Cu Kα radiation ($\lambda = 1.5405$ Å) was used to record the pattern and data in a wide range of Bragg angle θ ($20° \leqslant 2\theta \leqslant 80°$) at a scanning rate of 3 (°)/min. Using PVA (binder)-mixed calcined powders, the cylindrical pellets (10 mm in diameter and 1–2 mm in thickness) were fabricated under a uni-axial pressure of 4×10^6 N/m². The pellets were sintered at optimized temperature in the range of 820–850 ℃ in air atmosphere for 4 h to get mechanically stable, strong and high-density (>95% of theoretical density) samples. The polished sintered pellets were silvered with high-purity and high-quality silver paste, and dried at 160 ℃ for 8 h before taking dielectric and electrical measurements. The capacitance, dissipative factor and impedance parameters were measured as a function of temperature (25–400 ℃) in a wide range of frequency (1 kHz–1 MHz), using a computer-controlled LCR meter (PSM-4NL Model: 1735, UK) along with a laboratory-designed and fabricated sample holder and furnace. To record the temperature at small interval, a chromel–alumel thermo-couple and KUSAM MECO 108 digital milli-voltmeter were used.

3　Results and discussion

3.1　Structural analysis

The XRD patterns of $Bi_{1-x}Pr_xFeO_3$ ($x = 0$ and 0.1), recorded at room temperature on their powder samples, are compared in Fig. 1. The XRD patterns consist of many sharp and single peaks, which are different from those of the ingredients. Each XRD pattern clearly exhibits better homogeneity and crystallization of the sample, and thus confirms the formation of new

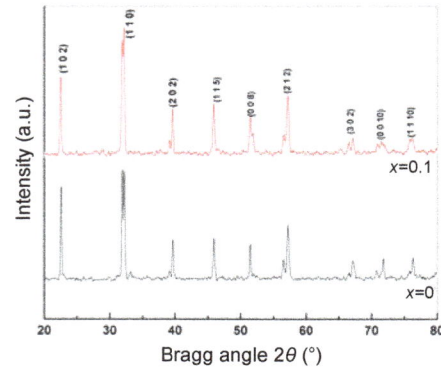

Fig. 1　XRD patterns of $Bi_{1-x}Pr_xFeO_3$ ($x = 0$ and 0.1).

compound in a single phase [10]. The reflection peaks are indexed in different unit cell configurations using standard computer software POWDMULT [11], and the lattice parameters of the BPFO ceramics as a function of Pr content are analyzed. From the best agreement between the observed (d_{obs}) and calculated (d_{cal}) inter-planner distances d ($\Sigma(d_{obs} - d_{cal}) =$ minimum), BPFO is found to be in the rhombohedral crystal system. The cell parameters have been compared in Table 1.

Table 1　Comparison of the cell dimensions of BFO and BPFO

	a (Å)	c (Å)	c/a	V (Å³)
BFO	5.5706(7)	13.8114(7)	2.4793	372.96
BPFO	5.575(8)	13.585(8)	2.4368	365.69

3.2　Dielectric study

The variation of relative dielectric constant (ε_r) and tangent loss (tanδ) as a function of temperature at some selected frequencies (10 kHz, 100 kHz and 1 MHz) is shown in Figs. 2(a) and 2(b). The magnitude of ε_r increases with rise in temperature. The increase in the value of ε_r can be ascribed as the electron–phonon interaction [12]. The increase in ε_r can be understood in terms of the thermally activated transport property and presence of space charges [13]. The dielectric anomaly (related to relaxation) is observed at 270 ℃ and it shifts towards higher temperature with increasing Pr substitution. This relaxation may be due to the anti-ferromagnetic transition [14]. The magnitude of ε_r in BPFO increases on increasing Pr content which is shown in Fig. 2(c) (at 100 kHz). This suggests that Pr-substitution has a significant effect on dielectric properties of BPFO. Therefore, substitution of Pr in BFO has enhanced the dielectric properties. The

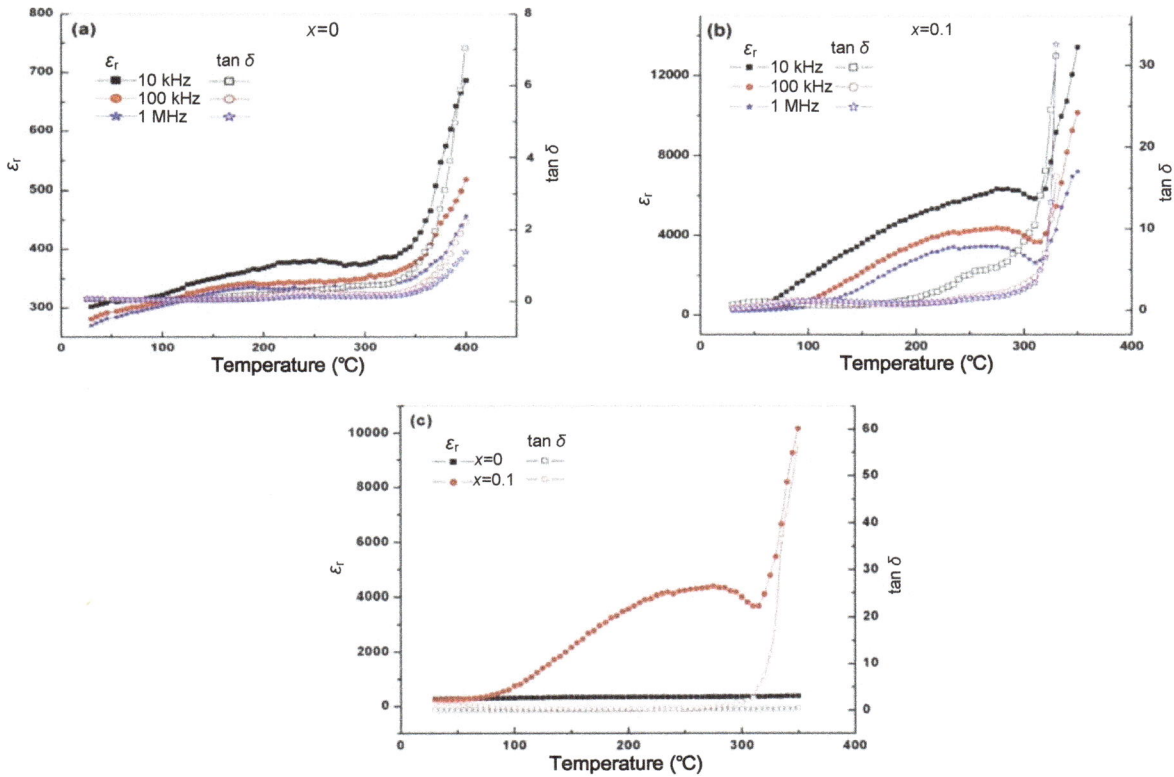

Fig. 2 Variation of ε_r and $\tan\delta$ of $Bi_{1-x}Pr_xFeO_3$ ($x = 0$ and 0.1) with temperature at different frequencies (10 kHz, 100 kHz and 1 MHz).

temperature dependence of $\tan\delta$ follows nearly the same trend as ε_r. However, at higher temperatures, $\tan\delta$ value increases very rapidly. This may be the result of (i) scattering of thermally activated charge carriers due to which the charges are deposited near the grain boundary region, that increases ε_r, and (ii) some inherent defects in the samples.

3.3 Impedance analysis

Complex impedance spectroscopy (CIS) is a non-destructive, unique and powerful technique to study electrical properties of polycrystalline and ionic materials over a wide range of frequency and temperature. This technique is very useful in separating the contributions of (i) bulk, (ii) grain boundary and (iii) electrode polarization in complex impedance and other related electrical parameters with different equivalent circuits.

For this technique, an AC signal is applied across the pellet sample, and the output response is measured. The impedance parameters of the material give us data having both real (resistive) and imaginary (reactive) components. The basic equations/formalism of impedance and electrical modulus has been used to

study impedance spectroscopic parameters of the samples [15].

Figure 3 shows the variation of real part (Z') and imaginary part (Z'') of impedance with frequency at different temperatures for different concentrations of Pr in $Bi_{1-x}Pr_xFeO_3$. The decrease in the magnitude of Z' with rise in both temperature and frequency indicates the increase in conductivity of the materials. At higher frequencies, the values of Z' coincide with each other, which is due to the possible release of space charges [16]. The frequency at which the magnitude of Z'' attains a maximum value, is known as relaxation frequency (f_r). The maximum value of Z'' decreases with rise in temperature and concentration of Pr, which indicates relaxation in the samples [17]. The broadening of peak with increase of temperature suggests the presence of temperature dependence of relaxation phenomenon in the materials. The relaxation process occurs due to the presence of immobile charges at lower temperatures and defects or vacancies at higher temperatures [18,19]. Figure 3(c) shows the effect of Pr concentration on real and imaginary parts of impedance at 300 ℃. It is observed that on increasing Pr content in BPFO, these real impedance

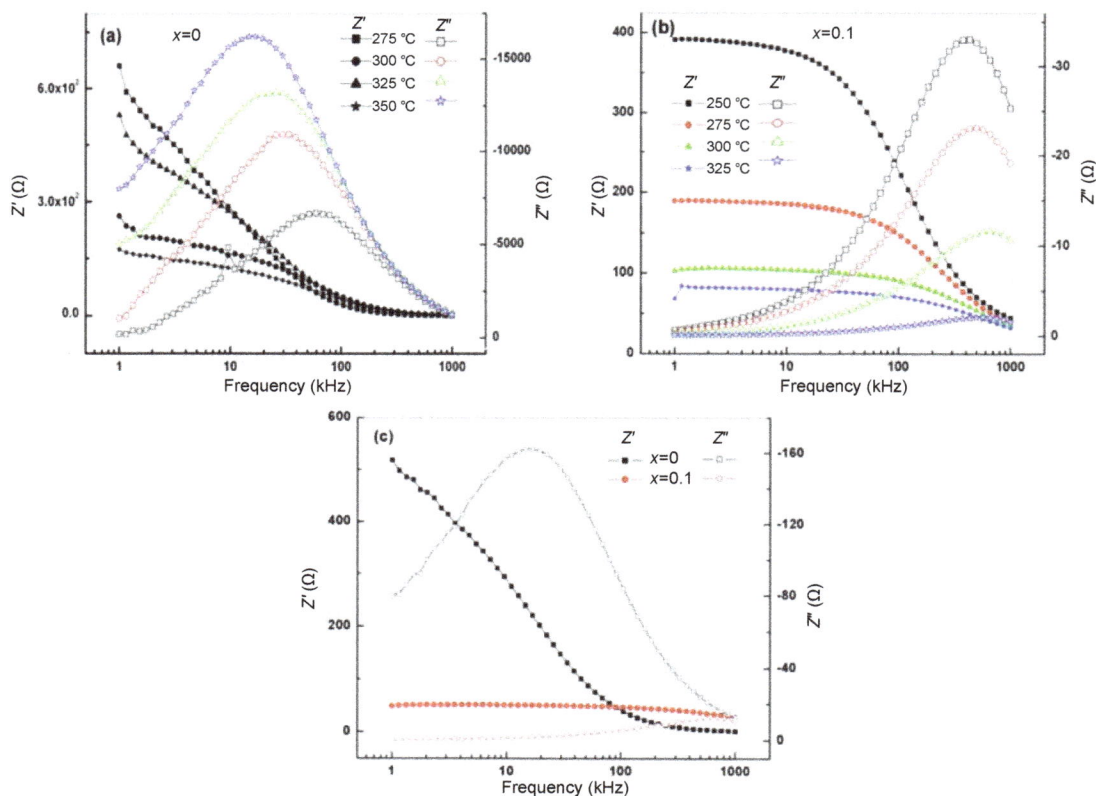

Fig. 3   Variation of Z' and Z'' of $Bi_{1-x}Pr_xFeO_3$ ($x = 0$ and 0.1) with frequency at different temperatures.

parameters significantly decrease, and the imaginary impedance component's peak shifts towards higher-frequency side.

Figure 4 shows the temperature dependence of complex impedance spectra (usually referred as Nyquist plot [20]) over a wide range of frequency (1 kHz–1 MHz) for $Bi_{1-x}Pr_xFeO_3$ ($x = 0$ and 0.1). At low temperatures, complex impedance plots are single semicircular arc (which is not shown in figure), whereas at higher temperatures (250–325 ℃) two semicircular arcs with center below the real axis are observed. The presence of a second semi-circle is due to grain boundary effect in the materials [21]. The plots of Fig. 4 show (real or practical) Debye-like response. An equivalent circuit consists of parallel combination of CQR and CR, where Q is called constant phase element (CPE). The presence of non-semicircle indicates the existence of non-Debye type of relaxation in the sample [22,23]. For Debye-type relaxation, a semicircle with its center at Z' axis is observed. As the value of Pr content increases, the bulk resistance (R_b) decreases (Fig. 4(c) at 300 ℃), which again indicates the increase in conductivity with the increase in Pr content. However, in the studied materials, we did not

get ideal Debye-type relaxation. Thus, depressed small semi-circles indicate non-Debye type of relaxation.

3.4   AC conductivity

Figure 5 shows the AC conductivity plots as a function of frequency at different temperatures. The values of σ_{AC} are calculated from dielectric data using an empirical relation: $\sigma_{AC} = \varepsilon_0 \varepsilon_r \omega \tan\delta$ [24], where ε_0, ω and $\tan\delta$ are dielectric permittivity in vacuum, angular frequency and tangent loss, respectively. In the low-frequency region, σ_{AC} remains almost constant (i.e., frequency-independent plateau), whereas the dispersion of conductivity is observed in the higher-frequency region. At higher temperatures, the conductivity curves show frequency-independent plateau in the low-frequency region, whereas at higher frequencies, $\sigma_{AC} \propto \omega^n$. As the Pr-substitution and temperature increase, all the curves tend to flatten at 300 ℃ in Fig. 5(c). The flattening is much clearer in low-frequency region, suggesting domination of DC conduction behavior. In the case of Debye model, the value of n is unity [25]. In another case, the motion of charge carriers is translational because of small value

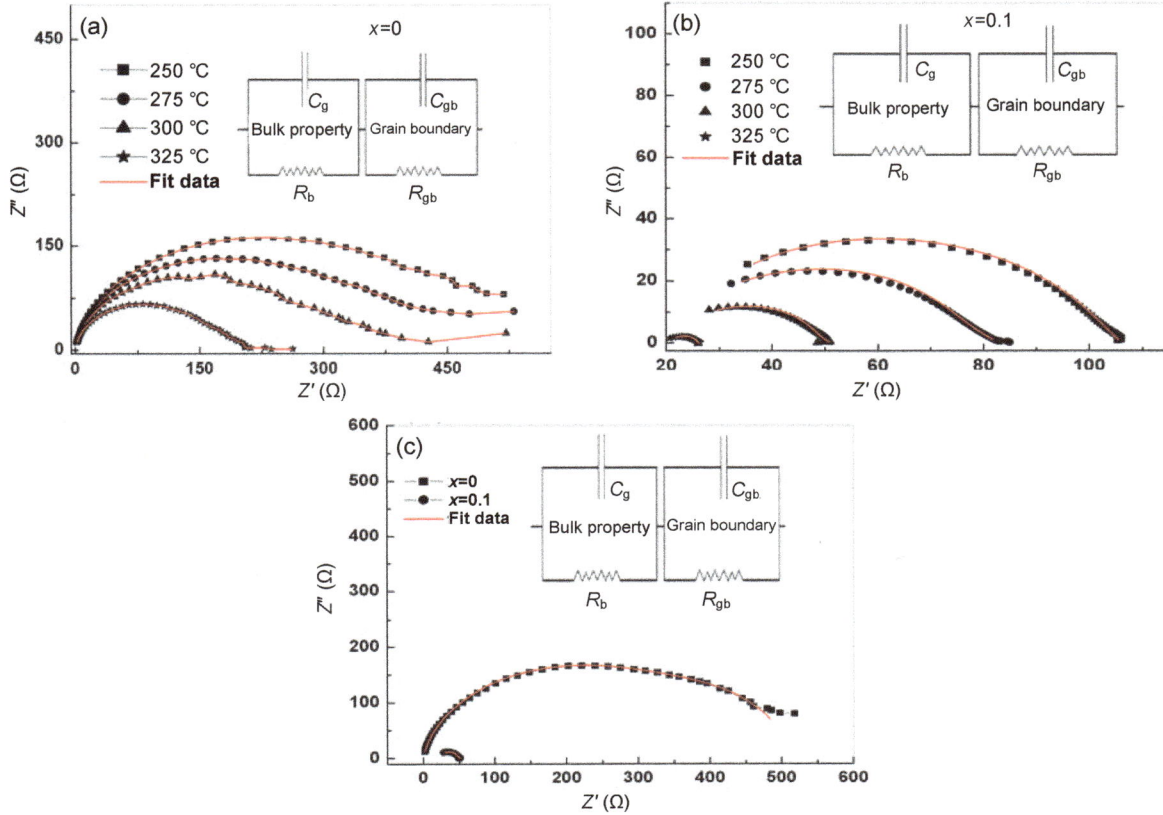

Fig. 4 Variation of Z'' with Z' of $Bi_{1-x}Pr_xFeO_3$ ($x = 0$ and 0.1) at different temperatures.

of n (<1). The value of $n<1$ signifies that the hopping motion involves a translational motion with a sudden hopping, whereas $n>1$ means that the motion involves localized hopping without the species leaving the neighborhood [13].

3.5 DC conductivity

Figure 6 shows the variation of DC conductivity (σ_{DC}) with inverse absolute temperature. It is calculated with the help of a mathematical relation: $\sigma_{DC} = t/(A \cdot R_b)$, where t is the thickness of the sample, A is the area of cross section and R_b is the bulk resistance. It is observed that DC conductivity increases with increase in temperature, which further supports the negative temperature coefficient of resistance (NTCR) behavior of the samples. The nature of the plots follows the Arrhenius relation: $\sigma_{DC} = \sigma_0 \exp[-E_a/(K \cdot T)]$ [25]. The slope of the graph gives the activation energy of the material. The calculated values of activation energy are found to be 0.78 eV and 0.69 eV for $x = 0$ and 0.1 of $Bi_{1-x}Pr_xFeO_3$, respectively. The conductivity increases considerably with increasing the Pr substitution.

4 Conclusions

The polycrystalline samples of $Bi_{1-x}Pr_xFeO_3$ ($x = 0$ and 0.1) were prepared by a standard mixed oxide method. Preliminary structural studies suggest that $Bi_{1-x}Pr_xFeO_3$ has rhombohedral structure. The dielectric constant is found to be increasing and the tangent loss decreasing at room temperature. The impedance studies exhibit the presence of grain (bulk) and grain boundary effects, and existence of NTCR behavior in the materials. The frequency-dependent electrical conductivity obeys Jonscher's power law, and is in good agreement with real parts of impedance versus frequency plots.

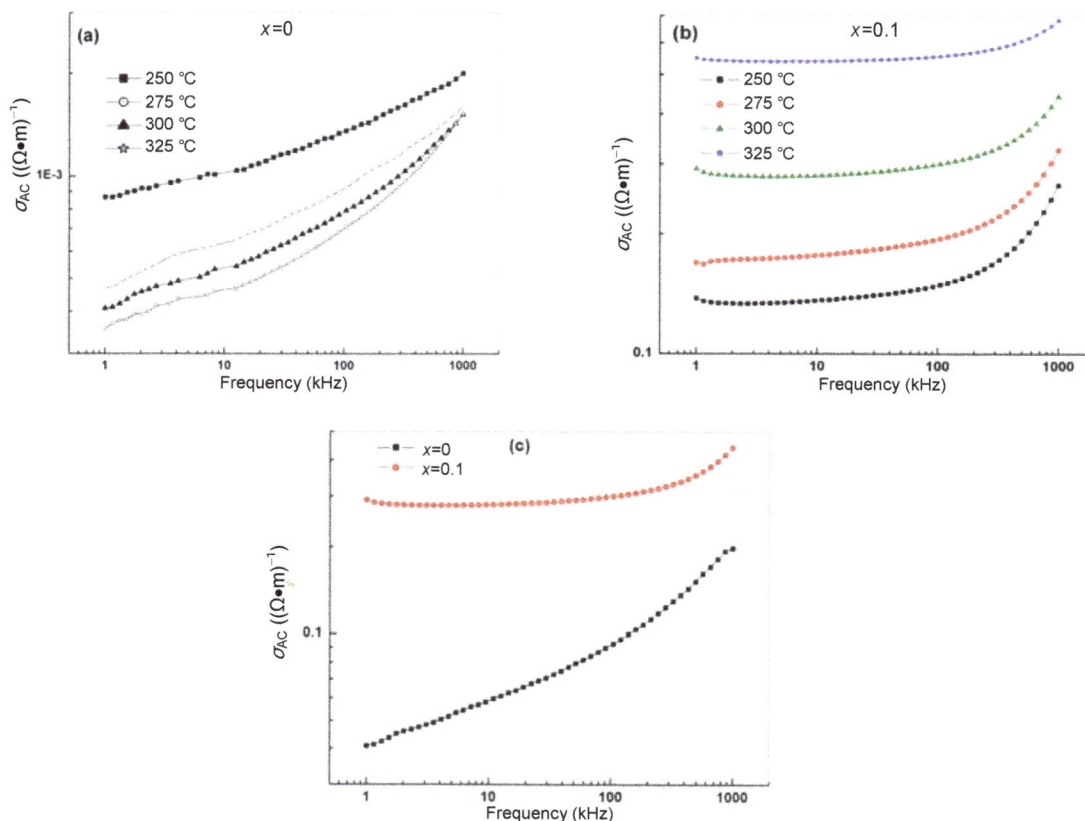

Fig. 5   Variation of AC conductivity (σ_{AC}) of $Bi_{1-x}Pr_xFeO_3$ ($x = 0$ and 0.1) with frequency at different temperatures.

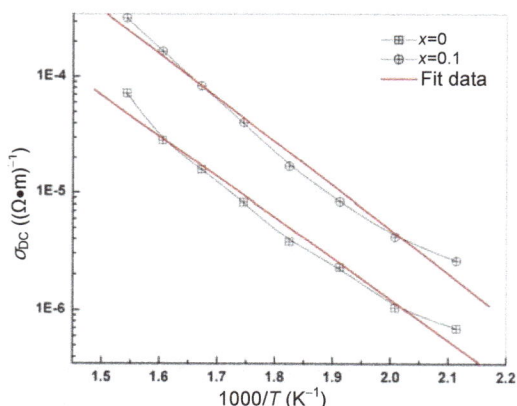

Fig. 6   Variation of DC conductivity of $Bi_{1-x}Pr_xFeO_3$ ($x = 0$ and 0.1) with temperature.

References

[1] Zhao T, Scholl A, Zavaliche F, *et al*. Electrical control of antiferromagnetic domains in multiferroic BiFeO$_3$ films at room temperature. *Nat Mater* 2006, **5**: 823–829.

[2] Ramesh R, Spaldin NA. Multiferroics: Progress and prospects in thin films. *Nat Mater* 2007, **6**: 21–29.

[3] Rehspringer J-L, Bursik J, Niznansky D, *et al*. Characterisation of bismuth-doped yttrium iron garnet layers prepared by sol–gel process. *J Magn Magn Mater* 2000, **211**: 291–295.

[4] Das R, Khan GG, Mandal K. Enhanced ferroelectric, magnetoelectric, and magnetic properties in Pr and Cr co-doped BiFeO$_3$ nanotubes fabricated by template assisted route. *J Appl Phys* 2012, **111**: 104115.

[5] Karimi S, Reaney IM, Han Y, *et al*. Crystal chemistry and domain structure of rare-earth doped BiFeO$_3$ ceramics. *J Mater Sci* 2009, **44**: 5102–5112.

[6] Michel C, Moreau J-M, Achenbach GD, *et al*. The atomic structure of BiFeO$_3$. *Solid State Commun* 1969, **7**: 701–704.

[7] Teague JR, Gerson R, James WJ. Dielectric hysteresis in single crystal BiFeO$_3$. *Solid State Commun* 1970, **8**: 1073–1074.

[8] Xu JM, Wang GM, Wang HX. Synthesis and weak ferromagnetism of Dy-doped BiFeO$_3$ powders. *Mater Lett* 2009, **63**: 855–857.

[9] Vopsaroiu M, Cain MG, Sreenivasulu G, *et al*. Multiferroic composite for combined detection of static and alternating magnetic fields. *Mater Lett* 2012, **66**: 282–284.

[10] Parida BN, Das PR, Padhee R, *et al*. Synthesis and

chracterization of a tungsten bronze ferroeletcric oxide. *Adv Mat Lett* 2012, **3**: 231–238.

[11] Wu E. POWDMULT: An interactive powder diffraction data interpretation and indexing program version 2.1. School of Physical Sciences, Flinders University of South Australia, Bradford Park, SA 5042, Australia.

[12] Parida BN, Das PR, Padhee R, *et al*. Phase transition and conduction mechanism of rare earth based tungsten–bronze compounds. *J Alloys Compd* 2012, **540**: 267–274.

[13] Barick BK, Mishra KK, Arora AK, *et al*. Impedance and Raman spectroscopic studies of $(Na_{0.5}Bi_{0.5})TiO_3$. *J Phys D: Appl Phys* 2011, **44**: 355402.

[14] Nalwa KS, Garg A, Upadhya A. Effect of samarium doping on the properties of solid-state synthesized multiferroic bismuth ferrite. *Mater Lett* 2008, **62**: 878–881.

[15] Parida BN, Das PR, Padhee R, *et al*. A new ferroelectric oxide $Li_2Pb_2Pr_2W_2Ti_4Nb_4O_{30}$: Synthesis and characterization. *J Phys Chem Solids* 2012, **73**: 713–719.

[16] Wieczorek W, Płocharski J, Przyłuski J. Impedance spectroscopy and phase structure of PEO NaI complexes. *Solid State Ionics* 1982, **28–30**: 1014–1017.

[17] Behera B, Nayak P, Choudhary RNP. Structural and impedance properties of $KBa_2V_5O_{15}$ ceramics. *Mater Res Bull* 2008, **43**: 401–410.

[18] Jonscher AK. The 'universal' dielectric response. *Nature* 1977, **267**: 673–679.

[19] Suman CK, Prasad K, Choudhary RNP. Complex impedance studies on tungsten–bronze electroceramic: $Pb_2Bi_3LaTi_5O_{18}$. *J Mater Sci* 2006, **41**: 369–375.

[20] Irvine JTS, Sinclair DC, West AR. Electroceramics: Characterization by impedance spectroscopy. *Adv Mater* 1990, **2**: 132–138.

[21] Macdonald JR. Note on the parameterization of the constant-phase admittance element. *Solid State Ionics* 1984, **13**: 147–149.

[22] Choudhary RNP, Perez K, Bhattachrya P, *et al*. Structural and electrical properties of $BiFeO_3$–$Pb(ZrTi)O_3$ composites. *Appl Phys A* 2007, **86**: 131–138.

[23] Pattanayak S, Parida BN, Das PR, *et al*. Impedance spectroscopy of Gd-doped $BiFeO_3$ multiferroics. *Appl Phys A* 2012, DOI: 10.1007/s00339-012-7412-6.

[24] Barick BK, Choudhary RNP, Pradhan DK. Phase transition and electrical properties of lanthanum-modified sodium bismuth titanate. *Mater Chem Phys* 2012, **132**: 1007–1014.

[25] Shukla A, Choudhary RNP. Study of electrical properties of La^{3+}/Mn^{4+}-modified $PbTiO_3$ nanoceramics. *J Mater Sci* 2012, **47**: 5074–5085.

21

Impedance spectroscopy study of $Na_2SmV_5O_{15}$ ceramics

P. S. DAS[a], P. K. CHAKRABORTY[a], Banarji BEHERA[b],
N. K. MOHANTY[b], R. N. P. CHOUDHARY[c,*]

[a]Department of Physics, Midnapore College, Midnapore 721 101, India
[b]School of Physics, Sambalpur University, Burla 768 019, Sambalpur, Odisha, India
[c]Department of Physics, ITER, Siksha "O" Anusandhan University, Bhubaneswar 751 030, Odisha, India

Abstract: The polycrystalline $Na_2SmV_5O_{15}$ (NSV), a new member of the tungsten bronze (TB) family, was prepared by a mixed-oxide technique. The room-temperature X-ray diffraction (XRD) confirmed the formation of single phase compound with orthorhombic crystal structure. The scanning electron microscopy (SEM) analysis indicated that the compound has homogeneous micrograph with a uniform distribution of small grains over the entire surface of the sample. The analysis of impedance spectra of NSV in a low-temperature range (-100 °C to 100 °C) at different frequencies exhibited interesting electrical properties like the contribution of bulk effect in conduction process. The study of imaginary part of the impedance at different temperatures showed existence of relaxation peak with its shift towards higher frequency on increasing temperature. This suggested the presence of frequency and temperature dependent relaxation process in the material. The loss peak spectra were found to abide by Arrhenius law with small activation energy of 0.12 eV. The temperature dependence of AC and DC electrical conductivity (σ_{AC} and σ_{DC}) was also obtained.

Keywords: solid-state reaction; microstructure; impedance; conductivity

1 Introduction

Due to the structural versatility of vanadium-containing compounds, some vanadates have attracted much attention of the researchers to work in such new compounds. They also have the ability to act as intercalation, ion-exchange, magnetism, cathode, and nonlinear optical (NLO) materials [1–7]. In vanadates, vanadium has been found in three-, four-, five-, and six-coordinate environments. Generally, the known ferroelectric oxides have perovskite, tungsten bronze (TB) or pyrochlore structures. Nowadays, there has been much interest to study the TB ferroelectric vanadates because they have useful applications. The TB structure with a general formula $(A_1)_2(A_2)_4(C)_4$-$(B_1)_2(B_2)_8O_{30}$ is a network of octahedral sharing of corners of a unit cell, and hence it would be easily said that this structural family stands at an intermediate position of perovskite and pyrochlore types of structures. A TB unit cell is composed of ten octahedral and six cages; the cage consists of four 15-coordinated sites and two 12-coordinated sites, surrounded by the octahedral. The octahedral sites are occupied by B ions and the remaining six sites by A ions, and the compound is generally represented by the formula

* Corresponding author.
E-mail: crnpfl@gmail.com

$A_6B_{10}O_{30}$ (called a filled TB), where either A or B sites are to be occupied by more than two kinds of ions. Accordingly, if the binary system is composed of A and B oxides, we could only find one $A_5B_{10}O_{30}$ TB type where six A sites are occupied by five ions; this is a typical TB structure. Though a lot of work on the niobates and vanadates of TB structural family have been reported in the past [8–14], we could not find any work on the titled compound. Giess *et al.* [13] reported the structural and dielectric properties of $KPb_2Nb_5O_{15}$ with Pb and K at the A site and Nb at the B site of TB structure. In view of the importance of materials within this family, we have attempted to study the synthesis, structure and electrical properties (impedance parameters) of $Na_2SmV_5O_{15}$ (NSV) ceramics by the substitutions of Na at the A site and V at the B site of the TB structure.

2 Experiment

Using a mixed-oxide method, the polycrystalline sample of NSV was prepared relatively at low temperature (450 ℃), with high-purity (>99%) ingredients: Na_2CO_3 (M/s. Sarabhai Chemicals, India), Sm_2O_3 (M/s. Loba Chem. Ltd, India) and V_2O_5 (M/s. Koch Light Ltd, England). The ingredients were thoroughly mixed in air atmosphere for 1 h and then in methanol atmosphere for 1 h. The mixed powder was calcined for 7 h at an optimum temperature (450 ℃) in an alumina crucible. The X-ray diffraction (XRD) analysis confirmed the formation of a new compound. The calcined powder was then cold-pressed into cylindrical pellets at a pressure of 4×10^6 N/m² using a hydraulic press. As a binder, polyvinyl alcohol was used to prepare pellets, and burnt out during the high-temperature sintering. The sample was sintered at the temperature of 475 ℃ (optimized) in an alumina crucible in air atmosphere for 7 h. The pellets were polished by fine emery papers to make the surfaces flat and parallel. The flat surfaces were coated with silver paint. After electroding, the pellets were dried at a temperature of 150 ℃ for 4 h to remove moisture and then brought to room temperature before taking electrical measurements. Structural data were recorded by X-ray diffractometer (Rigaku Miniflex, Japan) using Cu Kα radiation ($\lambda = 1.5405$ Å) in a wide range of Bragg angle 2θ ($20° \leqslant 2\theta \leqslant 80°$) with a scanning rate of 4 (°)/min. The electrical properties were measured by a computer-controlled LCR meter (HIOKI Model: 3532)

in a wide range of frequency (10^2–10^6 Hz) at different temperatures (–100 ℃ to 100 ℃).

3 Results and discussion

3.1 Structural and micro structural properties

The room-temperature XRD pattern of the calcined powder of NSV is shown in Fig. 1. It shows the formation of a new single-phase compound. All of the reflection peaks are indexed in different crystal systems and unit cell configuration. An orthorhombic unit cell is selected on the basis of the good agreement between observed and calculated d spacing ($\sum \Delta d = d_{obs} - d_{cal} = $ minimun). The lattice parameters of the selected unit cell were refined using the least-squares sub-routine of a standard computer program package "POWD" [15]. These are $a = 18.0487(33)$ Å, $b = 16.9000(33)$ Å, $c = 3.4986(33)$ Å, and $V = 1067.14$ Å³ (the number in parenthesis is estimated standard deviation). These unit cell parameters are in good agreement with those of reported ones [14,15] for the TB structure. However, with limited data, it is not possible to determine crystal structure and the space group of the compound. The average crystallite size (P) of NSV is determined from the broadening of a few XRD peaks using the Scherrer's equation [16]:

$$P = K\lambda / (\beta_{1/2} \cos\theta_{hkl})$$

where $K = 0.89$ (constant); $\lambda = 1.5405$ Å; $\beta_{1/2}$ is the peak width of the reflection at half intensity; and

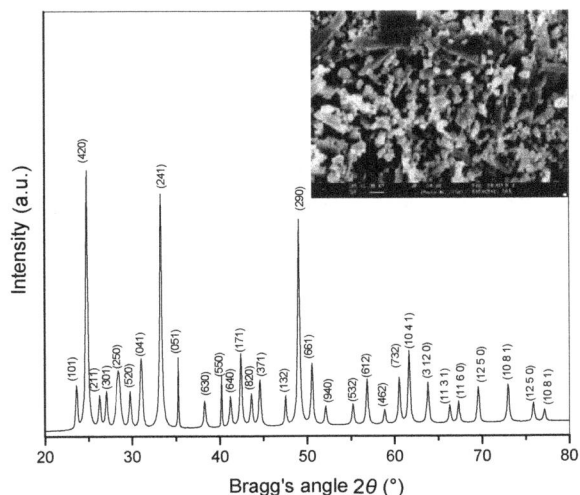

Fig. 1 Room-temperature XRD and SEM (inset) patterns of NSV.

θ_{hkl} is the Bragg angle. The average crystallite size of the compound is found to be 47 nm. The scanning electron microscopy (SEM) obtained using a scanning electron microscope (Model LEICA S440i) of the sample is given in Fig. 1 (inset), which clearly shows uniform distribution of the grains over the entire surface with certain degree of porosity.

3.2   Impedance analysis

Impedance spectroscopy is a powerful technique for the characterization of electrical behavior of electro-ceramic materials, where a sinusoidal perturbation is applied and the AC response is analyzed. Subsequently, impedance is calculated as a function of frequency of the perturbation. The output response appears in the form of a succession of semicircle representing electrical phenomena due to bulk, grain boundary effect and interfacial phenomena in the Argand plane [17]. Though polycrystalline materials usually show both grain and grain boundary effects with different time constants leading to two successive semicircles, in this present study, within the chosen temperature and frequency ranges, the second intercept corresponding to the low-frequency semicircle exhibiting grain boundary property is not observed. The frequency-dependent electrical properties of a material are represented in terms of complex impedance Z^*, as

$$Z^*(\omega) = Z' - jZ'' = R_s - 1/(j\omega C_s)$$

where Z' and Z'' are the real and imaginary components of impedance, respectively; $j = \sqrt{-1}$ is the imaginary factor; and R_s is the series resistance and C_s is the series capacitance.

The complex impedance plots of the material are obtained at different temperatures shown in Fig. 2. The presence of well-resolved semicircular arcs with rise in temperature has been attributed to the bulk properties of the material, and arises due to the parallel combination of bulk resistance R_b and bulk capacitance C_b. The low-frequency dispersion curve (second semicircular arc) not appearing in our selected temperature range can be assigned to the grain boundary. The intercept of the first semicircular arc on the real axis Z' gives the bulk resistance R_b (DC resistance). The DC resistance decreases on increasing temperature. This nature indicates that the bulk conductivity of the material increases with increase of temperature. Thus the material shows negative temperature coefficient of resistance (NTCR), which is

a property of semiconductors. The grain and grain boundary effects can be modeled as cascading of parallel RC combination in accordance with brick layer model [18], which has been represented in terms of the equivalent circuit given in Fig. 2 (inset). The value of the circuit elements obviously depends microscopically on the volume fraction of the grain; such a substantial decrease in the value of R_b has been observed with rise in temperature. All these curves start at almost the same R_b value, and do not coincide with the origin. So there would be a series of resistance that can be ascribed to the LCR circuit representation of the sample. The peak maxima of the plot decreases and shifts towards higher values with rise in temperature. This curve helps to determine the particle interaction like grain and grain boundary effects etc. This provides the information about the nature of the dielectric relaxation. For pure non-dispersive Debye process, one expects semicircular plots with the centre located on the Z' axis. For the poly-dispersive relaxation, these Argand plane [17] plots are close to the circular arcs with end points on the real axis, and the centre below it as observed in the present study. The complex impedance in such situation is described by Cole formalism:

$$Z^*(\omega) = Z' + iZ'' = R / [1 + (i\omega / \omega_0)^{1-\alpha}]$$

where α is the magnitude of the departure of the electrical response from an ideal condition and this can be determined from the location of the centre of the circles (not shown here); ω_0 is the Debye frequency. When α tends to zero, we have classical Debye

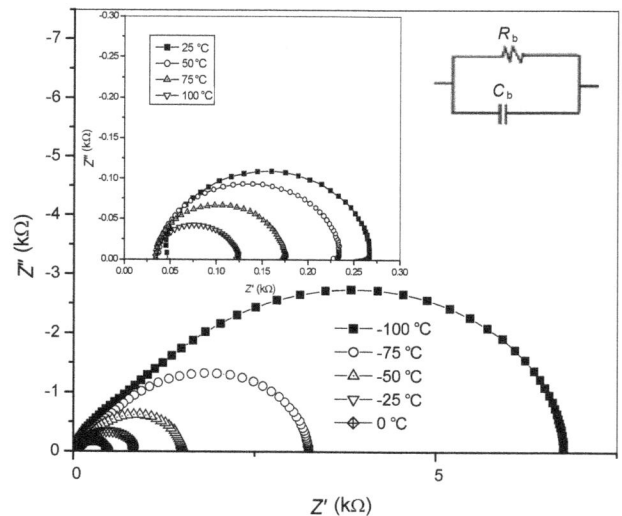

Fig. 2   Variation of Z'' and Z' at selected temperatures with equivalent circuit (inset).

formalism [19]. From these figures, it is clear that the dominance is only grain resistance. As the temperature increases, the bulk resistance which shows the NTCR behavior decreases like that of a semiconductor.

Figure 3 shows loss spectra (i.e., the variation of the imaginary part of the impedance Z'' with frequency) at different temperatures. This provides an insight into the process associated with the largest resistance. The loss spectra have a few important features: (i) typical peak broadening with rise in temperatures and (ii) asymmetric peak broadening. The presence of peak at unique frequency indicates the type and strength of electrical relaxation phenomenon in the material. The peak broadening with temperature indicates the temperature-dependent relaxation phenomenon in the material. At low temperatures, the relaxation phenomenon may be due to the immobile species/ electrons; at high temperatures, the relaxation may be due to the defects/vacancies. The decrease in the magnitude of Z'' with a shift in the peak frequency towards higher frequency arises possibly due to the presence of space charges at higher temperatures. The asymmetric broadening of the peaks suggests that multiple relaxation phenomenon is present in the material, with their own discrete relaxation time depending on the temperature. The relaxation peak in the loss spectrum coincides with the highly dispersive region in the Z' vs. frequency spectrum.

The variation of the real part of impedance Z' with frequency at several temperatures is shown in Fig. 4. It exhibits that Z' decreases with rise in frequency and temperature. The values of Z' merge after a certain frequency. This may be due to release of space charges

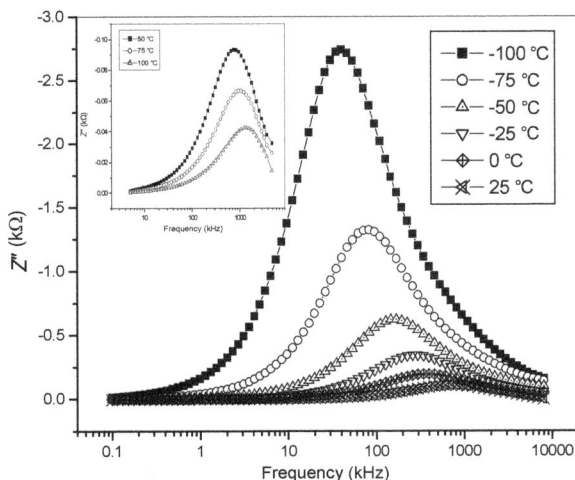

Fig. 4 Variation of Z' with frequency at a few selected temperatures.

at low temperatures. It indicates increment in AC conductivity with increment of temperature and frequency. As the temperature increases, the nature of variation of Z' as a function of frequency assumes a sigmoid type in the low-frequency region, and a plateau region beyond a fixed frequency. The magnitude of Z' (bulk resistance) decreases with increasing temperature, and appears to merge in the high-frequency region, irrespective of temperature. The behavior at high temperature may be attributed to the lowering of barrier properties of the sample with temperature, and then space charges would be emitted, which would enhance the conductivity and reduce the impedance properties. The modulus of Z' decreases with the rise in temperature, which means the sample exhibits NTCR behavior which is consistent with the previous results. This behavior is changed drastically in the high-frequency region. This particular frequency at which Z' value becomes constant is observed to shift towards the high-frequency side with rise of temperature. This shift in Z' plateau indicates the possibility of frequency relaxation process in the material.

4 Electrical conductivity

4.1 AC conductivity

The variation of AC conductivity (σ_{AC}) as a function of frequency (conductivity spectrum) at different temperatures is shown in Fig. 5. The conductivity

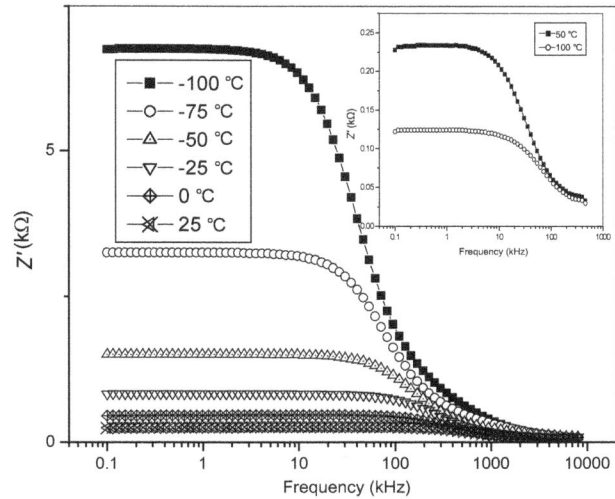

Fig. 3 Variation of Z'' with frequency at a few selected temperatures.

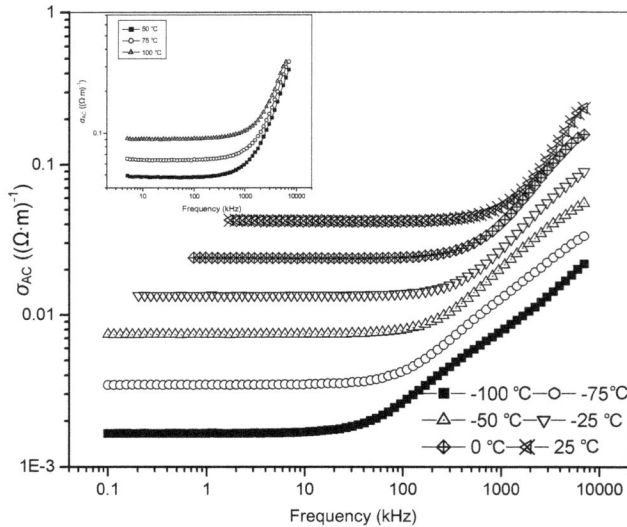

Fig. 5   Variation of σ_{AC} as a function of frequency at selected temperatures.

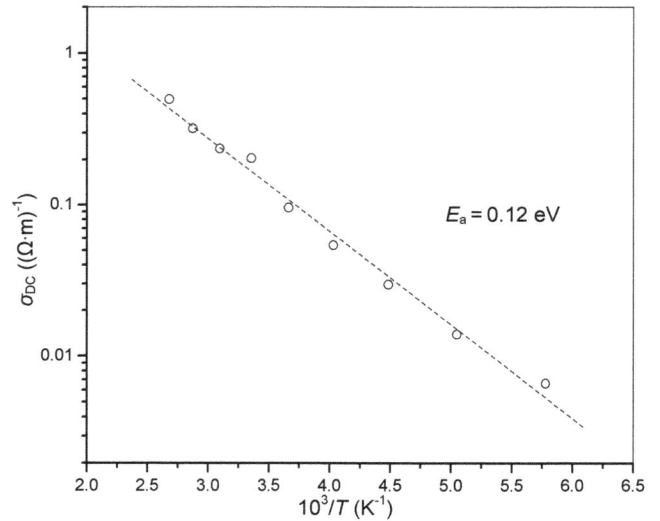

Fig. 6   Variation of σ_{DC} with $10^3/T$.

patterns can be divided into two parts. First at low temperatures, continuous dispersion curves of conductivity are observed in the lower-frequency region. Then the conductivity patterns approach close to each other in the high-frequency region for the same temperature region. This may be attributed to the space charge dependent region. The plateau region corresponds to the frequency-independent σ_{DC} conductivity. The rise in the conductivity value with temperature indicates that the electrical conduction is a thermally activated process.

4. 2   DC conductivity

The typical variation of σ_{DC} with temperature is a characteristic that the conductivity increases with the increment of temperature (Fig. 6). This is a typical Arrhenius type behavior:

$$\sigma_{DC} = \sigma_0 \exp[-E_a / (kT)]$$

where σ_{DC} is the DC conductivity; σ_0 is the pre-exponential factor; E_a is the activation energy; and k is the Boltzmann constant. The activation energy calculated from the slope of straight line is found to be 0.12 eV. The value of activation energy is found to be very less as compared to other TB and perovskite structure compounds [20–22]. This suggests that a small amount of energy is required to activate the electrons for electrical conduction.

The impedance data are used to evaluate the relaxation time (τ) of the electrical process in the material using the relation:

$$\omega_{max}\tau = \omega_{max}R_bC_b = 1$$

where ω_{max} is the maximum angular frequency; R_b and C_b are the bulk resistance and bulk capacitance respectively. It is independent of the sample geometry factors, but depends only on its intrinsic property (microstructure). The pattern shows a steady decrease in magnitude with rise in temperature (Fig. 7). This suggests the occurrence of temperature-dependent electrical relaxation phenomenon. This may be due to the migration of immobile species/defects. It is found to obey the Arrhenius law:

$$\tau = \tau_0\exp[-E_a / (kT)]$$

where τ is the relaxation time; τ_0 is the pre-exponential factor; E_a is the activation energy; and k is the Boltzmann constant. This expression

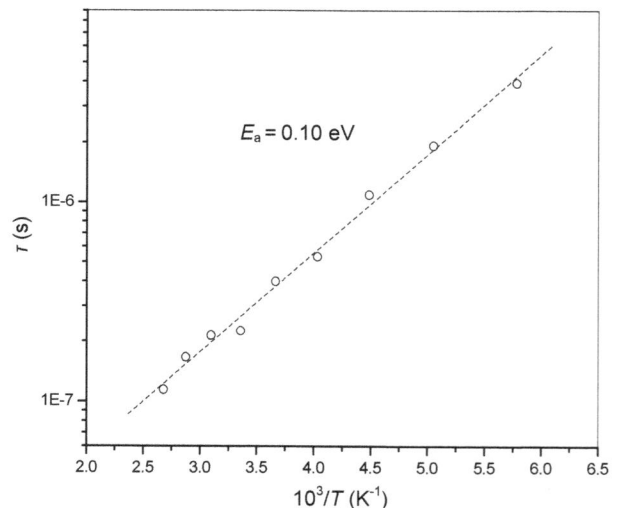

Fig. 7   Variation of relaxation time (τ) with $10^3/T$.

also estimates the activation energy which has a close resemblance to previously obtained E_a, and hence suggests the possibility of long-range mobility of charge carriers (ions) via hopping combination mechanism at higher temperatures.

5 Conclusions

The polycrystalline sample of NSV, a new member of TB structural family, was prepared by a mixed-oxide method at relatively low temperature (450 ℃). Preliminary X-ray analysis exhibited the orthorhombic crystal structure of the compound at room temperature. Impedance spectroscopy was used to characterize the electric properties of the material. The material showed the NTCR behavior which is the characteristics of a semiconductor. Study of the temperature dependence of DC conductivity provided activation energy with the value of 0.12 eV.

References

[1] Cavalli E, Belletti A, Mahiou R, *et al.* Luminescence properties of $Ba_2NaNb_5O_{15}$ crystals activated with Sm^{3+}, Eu^{3+}, Tb^{3+} or Dy^{3+} ions. *J Lumin* 2010, **130**: 733–736.

[2] Yin X, Shi L, Wei A, *et al.* Effect of structural packing on the luminescence properties in tungsten bronze compounds $M_2KNb_5O_{15}$ (M = Ca, Sr, Ba). *J Solid State Chem* 2012, **192**: 182–185.

[3] Szwagierczak D, Kulawik J. Thick film capacitors with relaxor dielectrics. *J Eur Ceram Soc* 2004, **24**: 1979–1985.

[4] Alaoui-Belghiti HE, Von der Mühll R, Simon A, *et al.* Relaxor or classical ferroelectric behavior in ceramics with composition $Sr_{2-x}A_{1+x}Nb_5O_{15-x}F_x$ (A = Na, K). *Mater Lett* 2002, **55**: 138–144.

[5] Ranga Raju MR, Choudhary RNP. Diffuse phase transition in $Sr_5RTi_3Nb_7O_{30}$ (R = La, Nd and Sm). *J Phys Chem Solids* 2003, **64**: 847–853.

[6] Rani R, Sharma S, Rai R, *et al.* Dielectric behavior and impedance analysis of lead-free CuO doped $(Na_{0.50}K_{0.50})_{0.95}(Li_{0.05}Sb_{0.05}Nb_{0.95})O_3$ ceramics. *Solid State Sci* 2013, **17**: 46–53.

[7] Ramam K, Chandramouli K. Ferroelectric and pyroelectric properties of Ce^{3+} modified tetragonal tungsten bronze structured lead barium niobate-55 ceramics. *J Phys Chem Solids* 2012, **73**: 1061–1065.

[8] Singh AK, Choudhary RNP. Diffuse ferroelectric phase transition in $Pb_5RTi_3Nb_7O_{30}$ (R = Eu and Gd). *Mater Lett* 2003, **57**: 3722–3728.

[9] Ravez J, Alaoui-Belghiti HE, Elaatmani M, *et al.* Relations between ionic order or disorder and classical or relaxor ferroelectric behaviour in two lead-free TKWB-type ceramics. *Mater Lett* 2001, **47**: 159–164.

[10] McCabe EE, West AR. New high permittivity tetragonal tungsten bronze dielectrics $Ba_2LaMNb_4O_{15}$: M = Mn, Fe. *J Solid State Chem* 2010, **183**: 624–630.

[11] Kathayat K, Panigrahi A, Pandey A, *et al.* Structural and electrical studies of $Ba_5RTi_3V_7O_{30}$ (R = Ho, Gd, La) compounds. *Physica B* 2012, **407**: 3753–3759.

[12] Parida BN, Das PR, Padhee R, *et al.* A new ferroelectric oxide $Li_2Pb_2Pr_2W_2Ti_4Nb_4O_{30}$: Synthesis and characterization. *J Phys Chem Solids* 2012, **73**: 713–719.

[13] Giess EA, Scott BA, Burns G, *et al.* Alkali strontium–barium–lead niobate systems with a tungsten bronze structure: Crystallographic properties and Curie points. *J Am Ceram Soc* 1969, **52**: 276–281.

[14] Mohanty NK, Satpathy SK, Behera B, *et al.* Complex impedance properties of $LiSr_2Nb_5O_{15}$ ceramic. *J Adv Ceram* 2012, **1**: 221–226.

[15] Wu E. *POWD*, an interactive program for powder diffraction data interpretation and indexing. *J Appl Cryst* 1989, **22**: 506–510.

[16] Patterson A. The Scherrer formula for X-ray particle size determination. *Phys Rev* 1939, **56**: 978–982.

[17] Shaw WT. *Complex Analysis with Mathematica.* Cambridge(UK): Cambridge University Press, 2006.

[18] Kidner NJ, Homrighaus ZJ, Ingram BJ, *et al.* Impedance/dielectric spectroscopy of electroceramics—Part 2: Grain shape effects and local properties of polycrystalline ceramics. *J Electroceram* 2005, **14**: 293–301.

[19] Hill RM, Dissado LA. Debye and non-Debye relaxation. *J Phys C: Solid State Phys* 1985, **18**: 3829–3836.

[20] Behera B, Nayak P, Choudhary RNP. Structural, dielectric and impedance properties of $NaCa_2V_5O_{15}$ ceramics. *Curr Appl Phys* 2009, **9**: 201–205.

[21] Behera B, Nayak P, Choudhary RNP. Impedance spectroscopy study of $NaBa_2V_5O_{15}$ ceramic. *J Alloys Compd* 2007, **436**: 226–232.

[22] Liu L, Huang Y, Su C, *et al.* Space-charge relaxation and electrical conduction in $K_{0.5}Na_{0.5}NbO_3$ at high temperatures. *Appl Phys A* 2011, **104**: 1047–1051.

22

Magnetic and structural properties of pure and Cr-doped haematite: α-Fe$_{2-x}$Cr$_x$O$_3$ (0 ≤ x ≤ 1)

Arvind YOGI[a,b], Dinesh VARSHNEY[a,*]

[a]*Materials Science Laboratory, School of Physics, Vigyan Bhawan, Devi Ahilya University, Khandwa Road Campus, Indore 452001, India*
[b]*School of Physics, Indian Institute of Science Education and Research, Thiruvananthapuram 695016, India*

Abstract: Solid-state ceramic technique route is used for synthesizing Cr-doped haematite α-Fe$_{2-x}$Cr$_x$O$_3$ (x = 0, 0.125, 0.50 and 1) samples. Single phase and corundum (Al$_2$O$_3$) type structure is revealed from the X-ray diffraction (XRD) patterns. The Raman spectra of α-Fe$_{2-x}$Cr$_x$O$_3$ illustrate seven phonon modes. On substitution of Cr (x = 0, 0.125, 0.50 and 1) at Fe site, all Raman active modes are shifted to higher wave numbers. The coercivity and remanence of Cr-doped haematites increase as x increases. The increased coercivity and remanence for Cr-doped samples can be attributed to their enhanced shape and magneto-crystalline anisotropy. The observed isomer shift δ values from room-temperature Mössbauer data clearly show the presence of ferric (Fe^{3+}) and Cr^{3+} ions illustrating strong ferromagnetic ordering up to x = 0.125 in α-Fe$_{2-x}$Cr$_x$O$_3$ haematite and weak ferromagnetic ordering for α-Fe$_{2-x}$Cr$_x$O$_3$ (x > 0.125) haematites.

Keywords: magnetic materials; X-ray diffraction (XRD); Raman spectra; Mössbauer spectroscopy

1 Introduction

Haematite, α-Fe$_2$O$_3$ is commonly described as rhombohedral or hexagonal with the space group $R\bar{3}c$ or D_{3d}^6, respectively. The structure of haematite is based on an arrangement of O^{2-} ions in a hexagonal close-packing, with Fe^{3+} ions being distributed in an ordered fashion in 2/3 of the octahedral sites. The crystal structure is the same as that of corundum, Al$_2$O$_3$. The framework of haematite is regarded as a set of O

and Fe layers arranged in normal to three-fold axis, and the chains of face-sharing octahedra are directed along the c-axis, while the Fe^{3+} ions within each chain form pairs separated by an empty interstitial site. The magnetic structure of α-haematite was the subject of considerable discussion and debate in the 1950s, because it appeared to be ferromagnetic with Curie temperature of around 1000 K but with an extremely tiny moment (0.002μ_B/Fe atom). Adding to the surprise was a transition to a phase with no net magnetic moment with a decrease in temperature at around 260 K (Morin transition) [1].

The substitution effect in α-Fe$_2$O$_3$ at Fe site influences the electrical, magnetic and other physical

---

* Corresponding author.
E-mail: vdinesh33@rediffmail.com

properties. The dopants used as substitution in α-Fe_2O_3 are Ti, Cr, Mn, Co, Ni, Cu, Zn, Al, Ga and Rh. The structural and magnetic properties of parent α-Fe_2O_3 are well established in the recent past; however, the doped α-Fe_2O_3 haematite leads to an increasing interest in the scientific community as the dopant like Zn ions substituting for Fe^{3+} in the corundum-related structure leads to formation of cationic and anionic vacancies [2–6]. The effect of substitution of Cr and Mn ions in $GaFeO_3$ manifests itself and can be delineated starting with suitable oxide precursors preferentially containing Cr (Mn) in the Ga or Fe site. It is noteworthy that Cr (Mn) substitution in the octahedral Fe (1,2) site of $GaFeO_3$ has marked effects on the structural parameters and magnetic properties, while substitution in the octahedral Ga 2 site has marginal effects [2–8]. The magnetic and structural properties of haematite are also affected by particle size [7,8], degree of crystallinity [9], pressure [10], doping [2–6], and mechanical alloying [11] due to its technological applications. It is noticed that depending upon milling conditions, structural phase transformations of nanostructured haematite to maghemite, magnetite and wüstite can be induced.

For identifying the vibrational phonon modes, Raman spectroscopy is a powerful probe. Raman measurements have been made at normal pressure and applying pressure for haematite (α-Fe_2O_3) as well [12]. The micro Raman spectra of Fe_2O_3 were measured at 62 GPa in a diamond anvil cell and at ambient conditions. The spectra depict all seven phonon (five E_g and two A_{1g}) modes predicted by group theory as $A_{1g}(1)$ (226 cm^{-1}), $E_g(1)$ (245 cm^{-1}), $E_g(2)$ (293 cm^{-1}), $E_g(3)$ (298 cm^{-1}), $E_g(4)$ (413 cm^{-1}), $A_{1g}(2)$ (500 cm^{-1}), and $E_g(5)$ (612 cm^{-1}) [13,14]. Apart from these modes, high-pressure spectra show an additional IR (infrared)-active E_u mode (~660 cm^{-1}) possibly induced by surface defects or stress. Furthermore, at high pressure, an additional Raman mode at 1320 cm^{-1} is observed and is attributed to a second order phonon–photon interaction. Earlier, McCarty has shown strong Raman features in $Fe_{2-x}Cr_xO_3$ at room temperature. The E_u mode (~660 cm^{-1}) of parent haematite varies linearly as a function of Cr doping and it is about 664 cm^{-1} for $Fe_{1.73}Cr_{0.27}O_3$ and 685 cm^{-1} for $Fe_{0.4}Cr_{1.6}O_3$, respectively [15,16]. McCarty and Boehme observed an additional band when chromium is substituted for iron in α-Fe_2O_3. The band ranges in Raman shift from 664 cm^{-1} for $Fe_{1.73}Cr_{0.27}O_3$ to 685 cm^{-1} for $Fe_{0.4}Cr_{1.6}O_3$, with the shift increasing nearly linearly with increasing chromium content [16].

The magnetic properties of haematite in bulk form and as spherical nanoparticles have been intensively studied because they have various applications in magnetic storage devices, spin electronic devices, drug delivery, tissue repair engineering, and magnetic resonance imaging [7,8,17–19]. In contrast to spherical nanoparticles, nanorods with their inherent one-dimensional (1-D) shape anisotropy may exhibit unique magnetic behavior, which is significantly different from that of the bulk material [20,21]. The magnetic properties of doped haematite are completely altered as compared to that of bulk haematite. However, few investigations of the magnetic properties of doped haematite have been reported [22,23].

Mössbauer spectroscopy is an impressive technique to probe the magnetic phases and to identify the magnetic ordering present in the structure as well. The parent haematite is paramagnetic above the Neel temperature ($T_N \approx 965$ K) and weakly ferromagnetic at room temperature, and at the Morin temperature ($T_M \approx 265$ K), it undergoes a phase transition to an antiferromagnetic state. The cation (Zn and Ni) substituted (5%) haematites show that T_M is essentially the same for both haematites and similar to that of undoped one. On the other hand, T_N is lower for the Ni-doped (5%) haematite ($\approx 883 \pm 2$ K) and Zn-doped (5%) haematite ($\approx 874 \pm 3$ K), and both are lower in comparison to T_N of single crystal haematite (965 K) [24]. Although the cation substitution suppresses T_N, an increase in T_M for Rh^{3+}, Ru^{3+} and Ir^{4+} is documented [25].

The room-temperature transmission Mössbauer spectra for parent haematite (α-Fe_2O_3) reveal magnetically resolved lines of haematite and are broadened progressively with increasing milling time (t_m). The static hyperfine field (B_{hf}) decreases from 51.5 T for non-milled sample down to 47 T for $t_m =$ 2 h [26]. As far as cation substitution is concerned, the magnetic cation Cr and non-magnetic cation Ti doped haematites (α-Fe_2O_3) are the most studied ones. The Mössbauer spectrum of Cr-substituted haematite fitted using only single-sextet model at 200 K infers anti-ferromagnetic (AF) ordering; however, weak-ferromagnetic (WF) ordering is noticed at room temperature. On the other hand, the spectra fitted through two-sextet model at about 245 K show the coexistence of both the phases (AF and WF). The magnetic hyperfine field (B_{hf}) for Cr-substituted Fe_2O_3 at AF (200 K) is $B_{hf} = 54.45 \pm 0.01$ T and it decreases ($B_{hf} = 53.70 \pm 0.01$ T) at WF (300 K) [27].

The near room-temperature Mössbauer spectra for $(Fe_{1-x}Ti_x)_2O_3$ ($x < 0.025$) fitted using single sextet of Lorentzian lines confirm the existence of WF phase. On the other hand, below 260 K, two sextets are essential for retracing the Mössbauer spectrum consisting of both WF and AF phases. The magnetic hyperfine field (B_{hf}) for Ti-doped haematite at 80 K results in $B_{hf} = 52.7$ T and $B_{hf} = 53.5$ T, respectively, while a reduced B_{hf} of about 51.0 T at 295 K with WF phase is documented [28]. On the other hand, non-magnetic cation as Zn-substituted haematites prepared by coprecipitation method is less studied and Zn^{2+}-doped haematites document an additional doublet pattern apart from single sextet and doublet. The hyperfine parameters are affected for Zn^{2+} substitution in haematites, and the decreasing magnetic ordering results in weak ferromagnetic nature [22].

In the present paper, we have prepared Cr^{3+}-doped α-$Fe_{2-x}Cr_xO_3$ haematites by solid-state ceramic technique route to seek the role of substitution (trivalent cation Cr^{3+} for trivalent Fe^{3+}) on the structural and magnetic properties following X-ray diffraction (XRD), Raman and Mössbauer spectroscopy. The immediate goal of the study is to see how the magnetic ordering changes with the impact of Cr^{3+} doping in α-Fe_2O_3.

2 Experimental techniques

2.1 Synthesis of samples

The polycrystalline samples with the composition α-$Fe_{2-x}Cr_xO_3$ ($x = 0$, 0.125, 0.50 and 1) were prepared by the solid-state ceramic technique route as described earlier [29]. Stoichiometry amounts for Fe_2O_3 and Cr_2O_3 were mixed and heated (calcinations) at different temperatures (850 °C, 950 °C and 1050 °C) in air for 24 h. After calcinations, we performed oxygen annealing at temperatures of 950 °C and 1050 °C for 24 h with intermediate grindings. The pellets were finally sintered and annealed at 1050 °C in oxygen atmosphere. All the samples were studied on a microscope slide as prepared and their structure were checked by XRD.

2.2 Characterization

2.2.1 X-ray diffraction

The XRD measurements were carried out with Cu Kα radiation using a Rigaku powder diffractometer which was equipped with a rotating anode scanning (0.01 (°)/step in 2θ) over the angular range of 10°–80° at room temperature and generating X-ray by 40 kV and 100 mA power settings. Monochromatic X-ray of $\lambda = 1.5406$ Å $K_{\alpha 1}$ line from a Cu target was made to fall on the prepared samples. The diffraction patterns were obtained by varying the scattering angle 2θ from 10° to 80° in step size of 0.01°.

2.2.2 Raman spectroscopy

The Raman equipment was a Jobin-Yovn Horiba Labram (System HR800) consisting of a single spectrograph (focal length 0.25 m) containing a holographic grating filter (1800 grooves/mm), and a Peltier-cooled CCD detector (1024×256 pixels of 26 µm). The spectra were excited with 632.8 nm radiation (1.95 eV) from a 19-mW air-cooled He–Ne laser (max laser power 19 mW) and the laser beam was focused on the sample by a 50× lens to give a spot size of 1 µm; the resolution was better than 2 cm^{-1}. The laser power was always kept on 5 mW at the samples to avoid sample degradation, except in the laser power dependence experiments. After each spectrum had been recorded, a careful visual inspection was performed using white light illumination on the microscope stage in order to detect any change that could have been caused by the laser.

2.2.3 Magnetization

The magnetization measurements on all the samples under investigation were performed using a vibrating sample magnetometer (VSM). Magnetic measurements for the samples in the powder form were carried out at room temperature using a VSM (Lakeshore 7300 model, USA). The hysteresis curves were obtained for the samples with different compositions. The applied field was varied from 0 to $\pm 10\,000$ Oe, and from hysteresis curves representing the magnetization process at room temperature (300 K), the saturation magnetization (M_{sat}), remanent magnetization (M_r) and coercive force (H_c) have been determined.

2.2.4 Mössbauer spectroscopy

^{57}Fe Mössbauer measurements were performed in transmission mode with a ^{57}Co radioactive source in constant acceleration mode using a standard PC-based Mössbauer spectrometer equipped with a WissEl velocity drive. Velocity calibration of the spectrometer was done with a natural iron absorber at room temperature. High magnetic field ^{57}Fe Mössbauer measurements were carried out using a Janis superconducting magnet. For the high magnetic field measurements, the external field was applied parallel to the γ rays (i.e., longitudinal geometry). Absorbers were carefully prepared to optimize both a good compactness and an adequate heat transmission. Analysis of the spectra was performed using the NORMOS [30] least-squares fitting program. Doublets and sextets of Lorentzian lines were used, doublets in the paramagnetic region and sextets in the magnetic region. The spectra were calibrated referring to α-Fe.

3 Results and discussion

The XRD patterns have been collected on the surfaces of the disks with a Rigaku diffractometer using Cu Kα radiation. Figure 1 shows the representative 2θ scans of XRD for the parent and Cr^{3+}-substituted α-Fe$_{2-x}$Cr$_x$O$_3$ ($x = 0$, 0.125, 0.50 and 1) haematites. Comparing the observed spectra with the standard diffraction pattern (JCPDS No. 86-0550), all Bragg peaks of the XRD patterns for the synthesized haematites in the doping range ($0 \leqslant x \leqslant 1$) are assigned to the presence of α-Fe$_2$O$_3$ phase and no other phase is identified. The XRD patterns for the α-Fe$_{2-x}$Cr$_x$O$_3$ are indexed as that of corundum type (α-Al$_2$O$_3$) structure at room temperature consistent with the earlier report [31]. In the XRD pattern for higher Cr doping, an additional Bragg peak is appeared

at 2θ value of around 25.48°. The argument lies in a fact that iron oxides show multiple phases like haematite (α-Fe$_2$O$_3$), maghemite (γ-Fe$_2$O$_3$) and magnetite (Fe$_3$O$_4$). However, haematite (α-Fe$_2$O$_3$) is very sensitive with doping. For small Cr doping concentration, T_M changes and suppresses abruptly, and at higher Cr doping concentration, it completely disappears. The above is due to the variation of the charge concentration on Fe site [2–6]. Since powder XRD gives the average structural information and it is not a powerful tool for low doping concentration, we expect a change in XRD pattern. Thus, we observe an additional appearance of Bragg peak at around 25.48° in XRD pattern.

The Raman spectra of the Cr-doped α-Fe$_{2-x}$Cr$_x$O$_3$ ($x = 0$, 0.125, 0.50 and 1) haematites in the range of 100 cm^{-1} to 800 cm^{-1} at room temperature and normal pressure are illustrated in Figs. 2 and 3. Figure 2 shows the Raman spectra of parent α-Fe$_2$O$_3$ and Fe$_{1.875}$Cr$_{0.125}$O$_3$ with seven phonon modes (two A$_{1g}$ modes, five E$_g$ modes and one Raman inactive or disorder E$_u$ mode) predicted from the group analysis (Table 1) confirming corundum structure. For haematite α-Fe$_2$O$_3$, the phonon modes predicted by

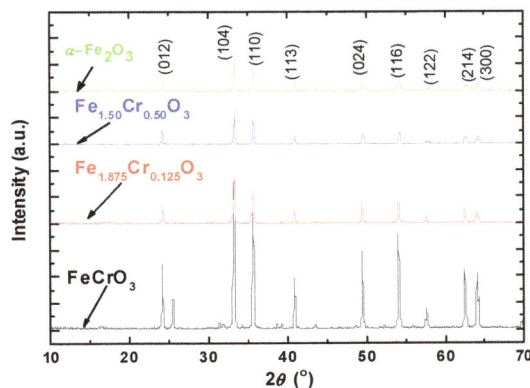

Fig. 1 The XRD patterns for α-Fe$_{2-x}$Cr$_x$O$_3$ ($x = 0$, 0.125, 0.50 and 1).

Table 1 Raman shift (ν) and full width at half maximum (FWHM) (Γ) for haematite at a laser power of 5 mW

No.	Active mode	α-Fe$_2$O$_3$		Fe$_{1.875}$Cr$_{0.125}$O$_3$		Fe$_{1.50}$Cr$_{0.50}$O$_3$		FeCrO$_3$	
-----	-------------	ν(cm^{-1})	Γ(cm^{-1})	ν(cm^{-1})	Γ(cm^{-1})	ν(cm^{-1})	Γ(cm^{-1})	ν(cm^{-1})	Γ(cm^{-1})
1	A$_{1g}$(1)	225	5.6	226	5.7	227	5.9	229	6.1
2	E$_g$(1)	244	6.8	245	6.9	246	6.9	—	—
3	E$_g$(2)	292	7.6	292	7.6	293	8.1	285	28.2
4	E$_g$(3)	298	7.9	298	7.9	—	—	—	—
5	E$_g$(4)	410	13.1	412	13.6	413	15.1	401	18.1
6	A$_{1g}$(2)	496	20.6	498	20.8	500	21.1	504	25.3
7	E$_g$(5)	611	14.1	612	14.4	612		—	—
	E$_u$(disorder)	660	—	659	25.2	656	29.1	654	66.9

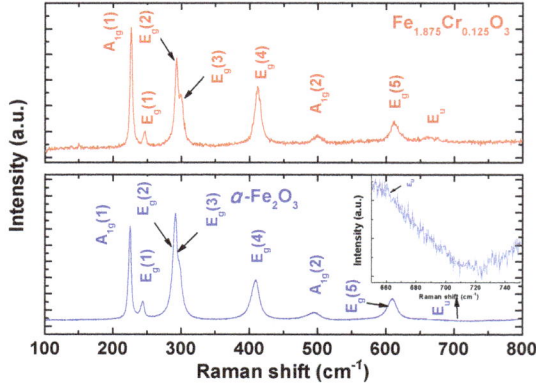

Fig. 2   Raman scattering intensities shown as a function of Raman shift (or wave number) for Fe_2O_3 and $Fe_{1.875}Cr_{0.125}O_3$. Inset shows the E_u mode for Fe_2O_3.

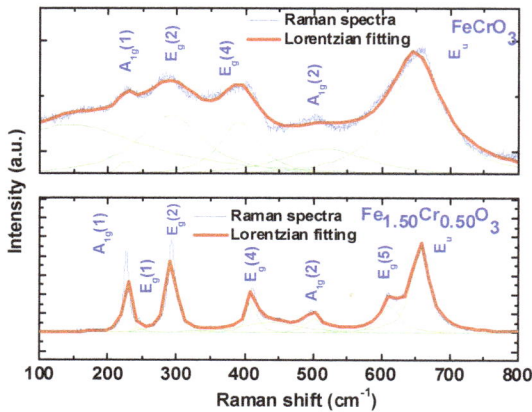

Fig. 3   Raman scattering intensities shown as a function of Raman shift (or wave number) for $Fe_{1.50}Cr_{0.50}O_3$ and $FeCrO_3$.

group theory are observed at $A_{1g}(1) \approx 225$ cm^{-1}, $E_g(1) \approx 244$ cm^{-1}, $E_g(2) \approx 292$ cm^{-1}, $E_g(3) \approx 298$ cm^{-1}, $E_g(4) \approx 410$ cm^{-1}, $A_{1g}(2) \approx 496$ cm^{-1}, and $E_g(5) \approx 611$ cm^{-1} consistent with the reported values in literatures [13,14]. It is noticed that the Raman spectrum in the range of 620–750 cm^{-1} shows an additional feature illustrating the Raman-inactive and IR-active E_u (LO) at about 660 cm^{-1} (the inset of Fig. 2).

On substitution of Cr ($x = 0$, 0.125) at Fe site, all seven phonon modes are observed and a small shift to higher wave numbers or Raman shift is seen in Fig. 2. The IR-active E_u (LO) (≈ 660 cm^{-1} for haematite (α-Fe_2O_3)) becomes sharp on Cr substitution and the FWHM (≈ 26 cm^{-1}) increases on enhancing the doping concentration. We note that IR-active E_u (LO) mode at 659 cm^{-1} is not group theoretically allowed in Raman

spectrum. The sharp feature of E_u (LO) mode ($0.125 \leqslant x \leqslant 1$) at about 660 cm^{-1} is attributed to the possible disorder induced by Cr surface defects or stress. The E_u mode of parent haematite at 660 cm^{-1} decreases linearly as a function of Cr doping and it reaches at about 654 cm^{-1} for $x = 1$. On the other hand, the FWHM increases from 26 cm^{-1} to 67 cm^{-1} consistent with the earlier measurements [15].

The increase in FWHM (i.e., bandwidth, as shown in Table 1) might also arise from local heating due to the relatively high laser power (≈ 5 mW), which enhances anharmonic interactions. In present, the laser power is fixed to 5 mW (for all the samples) causing the bands to broaden and undergo a small shift to higher wave numbers or Raman shift, as illustrated in Table 1. Beattie and Gilbson argued that in α-Fe_2O_3 the bandwidth decreases and Fe band positions shift slightly to higher wave numbers at low temperature (77 K) [14]. This behavior seems to be fully reversible, as turning down the laser power causes the bandwidth to decrease.

In passing, we may refer to the work of McCarty and Boehme who observed an additional band when chromium is substituted for iron in α-Fe_2O_3. The band ranges in Raman shift from 664 cm^{-1} for $Fe_{1.73}Cr_{0.27}O_3$ to 685 cm^{-1} for $Fe_{0.4}Cr_{1.6}O_3$, with the shift increasing nearly linearly with increasing Cr content [16]. The present Raman spectrum changes gradually for Cr doping in α-$Fe_{2-x}Cr_xO_3$ ($x = 0$, 0.125, 0.50 and 1) which is attributed to the strong electron–phonon interaction in this system. The $E_g(3)$ mode is more pronounced for lower doping ($x = 0.125$), but for higher doping ($x > 0.125$), it does not document in the Raman spectra attributed to the fact that the non-magnetic ion Cr doping leads to structural changes in α-$Fe_{2-x}Cr_xO_3$ system. The 100% Cr^{3+} replacement by Fe ($FeCrO_3$) does not show three E_g modes in the Raman spectrum ($E_g(1)$, $E_g(3)$ or $E_g(5)$), while that the FWHM increases as Cr doping increases represented in Table 1 is consistent with earlier reported data [15,16].

Magnetic hysteresis measurements for Cr-doped bulk haematite α-$Fe_{2-x}Cr_xO_3$ ($x = 0$, 0.125, 0.50 and 1) were carried out in an applied magnetic field (H) at room temperature (300 K) as present in Figs. 4–7. The hysteresis loops of the Cr-doped α-Fe_2O_3 show a ferromagnetic behavior for all the prepared samples. Moreover, the magnetization measurements of parent and Cr-doped bulk haematites α-$Fe_{2-x}Cr_xO_3$ ($x = 0$, 0.125, 0.50 and 1) exhibit hysteretic features with coercivity being determined to be 995 Oe, 1909 Oe, 2198 Oe and 3725 Oe, respectively (Figs. 4–7). The coercivity is an extrinsic property of a magnet, which depends not only on the spin carrier but also on the shape or size of the magnets [31,32]. The larger coercivity and remanence for Cr-doped samples can be attributed to their enhanced shape and magneto-crystalline anisotropy. The parent α-Fe$_2$O$_3$ shows the magnetic hysteresis loop at 300 K. We fail to observe saturation of the magnetization even under the maximum applied magnetic field of 10 kOe for all the prepared samples (Figs. 4–7). The parent α-Fe$_2$O$_3$ shows weak ferromagnetic behavior with the remanent magnetization (M_r) of 0.434 emu/g and coercivity (H_c) of 995 Oe. As increasing Cr doping

concentration x in α-Fe$_{2-x}$Cr$_x$O$_3$ system, the coercivity and the remanence increase abruptly, which show ferromagnetic nature. These results are summarized in Table 2.

Hill et $al.$ [21] observed the magnetic properties of bulk and mesoporous haematite (α-Fe$_2$O$_3$) at room temperature as well as at low temperature. For bulk α-Fe$_2$O$_3$, no hysteresis loop is observed at 12 K, in line with what is expected for a simple antiferromagnet. At 295 K, hysteresis is observed due to the weak ferromagnetic state of the material. Extrapolation of the high-field region of the $M(H)$ curve (where saturation occurs) provides a spontaneous magnetization, M_r, of 0.28 emu/g. At room temperature, however, the bulk material is a harder magnet than the

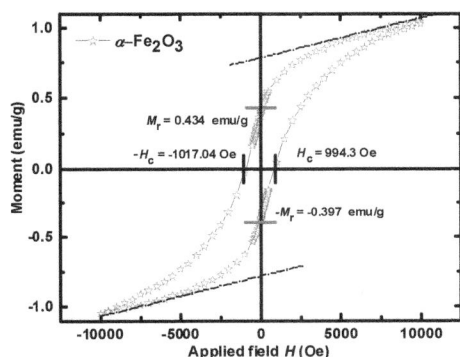

Fig. 4   Hysteresis loop for parent α-Fe$_2$O$_3$ sample at room temperature (300 K).

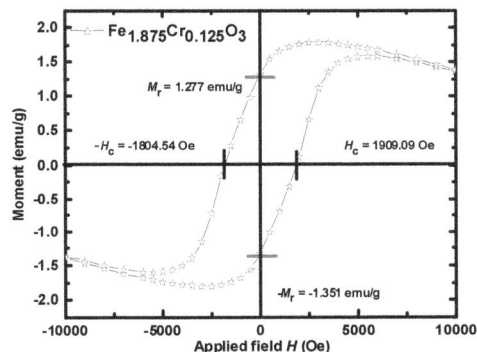

Fig. 5   Hysteresis loop for Cr-doped haematite α-Fe$_{1.875}$Cr$_{0.125}$O$_3$ sample at room temperature (300 K).

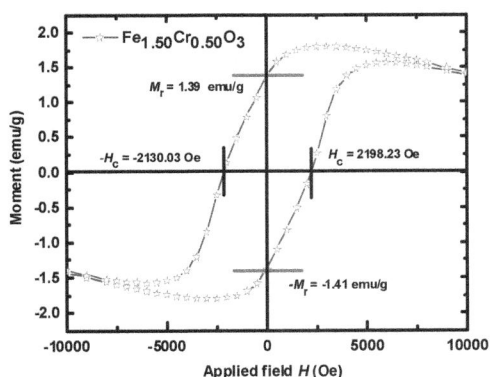

Fig. 6   Hysteresis loop for Cr-doped haematite α-Fe$_{1.50}$Cr$_{0.50}$O$_3$ sample at room temperature (300 K).

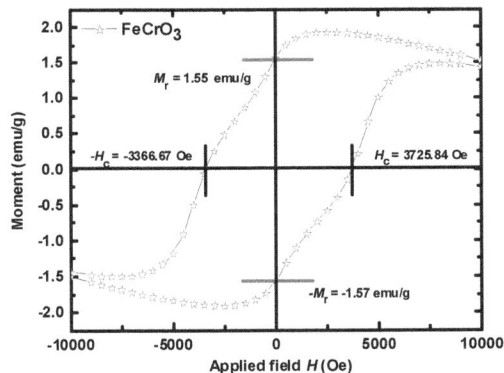

Fig. 7   Hysteresis loop for Cr-doped haematite α-FeCrO$_3$ sample at room temperature (300 K).

Table 2   Remanent magnetization (M_r) and coercive force (H_c) for Cr-doped α-Fe$_{2-x}$Cr$_x$O$_3$

Sample	Remnant magnetization, M_r (emu/g)	Coercive force, H_c (Oe)
Fe$_2$O$_3$	0.434	994.3
Fe$_{1.875}$Cr$_{0.125}$O$_3$	1.277	1909.1
Fe$_{1.50}$Cr$_{0.50}$O$_3$	1.390	2198.2
FeCrO$_3$	1.550	3725.8

mesoporous material, $H_c = \sim 0.17$ T. These results are higher in magnitude as compared to our results for parent α-Fe$_2$O$_3$, which may be due to the morphological difference. The effect of Co-coated haematite is studied and noticed that magnetic properties are changed with Co coating. According to the report, the coercivity of Co-coated haematite is higher than that of pure haematite by about 400 Oe [23]. In the present study of magnetization, the hysteretic features are changed gradually for Cr doping in α-Fe$_{2-x}$Cr$_x$O$_3$ ($x = 0.125$, 0.50 and 1) because magnetic properties are sensitive for doping concentration. It is well known that the magnetization of ferromagnetic materials is very sensitive to the morphology and structure of the as-synthesized sample. So the higher coercivity and remanent magnetization are due to the shape anisotropy of α-Fe$_{2-x}$Cr$_x$O$_3$ superstructures, which prevent them from magnetizing in directions other than along their easy magnetic axes, hence leading to the higher remanent magnetization and higher coercivity.

The ^{57}Fe Mössbauer spectra recorded at 300 K for Cr doping at Fe site α-Fe$_{2-x}$Cr$_x$O$_3$ ($x = 0$, 0.125, 0.50 and 1) are illustrated from Fig. 8 and Fig. 9, respectively. The Mössbauer spectra are analysed by considering single symmetric Lorentzian-shaped sextet model for parent (doped) haematites. The room-temperature Mössbauer spectra are fitted with NORMOS-SITE program [30] and all samples show strong magnetic ordering. The obtained value of chi-2 (χ^2) is minimum and NORMOS-SITE program is well fitted to the experimental data. The ^{57}Fe fitted Mössbauer parameters at room temperature are listed in Table 3. The Mössbauer parameters are derived from the spectra using the single symmetric Lorentzian-

shaped sextet model, and an additional quadruple model is required for Cr doping ($x > 0.50$). The obtained hyperfine parameters are with respect to natural iron consistent with the earlier experimental data [22,26–28].

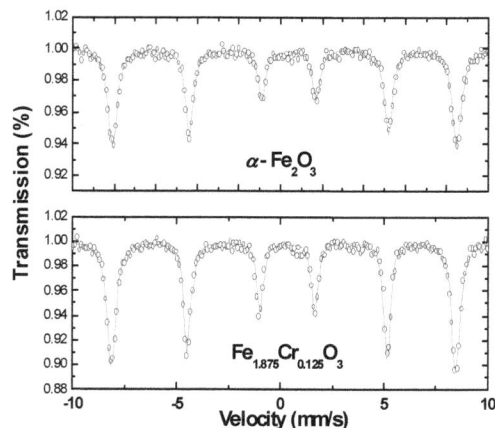

Fig. 8   Room-temperature Mössbauer spectra of the Fe$_2$O$_3$ and Fe$_{1.875}$Cr$_{0.125}$O$_3$ haematites.

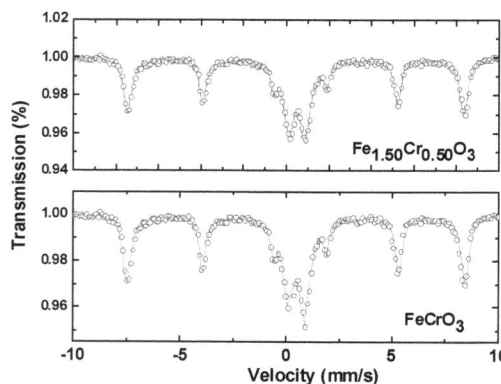

Fig. 9   Room-temperature Mössbauer spectra of the Fe$_{1.50}$Cr$_{0.50}$O$_3$ and FeCrO$_3$ haematites.

Table 3 ^{57}Fe Mössbauer parameters for Cr-doped α-Fe$_2$O$_3$

Sample	Assignment	δ (mm/s)	ΔE_Q (mm/s)	B_{hf} (T)	Relative area (%)
Fe$_2$O$_3$	Sextet	0.30	−0.15	51.18	100
Fe$_{1.875}$Cr$_{0.125}$O$_3$	Sextet	0.32	−0.19	52.0	100
Fe$_{1.50}$Cr$_{0.50}$O$_3$	Sextet	0.27	−0.21	50.9	76.93
	Doublet	0.24	0.38	—	23.07
FeCrO$_3$	Sextet	0.29	−0.23	49.7	75.18
	Doublet	0.26	0.36	—	24.82

Though α-Fe$_2$O$_3$ is a very common iron oxide in solids, it is particularly interesting due to its two magnetic phases namely week ferromagnetism (WF, at ambient condition, $\sim$300 K) and antiferromagnetic (AF, at non-ambient condition, $\sim$265 K). Due to its high Neel temperature ($T_N = 965$ K) and possibly a high

effective anisotropy constant, the Mössbauer spectrum at room temperature appears as a sextet with a hyperfine field of about 51 T and a quadruple shift of about 0.2 mm/s. We have calculated these parameters using NORMOS-SITE program for all prepared samples, which are in good agreement with reported results [22,24]. From these two results of α-Fe$_2$O$_3$ Mössbauer spectrum, one can be readily distinguished even at room temperature from other iron oxides, making again Mössbauer spectroscopy to be a powerful method for identification purposes.

From Fig. 8, the transmission Mössbauer spectroscopy of α-Fe$_2$O$_3$ reveals only single sextet indicating magnetically ordered state. The peaks with high hyperfine field values are related to the irons involved in the electron delocalization process, without the presence of Cr as the nearest neighbors. The transmission Mössbauer spectroscopy of α-Fe$_{2-x}$Cr$_x$O$_3$ $(0.50 \leqslant x \leqslant 1)$ shows an additional peak corresponding to those iron atoms surrounded by Cr neighbors as depicted in Fig. 9. The additional feature is more pronounced for $x = 1$. It is noticed that the Fe site distributions become broader and occupy more area as the amount of Cr increases and the relative area decreases. The observed isomer shift values from room-temperature Mössbauer data clearly show the presence of ferric (Fe^{3+}) state and Cr is in Cr^{3+}. The negative values of the quadruple coupling constants (Table 3) for all Cr-doped haematites indicate that the material is weakly ferromagnetic at room temperature. This is mainly because the occupation of interstitial sites by Cr^{3+} ions replaces the vacancies on Fe^{3+} sites.

For undoped α-Fe$_2$O$_3$ we refer to our earlier work, where hyperfine field of about 51.18 T and a quadruple shift of about -0.15 mm/s are obtained [33]. Since Cr^{3+} shows low spin value ($S = 3/2$) and Fe^{3+} shows high spin value ($S = 5/2$), the difference in spin values of Cr^{3+} and Fe^{3+} ground state quantum fluctuations might be enhanced, which motivates us to perform Cr doping on this system. Because of the replacing of huge difference spin values, the Cr doping leads to the hyperfine field changing from 51.18 T to 49.7 T and the quadruple shift from -0.15 mm/s to -0.23 mm/s as a function of Cr doping at Fe site.

The room-temperature Mössbauer spectra of α-Fe$_{2-x}$Cr$_x$O$_3$ $(0.50 \leqslant x \leqslant 1)$ haematites show an additional presence of a quadruple-split doublet [22,24]. The ^{57}Fe Mössbauer spectrum of FeCrO$_3$ recorded at 300 K (Fig. 9) shows the clear sextet pattern and a broad doublet. The doublet component value of isomer shift is $\delta = 0.26$ mm/s and the quadrupole is $\Delta E_Q = 0.36$ mm/s at 300 K. The NORMOS-SITE program fitting for Mössbauer spectrum of the magnetically split component of the FeCrO$_3$ spectrum to a single sextet yields a hyperfine

magnetic field ($B_{hf} = 49.7 \pm 0.3$ T) smaller to that of pure α-Fe$_2$O$_3$ ($B_{hf} = 51.18 \pm 0.3$ T).

Furthermore, the linewidth of FeCrO$_3$ (0.32 mm/s) is an indicative of a distribution of sites originating from the presence of the dopant Cr^{3+} within the α-Fe$_2$O$_3$ structure. The spectrum is therefore fitted with sextet pattern and the field values (B_{hf}) are smaller for all Cr-doped haematites α-Fe$_{2-x}$Cr$_x$O$_3$. The above feature indicates that α-Fe$_{2-x}$Cr$_x$O$_3$ $(0.50 \leqslant x \leqslant 1)$ is weakly ferromagnetic at room temperature as compared to the parent α-Fe$_2$O$_3$. In the α-Fe$_2$O$_3$ pattern the larger field ($B_{hf} = 51.18 \pm 0.3$ T) can be associated with Fe^{3+} ions which are not influenced by the presence of Cr, while in α-Fe$_{2-x}$Cr$_x$O$_3$ $(0.50 \leqslant x \leqslant 1)$ the smaller field ($B_{hf} = 49.7 \pm 0.3$ T) can be assigned to Fe^{3+} ions that have Cr^{3+} ions in close proximity. The relative population of each magnetic component is varied with assumed linewidth without any significant change to the value of χ^2, and the percentage of Fe^{3+} influenced by Cr is therefore difficult to quantify.

4 Conclusions

The structural and magnetic ordering effects in polycrystalline α-Fe$_{2-x}$Cr$_x$O$_3$ $(0 \leqslant x \leqslant 1)$ are studied by XRD, Raman spectra and Mössbauer spectroscopy. The crystal structure of the samples is studied using XRD. The XRD patterns identify the single phase and corundum or Al$_2$O$_3$ type structure of all the polycrystalline samples. The Raman spectra reveal seven Raman active modes (two A$_{1g}$ modes, five E$_g$ modes) and an additional Raman inactive E$_u$ mode. The IR-active E$_u$ (LO) (≈ 660 cm^{-1} for haematite (α-Fe$_2$O$_3$)) becomes sharp on Cr substitution, and the FWHM increases on enhancing the doping concentration (for 100% Cr doping, $\Gamma \approx 67$ cm^{-1}). The E$_g$(3) mode is more pronounced for lower doping ($x = 0.125$), but for higher doping ($x > 0.125$) it does not document in the Raman spectra attributed to the fact that the non-magnetic ion Cr doping leads to structural changes in α-Fe$_{2-x}$Cr$_x$O$_3$ system. The 100% Cr^{3+} replacement of Fe (FeCrO$_3$) does not show three E$_g$ modes in the Raman spectra (E$_g$(1), E$_g$(3) or E$_g$(5)) while the FWHM increases as Cr doping increases. Increase in Γ with Cr doping concentration is attributed to enhanced α-Fe$_2$O$_3$ bandwidth due to presence of Cr ion in the lattice and local heating by the relatively high laser power as well, which enhances anharmonic interactions.

The coercivity and remanence of Cr-doped haematites increase as x increases. The coercivity H_c (remanence M_r) for $x = 0$, 0.125, 0.50 and 1 is determined to be 995 Oe (0.44 emu/g), 1909 Oe (1.27 emu/g), 2198 Oe (1.39 emu/g) and 3725 Oe (1.55 emu/g), respectively. The larger coercivity and remanence for Cr-doped samples can be attributed to their enhanced shape and magneto-crystalline anisotropy. The room-temperature ^{57}Fe transmission Mössbauer spectroscopy of α-Fe$_2$O$_3$ reveals only single sextet indicating magnetically ordered state; however, a quadruple-split doublet is used for doped α-Fe$_{2-x}$Cr$_x$O$_3$ ($0.50 \leqslant x \leqslant 1$) implying weak magnetic ordering. The observed isomer shift δ values from room-temperature Mössbauer data clearly show the presence of ferric (Fe^{3+}) and Cr^{3+} ions illustrating strong ferromagnetic ordering up to $x = 0.125$ in α-Fe$_{2-x}$Cr$_x$O$_3$ haematite and weak ferromagnetic ordering for α-Fe$_{2-x}$Cr$_x$O$_3$ ($x > 0.125$) haematites.

Acknowledgements

Authors are thankful to UGC New Delhi for financial assistance and to UGC-DAE CSR, Indore for providing characterization facilities. Useful discussions with Dr. D. M. Phase, Dr. R. J. Choudhary, Dr. V. Sathe and Dr. V. R. Reddy are gratefully acknowledged.

References

[1] Saragovi C, Arpe J, Sileo E, et al. Changes in the structural and magnetic properties of Ni-substituted hematite prepared from metal oxinates. Phys Chem Miner 2004, 31: 625–632.

[2] Van San E, De Grave E, Vandenberghe RE, et al. Study of Al-substituted hematites, prepared from thermal treatment of lepidocrocite. Phys Chem Miner 2001, 28: 488–497.

[3] Kanai H, Mizutani H, Tanaka T, et al. X-ray absorption study on the local structures of fine particles of α-Fe$_2$O$_3$–SnO$_2$ gas sensors. J Mater Chem 1992, 2: 703–707.

[4] Owoc D, Przewoznik J, Kozlowski A, et al. X-ray studies of Fe$_{3-x}$Me$_x$O$_4$, Me = Zn, Ti and Al; the impact of doping on the Verwey transition. Physica B 2005, 359–361: 1339–1341.

[5] Busca G, Ramis G, Prieto MC, et al. Preparation and characterization of Fe$_{2-x}$Cr$_x$O$_3$ mixed oxide powders. J Mater Chem 1993, 3: 665–673.

[6] Musić S, Lenglet M, Popović S, et al. Formation and characterization of the solid solutions (Cr$_x$Fe$_{1-x}$)$_2$O$_3$, $0 \leqslant x \leqslant 1$. J Mater Sci 1996, 31: 4067–4076.

[7] Zysler RD, Fiorani D, Testa AM, et al. Size dependence of the spin-flop transition in hematite nanoparticles. Phys Rev B 2003, 68: 212408.

[8] López JL, Pfannes H-D, Paniago R, et al. Investigation of the static and dynamic magnetic properties of CoFe$_2$O$_4$ nanoparticles. J Magn Magn Mater 2008, 320: e327–e330.

[9] Dang M-Z, Rancourt DG, Dutrizac JE, et al. Interplay of surface conditions, particle size, stoichiometry, cell parameters, and magnetism in synthetic hematite-like materials. Hyperfine Interact 1998, 117: 271–319.

[10] Bruzzone CL, Ingalls R. Mössbauer-effect study of the Morin transition and atomic positions in hematite under pressure. Phys Rev B 1983, 28: 2430–2440.

[11] Hofmann M, Campbell SJ, Kaczmarek WA, et al. Mechanochemical transformation of α-Fe$_2$O$_3$ to Fe$_{3-x}$O$_4$—Microstructural investigation. J Alloys Compd 2003, 348: 278–284.

[12] Bersani D, Lottici PP, Montenero A. Micro-Raman investigation of iron oxide films and powders produced by sol–gel syntheses. J Raman Spectrosc 1999, 30: 355–360.

[13] Shim S-H, Duffy TS. Raman spectroscopy of Fe$_2$O$_3$ to 62 GPa. Am Mineral 2001, 87: 318–326.

[14] Beattie IR, Gilson TR. The single-crystal Raman spectra of nearly opaque materials. Iron(III) oxide and chromium(III) oxide. J Chem Soc A 1970: 980–986.

[15] McCarty KF. Inelastic light scattering in α-Fe$_2$O$_3$: Phonon vs magnon scattering. Solid State Commun 1988, 68: 799–802.

[16] McCarty KF, Boehme DR. A Raman study of the systems Fe$_{3-x}$Cr$_x$O$_4$ and Fe$_{2-x}$Cr$_x$O$_3$. J Solid State Chem 1989, 79: 19–27.

[17] Zhao Y, Dunnill CW, Zhu Y, et al. Low-temperature magnetic properties of hematite nanorods. Chem Mater 2007, 19: 916–921.

[18] Bødker F, Hansen MF, Koch CB, et al. Magnetic properties of hematite nanoparticles. Phys Rev B 2000, 61: 6826–6838.

[19] Rath C, Sahu KK, Kulkarmi SD, et al.

Microstructure-dependent coercivity in mono-dispersed hematite particles. *Appl Phys Lett* 1999, **75**: 4171.

[20] Gregg KA, Perera SC, Lawes G, *et al.* Controlled synthesis of MnP nanorods: Effect of shape anisotropy on magnetization. *Chem Mater* 2006, **18**: 879–886.

[21] Hill AH, Jiao F, Bruce PG, *et al.* Neutron diffraction study of mesoporous and bulk hematite, α-Fe$_2$O$_3$. *Chem Mater* 2008, **20**: 4891–4899.

[22] Ayub I, Berry FJ, Bilsborrow RL, *et al.* Influence of zinc doping on the structural and magnetic properties of α-Fe$_2$O$_3$. *J Solid State Chem* 2001, **156**: 408–414.

[23] He T, Luo H-L, Li S. Effect of cobalt on the morin transition of hematite. *J Magn Magn Mater* 1988, **71**: 323–328.

[24] Barrero CA, Arpe J, Sileo E, *et al.* Ni- and Zn-doped hematite obtained by combustion of mixed metal oxinates. *Physica B* 2004, **354**: 27–34.

[25] Morrish AH. *Canted Antiferromagnetism: Hematite.* Singapore: World Scientific Publishing Company, 1994.

[26] Stewart SJ, Borzi RA, Cabanillas ED, *et al.* Effects of milling-induced disorder on the lattice parameters and magnetic properties of hematite. *J Magn Magn Mater* 2003, **260**: 447–454.

[27] Sileo EE, Daroca DP, Barrero CA, *et al.* Influence of the genesis on the structural and hyperfine properties of Cr-substituted hematites. *Chem Geol* 2007, **238**: 84–93.

[28] Ericsson T, Krisnhamurthy A, Srivastava BK. Morin-transition in Ti-substituted hematite: A Mössbauer study. *Phys Scr* 1986, **33**: 88.

[29] Varshney D, Yogi A. Structural and transport properties of stoichiometric and Cu^{2+}-doped magnetite: Fe$_{3-x}$Cu$_x$O$_4$. *Mater Chem Phys* 2010, **123**: 434–438.

[30] Ruebenbauer K, Birchall T. A computer programme for the evaluation of Mössbauer data. *Hyperfine Interact* 1979, **7**: 125–133.

[31] Li L, Chu Y, Liu Y. Synthesis and characterization of ring-like α-Fe$_2$O$_3$. *Nanotechnology* 2007, **18**: 105603.

[32] Jacob J, Abdul Khadar M. VSM and Mössbauer study of nanostructured hematite. *J Magn Magn Mater* 2010, **322**: 614–621.

[33] Varshney D, Yogi A. Structural and electrical conductivity of Mn doped hematite (α-Fe$_2$O$_3$) phase. *J Mol Struct* 2011, **995**: 157–162.

23

Edge chipping resistance of ceramics: Problems of test method

George A. GOGOTSI[*]

Pisarenko Institute for Problems of Strength, 2, Timiryazevskaya Str., 01014, Kiev, Ukraine

Abstract: An unconventional method for determining the fracture resistance of brittle materials is discussed. This method employs a conical indenter to chip the rectangular edge of the specimen. Particular features of the method are the use of small specimens and the evaluation of the resistance of materials to the nucleation, initiation and propagation of a crack. It is shown that this method is somewhat similar to the Hertzian fracture method and to the way that early man selected stones to make tools and weapons. Measured data of the fracture resistance of ceramics is presented. It is confirmed that if a ceramic material is similar to the model material of linear elastic fracture mechanics (LEFM), then those fracture resistance values are directly proportional to the critical stress intensity factors (baseline). For elastic and inelastic ceramics, R-lines characterizing the fracture resistance to crack growth are plotted. It is shown that proportionality lines (edge chipping resistance versus critical stress intensity factor) may be straight lines for ceramics with similar structure (such as Y-TZP and Mg-PSZ). The effect of rounding of the conical indenter tip (10–800 μm) on chip scar shape is indicated. Other aspects in the fracture behavior of ceramics during edge chipping are also analyzed. The advantages and disadvantages of the method are discussed. Further studies in this mechanico-physical research area are suggested.

Keywords: mechanical characterization; micromechanics; indentation; phase transformation; edge fracture (EF) method

1 Introduction

Modern evaluations of the fracture resistance of ceramics are usually based on linear elastic fracture mechanics (LEFM) [1]. In these evaluations, it is assumed that the material is linear elastic and isotropic and may be inelastic only at the crack tip [2]. Such materials include aluminum oxide (Al_2O_3), silicon nitride (Si_3N_4) and other ceramics. However, many materials differ in structure and are inelastic, which is

disregarded even in the relevant ASTM, CEN and ISO standards [3], though it would be more reasonable, for instance, to use elastic-plastic fracture mechanics [1] and *J*-integral as a fracture criterion [4] for the reliable evaluation of inelastic ceramics. A telling illustration is the expansive attempt made to create high-toughness ceramics of which the first was "ceramic steel" (Mg-PSZ) [5]. However, this lesson has not been learnt and the critical stress intensity factor K_{Ic} is still considered applicable to the evaluation of the fracture resistance of ceramics that differ in structure and mechanical behavior. This hinders the improvement of ceramics and affects the reliability of ceramic products

---
* Corresponding author.
E-mail: gogotsi@ipp.kiev.ua

in engineering and restorative medicine. A shortcoming of the conventional fracture toughness methods is the use of relatively large specimens, which may be much bigger than the samples that will be applied in dental and other products. Moreover, these methods do not involve the determination of the crack-growth resistance of materials—a characteristic that may play an important role in evaluating the performance of ceramic products. These problems may possibly be resolved with the edge-chipping method, which is currently far from being perfect. Therefore, it is reasonable to optimize this method and to improve the test data treatment procedure. This problem is discussed below.

2   Energy-based evaluation of the fracture resistance of ceramics

In the infancy of mankind (more than half a million years B.C. [6,7]), there was no need for any hypotheses or models to successfully select a more crack-resistant stone for making weapons and tools. Two stones were stricken against each other (equal energy was supplied to each of them) to choose the one that displayed less damage. In his classical study, which opened the road to the development of fracture mechanics, Griffith [8] analyzed the energy needed to fracture glass and proved that its capability of resisting fracture is characterized by the energy γ_s required to create new surfaces when a crack propagates.

At the National Physical Laboratory (United Kingdom) [9], early man's core stone was replaced by a specimen with a sharp rectangular edge, while the hammer stone with a standard Rockwell indenter[1] for chipping this edge (Fig. 1(a)). It went, however, unnoticed that this is an attempt to experimentally solve a fracture problem somewhat similar to the classical Hertzian problem [11]. A test fixture with a movable microscope for precision edge chipping was created. The ratio of the fracture load p_f to the distance d from the point of its application to the edge [12] (or the slope angle of p_f vs. d curve [13]) was used as a fracture characteristic (M). Since its

---

[1] In Ref. [9], the Rockwell indenter was chosen without justification, though a spherical tip with a radius of 200 μm is known to be minimum for elastic contact between the indentor and the material [10].

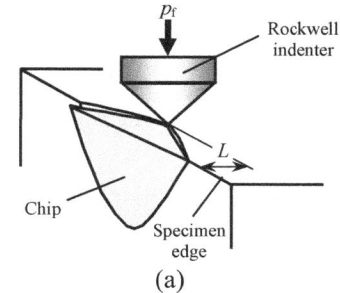

(a)

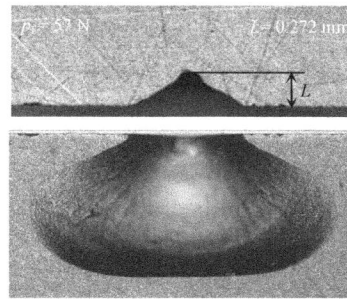

(b)

Fig. 1   (a) Schematic of edge-chipping method and (b) chip scars of $La_{0.8}Ca_{0.2}CoO_3$ perovskite ceramics.

unit is N/m, i.e., the same as the unit of fracture energy, this characteristic is named edge toughness. In Refs. [12] and [13], it is shown that for hard metals (and some other materials), this characteristic is in a nonlinear relationship with the critical strain energy release rate G_{Ic} [1]. However, such a relationship is not observed for all brittle materials [14], probably because chip scars (and the corresponding fracture surfaces) on specimen edges can differ considerably among tested materials[2] (Fig. 2). It should be noted that the edge-chipping method is the only method to determine the fracture resistance of brittle materials without the need to control the shape of fracture surfaces.

3   Description of tests

In the tests described in Ref. [15], a specimen bonded to a photographic glass was placed on the two-axis table of a CeramTest unit (Gobor Ltd.) mounted on a universal testing machine. To chip the edge, a standard diamond Rockwell indenter (Gilmore Diamond Tools, Inc.) with spherical tip of radius of 200 μm fixed in the load rod of the test unit was used. The indentation

---

[2] In Ref. [9], however, it is stated that "flake geometries are remarkably similar and independent of the distance, force, or material."

(a)   (b)

(c)   (d)

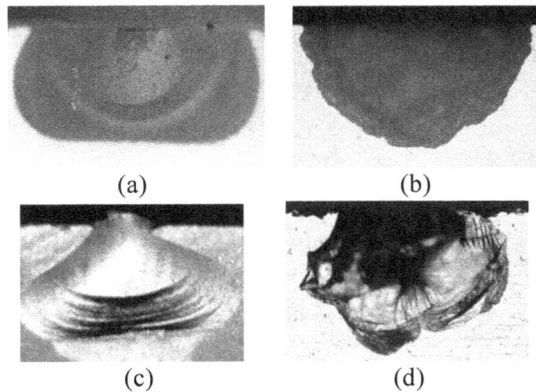

Fig. 2   Chip scars of (a) elastic Y-TZP and (b) inelastic Mg-PSZ zirconia ceramics as well as (c) silicon carbide EKasic® F ceramics and (d) polycrystalline silicon.

point was selected visually, with a magnifier. The indentation rate (crosshead speed) was constant and equal to 0.5 mm/min, as in standard tests on ceramics. During the tests, a computer recorded the time dependence of the displacement of the crosshead, which was stopped upon a decrease in the indenter load. The maximum load on the indenter is considered to be the fracture load p_f (Fig. 1). Then a BX51M Olympus (50–1000×) binocular microscope with QuickPhoroMicro 2.3 software was used to photograph the chip scars and to measure the distance (called fracture distance L) from the specimen edge to the extreme point of the chip scar (Fig. 1). These tests employed fragments of specimens that had been used to measure fracture toughness (K_{Ic} values [16]). A load support was installed in the CeramTest unit to provide free displacement of a specimen with a V-shaped notch with a tip radius of no greater than 10 μm (SEVNB method). The critical stress intensity factor (K_{Ic}) was determined using the following formula:

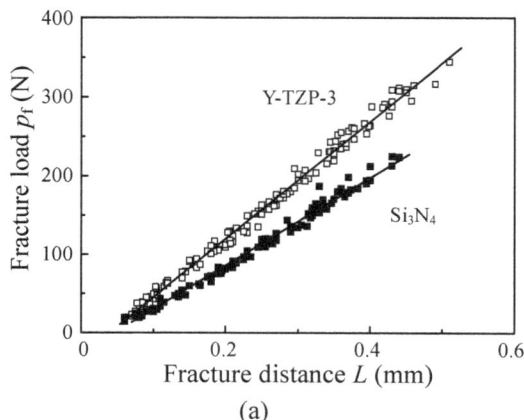

$$K_{Ic} = \frac{F}{B\sqrt{W}} \cdot \frac{S}{W} \cdot \frac{3\sqrt{\alpha}}{2} \cdot Y^*$$

where

$$Y^* = \frac{1.99 - \alpha(1-\alpha)(2.15 - 3.93\alpha + 2.7\alpha^2)}{(1+2\alpha)(1-\alpha)^{3/2}}$$

F is the fracture load; B and W are the width and height of the specimen, respectively; S is the distance between the support rollers; a and $\alpha = a/W$ are the absolute and relative depths of the notch, respectively.

The specimens were mirror-polished by 4 mm × 3 mm rectangular beams with sharp edges rounded to a radius of no greater than 10–15 μm. Fracture toughness and fracture resistance were measured on the same specimen to avoid any doubt that the materials tested under different conditions were different. Based on a great amount of statistically reliable test data, a method for edge fracture (EF) tests with a Rockwell indenter was developed [15]. Its major feature is an original analysis of edge-chipping data and a mechanico-physical approach to the problem. Emphasis in the method is placed on the fact that dissimilar brittle materials have very different chip scars; therefore, their surfaces are photographed in the indentation and perpendicular directions for further examination. After the completion of tests, fracture diagrams (p_f versus L) are plotted. They are linear for elastic materials and nonlinear for inelastic materials (Fig. 3). Fracture resistance value is calculated as follows [15]:

$$F_r = \frac{1}{n}\sum_{i=1}^{n}\frac{p_{fi}}{L_i}$$

where p_{fi} and L_i are the current fracture load and fracture distance, respectively.

The collected statistically reliable test data were

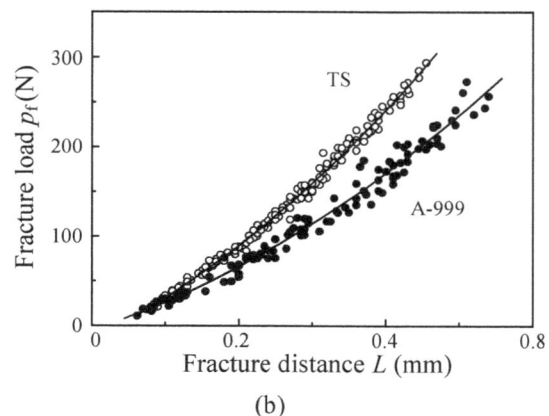

(a)   (b)

Fig. 3   Fracture diagrams of (a) elastic Y-TZP and silicon nitride Si_3N_4 as well as (b) inelastic TS and alumina A-999 ceramics.

used to compare fracture resistance, F_r, and fracture toughness K_{Ic} values, determined on the same specimens of ordinary homogeneous single-phase ceramics similar in mechanical behavior to the LEFM model material. It appears that these quantities are in a linear relationship (Fig. 4), called the baseline [15]. This line indicates that the two essentially different test methods make equivalent assessments of fracture resistance. It can be used to determine how similar the tested ceramic material is to the LEFM model material and, hence, to validate the K_{Ic} values found by any LEFM-based test method. Therefore, it was proposed to add the EF method to the fracture toughness test standards for ceramics [17].

4   Analysis of test data

An analysis of the test data reveals that the relationship between edge-chipping fracture characteristic and fracture toughness is linear for ceramics with similar mechanical behavior (such as inelastic Mg-PSZ and elastic Y-TZP ceramics; Fig. 5). Direct proportionality (similar to the baseline) is also observed in tests with other indenters such as a Vickers indenter [18]. When plotted on the base diagram, such straight lines (which may be called proportionality lines) have a smaller slope angle than that of the baseline (Fig. 5) and therefore, they are less effective for comparison of tested ceramics.

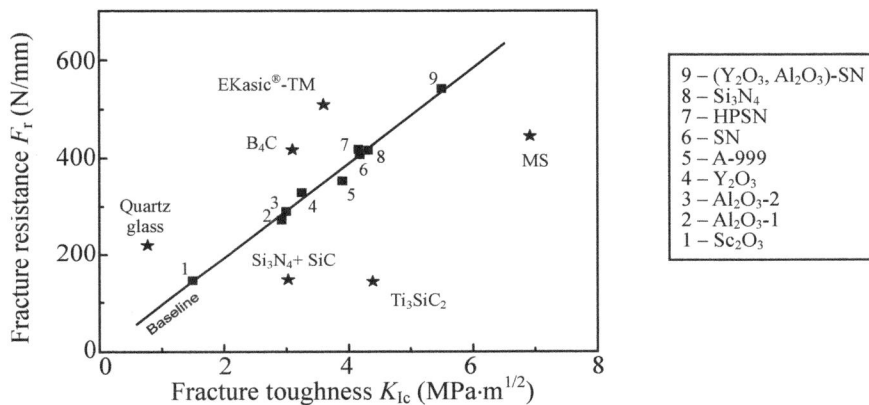

Fig. 4   Base diagram with baseline and data points for various ceramics and glasses.

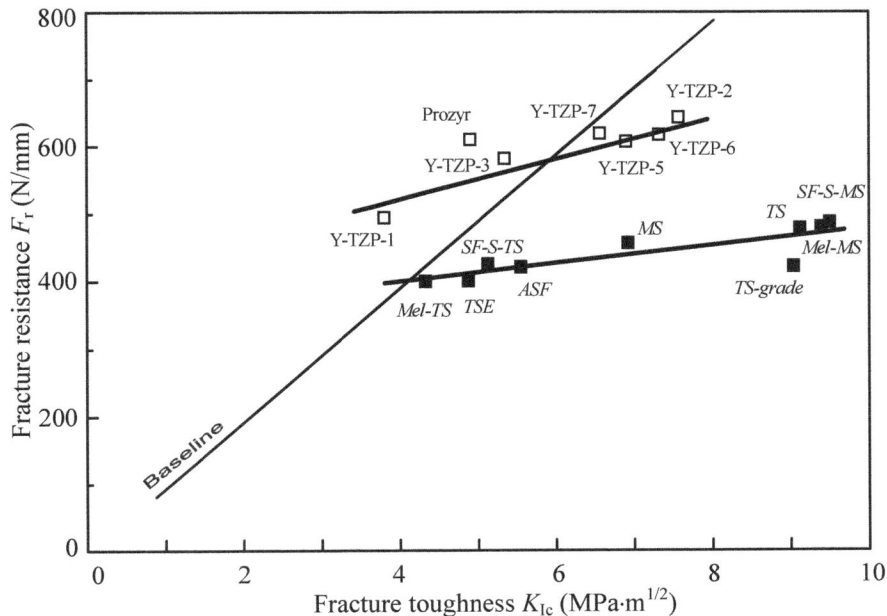

Fig. 5   Proportionality lines for elastic (Y-TZP) and inelastic (Mg-PSZ) zirconia ceramics.

An analysis of the test data shows that both the fracture load and the surface area of the chip scar increase with the distance L. In Ref. [15], it was proposed to describe this effect by plotting fracture resistance F_r versus fracture distance L (R-lines shown in Fig. 6). Note that R-lines are qualitatively similar to R-curves in LEFM [1]: in both cases, the fracture resistance of ceramics increases, which, in fact, corresponds to the increase in the area of the fracture surface (LEFM methods deal only with the length of the propagating crack). The correspondence of nonlinearly rising R-line and R-curve was analyzed in Ref. [15], and it was established that the respective evaluations of the fracture resistance of ceramics are similar. This is important because the R-line is obtained from small specimens tested with relatively simple test equipment. Contrastingly, plotting the R-curve is quite difficult and possible only in a specialized mechanical laboratory.

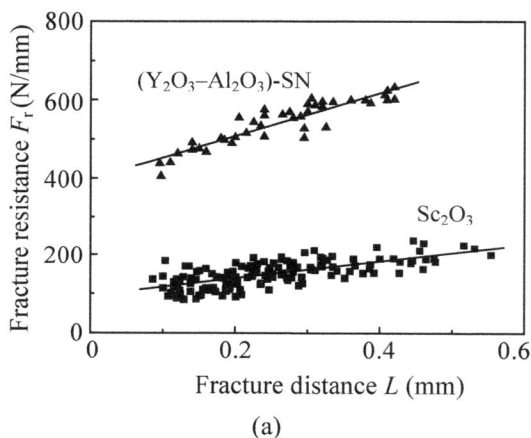

(a)

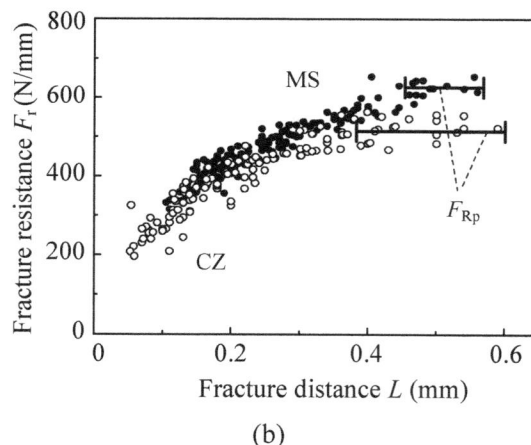

(b)

Fig. 6   R-lines of (a) elastic silicon nitride $(Y_2O_3-Al_2O_3)$-SN and scandia oxide Sc_2O_3 ceramics, and (b) inelastic zirconia MS and composites alumina+zirconia CZ ceramics.

Summing up the aforementioned, it should be pointed out that the EF method is promising for the comparative fracture resistance evaluations of ceramics with similar chip scars and for the study of their fracture resistance. In this case, not only R-lines are plotted and F_r is determined, but also a new fracture characteristic F_{Rp} of ceramics corresponding to the plateau of R-lines (Fig. 6(b)) [17] can be determined. Note that a similar characteristic was proposed earlier in analyzing an R-curve [19], but it was not accepted probably because of the technical difficulties of obtaining it, which is not so when using the EF method. This characteristic may be of practical interest. It should also be pointed out that all methods for determining the edge chipping resistance of ceramics (as well as ordinary LEFM methods), unlike the prehistoric method, are actually unsuitable for the reliable evaluation of dissimilar ceramics and other brittle materials, i.e., are not "universal" methods.

In studying the fracture of materials with the method in question, it is possible, following Ref. [20], to examine a chip scar on a quartz glass specimen (Fig. 7) made with a Rockwell indenter (spherical tip). At the first stage of fracture, tensile stresses cause a microdefect that is always present on the surface to grow to a critical size, thus nucleating a crack, which is a critical event leading to fracture [21]. At the second stage of fracture, an incipient surface crack (a part of a ring crack) develops[3]. The third stage is the initiation and propagation of a pseudo-conical crack. This pattern is somewhat similar to the formation of a Hertzian cone crack [12], which is studied using compound specimens (two pieces glued together) separated after the test [24] to examine the fracture surfaces.

It should be noted that the second stage of fracture is controlled by the properties of the subsurface layer, which is, unfortunately, disregarded. Therefore, the most complicated is the second stage of fracture, which corresponds to the fracture onset barrier [25], typical for the materials under consideration. In Ref. [15], it was shown that brittle materials may have barriers of three types. A low fracture barrier (mentioned in Ref. [26] with reference to inelastic and composite ceramics)

---

[3]   At large values of L, not only parts of a ring crack, but also radial cracks may propagate because the stress–strain state in the contact region between the indenter and the specimen becomes symmetric [22] (Hertzian problem [11]). This is disregarded in Ref. [23].

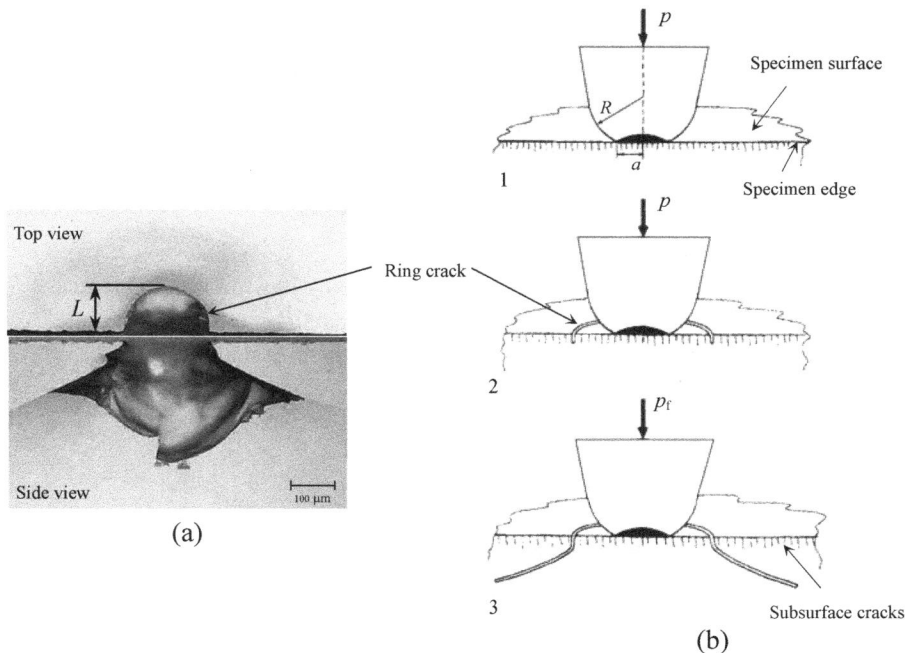

Fig. 7   (a) Chip scar on quartz glass specimen and (b) fracture development scheme: 1, 2 and 3 denote stages of fracture; R is the indenter tip radius; a is the radius of the contact area between the indenter and the specimen.

is observed when data points lie below the baseline (Figs. 4 and 5). It was established that it is due to the presence of microcracks, weak bonds between grains, phase transitions, and other effects in the subsurface layer. Contrastingly, the "normal" fracture barrier does not affect fracture resistance (it is typical for ceramics with data points on the baseline). Of interest is a high fracture barrier observed in glass and in ceramics used to manufacture cutting tools and armor: data points lie above the baseline (Fig. 4). Its cause is unclear. It is only known [15] that the fracture zones caused by an indenter in such materials are very different from those in materials with data points on the baseline (Fig. 8). This is indicative of the fact, known in contact mechanics, that each material resists local fracture in its own way, which was earlier disregarded. An attempt was made in Ref. [18] to analyze a high fracture barrier based on edge-chipping data obtained with a Vickers indenter, which causes elastic-plastic fracture of the material. In this case, however, the concept of fracture onset barrier loses its meaning. At the beginning of our studies, following Ref. [27], we also tried to understand this effect by scratching the specimen with a Rockwell indenter until chipping [28]. Actually, that is meaningless because a scratch not only partially excludes the first stage of fracture, but also forms an additional uncontrollable stress concentrator that depends on the material and the load applied to the indenter. Thus, the high fracture onset barrier in ceramics is yet to be evaluated and understood, and it is still unclear whether the descriptions of the edge-chipping fracture of many brittle materials are reliable. This is one of the important reasons why edge-chipping data cannot often be used to compare the fracture resistance of different brittle materials. A shortcoming of ordinary edge-chipping tests is the use of a relatively sharp Rockwell indenter, which causes both quasi-cone chip scars (elastic fracture) and scallop-like chip scars (elastic-plastic fracture) on fused quartz specimens [29]. This is why the fracture resistance of this material may be assigned different values. The problem can be resolved with an indenter with a tip radius of 400 μm (Fig. 9). Therefore, conical diamond indenters with different tip radii (Gilmore Diamond Tools, Inc., USA) were used in edge chipping of soda lime glass (often considered as a model material). It was established that an indenter with a tip radius of 10 μm does not produce pseudo-conical chip scars on glass specimens. There are 38% of such scars if the tip radius is 100 μm, 68% if it is 200 μm, and 100% if it is 400 μm. An 800-μm indenter leaves only pseudo-conical scars. It was noticed that a steel indenter with a tip radius of 2000 μm makes a chip scar of different shape on the edge of Y-TZP ceramics specimens (see Figs. 2 and 10).

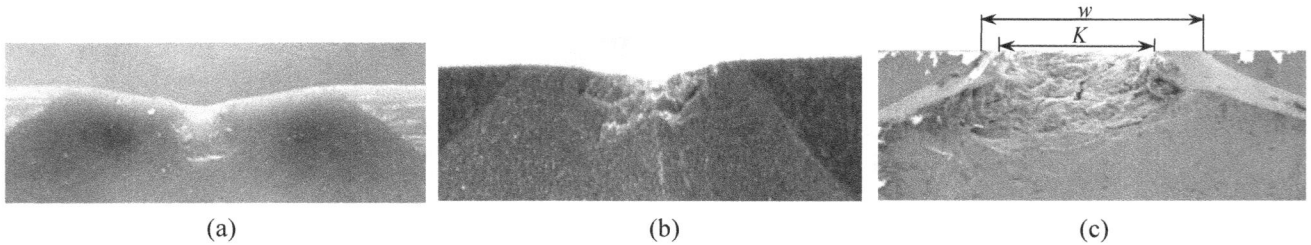

(a)   (b)   (c)

Fig. 8   (a) Chip scar of silicon nitride GPSSN ceramics with data points on the baseline and (b) chip scars of silicon carbide EKasic® TM ceramics and (c) light crown glass whose data points are above the baseline: K is the width of the zone damaged by the indenter and w is the chip scar size in the indentation direction.

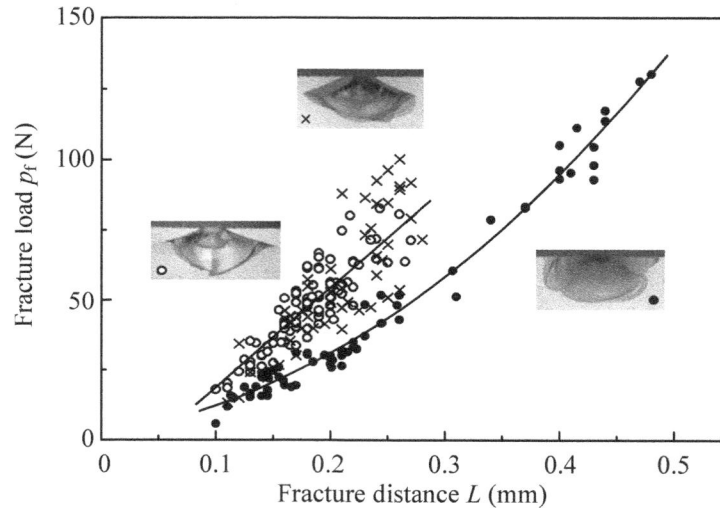

Fig. 9   Fracture diagrams of fused silica: data points for quasi-cone chip scars (($\bigcirc$) Rockwell and ($\times$) conical indenters) and for scallop-like chip scars (($\bullet$) Rockwell indenter).

Fig. 10   Chip scar on Y-TZP ceramics (indenter tip radius of 2000 μm).

be used only for the comparative evaluations of the fracture resistance of brittle materials that show similar local fracture behavior. Important tasks for subsequent research is to find the optimal radius of the indenter tip and to study the fracture behavior of the surface and subsurface layer of specimens to improve the reliability of evaluations of the fracture resistance of brittle materials made by this method.

Acknowledgements

The author is grateful to V. Galenko for technical assistance.

5   Conclusions

The edge-chipping method (with Rockwell indenter) is still a long way from perfection and can successfully

References

[1] Broek D. *Elementary Engineering Fracture Mechanics*. Dordrecht: Kluwer Academic Publishers, 1986.

[2] Irwin GR. Analysis of stresses and strains near the end of a crack traversing a plate. *J Appl Mech* 1957, **24**: 361–364.

[3] ISO 23146:2008. Fine ceramics (advanced ceramics, advanced technical ceramics)—Test methods for fracture toughness of monolithic ceramics—Single-edge V-notch beam (SEVNB) method. Gevena: ISO, 2008.

[4] Evans AG, Faber KT. Crack-growth resistance of microcracking brittle materials. *J Am Ceram Soc* 1984, **67**: 255–260.

[5] Garvie RC, Hannink RH, Pascoe RT. Ceramic steel? *Nature* 1957, **258**: 703–704.

[6] Brumm A, Aziz F, van den Bergh GD, *et al*. Early stone technology on Flores and its implications for *Homo floresiensis*. *Nature* 2006, **441**: 624–628.

[7] Balter M. New light on revolutions that weren't. *Science* 2012, **336**: 530–531.

[8] Griffith A. The phenomena of rupture and flow in solids. *Phil Trans R Soc Lond A* 1921, **221**: 163–198.

[9] Almond EA, McCormick NJ. Constant-geometry edge-flaking of brittle materials. *Nature* 1986, **321**: 53–55.

[10] Evans AG, Wilshaw TR. Quasi-static solid particle damage in brittle solids—I. Observations analysis and implications. *Acta Metall* 1976, **24**: 939–956.

[11] Fischer-Cripps AC. *Introduction to Contact Mechanics*, 2nd edn. New York: Springer, 2007.

[12] McCormick NJ, Almond EA. Edge flaking of brittle materials. *J Hard Mater* 1990, **1**: 25–51.

[13] Morrell R, Gant AJ. Edge chipping of hard materials. *Int J Refract Met H* 2001, **19**: 293–301.

[14] Gogotsi GA, Galenko VI, Mudrik SP, *et al*. Fracture resistance estimation of elastic ceramics in edge flaking: EF baseline. *J Eur Ceram Soc* 2010, **30**: 1223–1228.

[15] Gogotsi GA. Fracture resistance of ceramics: Base diagram and *R*-line. *Strength Mater* 2006, **38**: 261–270.

[16] Gogotsi GA. Fracture toughness of ceramics and ceramic composites. *Ceram Int* 2003, **29**: 777–784.

[17] Gogotsi GA. Criteria of ceramics fracture (edge chipping and fracture toughness tests). *Ceram Int* 2013, **39**: 3293–3300.

[18] Gogotsi GA, Mudrik SP. Fracture barrier estimation by the edge fracture test method. *Ceram Int* 2009, **35**: 1871–1875.

[19] Readey MJ, Heuer AH, Steinbrech RW. Crack propagation in Mg-PSZ. *MRS Proc* 1986, **78**: 107–120.

[20] Wilshaw TR. The Hertzian fracture test. *J Phys D: Appl Phys* 1971, **4**: 1567–1581.

[21] Knott JF. Micromechanisms of fracture and the fracture toughness of engineering alloys. In *Fracture*. Taplin DMR, Ed. Canada: University of Waterloo Press, 1977: 61–92.

[22] Batanova OA, Gogotsi GA, Matvienko YuG. Numerical analysis of edge chipping data. *Ind Lab Diag Mater* 2011, **77**: 53–56. (in Russian)

[23] Chai H. On the mechanics of edge chipping from spherical indentation. *Int J Fracture* 2011, **169**: 85–95.

[24] Guiberteau F, Padture NP, Lawn BR. Effect of grain size on Hertzian contact damage in alumina. *J Am Ceram Soc* 1994, **77**: 1825–1831.

[25] Drucker DC. Macroscopic fundamentals in brittle fracture. In *Treatise on Fracture*. Liebowitz H, Ed. New York: Academic Press, 1968: 473–531.

[26] Gogotsi GA. The problem of the classification of low-deformation materials based on the features of their behavior under load. *Strength Mater* 1977, **9**: 77–83.

[27] Quinn J, Su L, Flanders L, *et al*. "Edge toughness" and material properties related to the machining of dental ceramics. *Mach Sci Technol* 2000, **4**: 291–304.

[28] Gogotsi GA, Mudrik SP, Quinn J. Edge toughness of silicon nitride ceramics: Method and results. In *Proceedings of the International Conference on Novel Technologies in Power Metallurgy and Ceramics*. Kiev: IPMS, 2003: 375–376.

[29] Gogotsi GA, Mudrik SP. Glasses: New approach to fracture behavior analysis. *J Non-Cryst Solids* 2010, **356**: 1021–1026.

Spark plasma sintering of damage tolerant and machinable YAM ceramics

Qian WANG[a,b], Salvatore GRASSO[c], Chunfeng HU[a,*], Haibin ZHANG[d],
Shu CAI[b,*], Yoshio SAKKA[e], Qing HUANG[a,*]

[a]Ningbo Institute of Materials Technology and Engineering (NIMTE), Chinese Academy of Sciences (CAS), Ningbo 315201, China
[b]Key Laboratory for Advanced Ceramics and Machining Technology of Ministry of Education, Tianjin University, Tianjin 300072, China
[c]School of Engineering & Materials Science and Nanoforce Technology Ltd., Queen Mary University of London, Mile End Road, London E1 4NS, UK
[d]China Academy of Engineering Physics, Mianyang 621900, China
[e]National Institute for Materials Science (NIMS), Tsukuba, Ibaraki 305-0047, Japan

Abstract: Single-phase $Y_4Al_2O_9$ (YAM) powders were synthesized via solid-state reaction starting from nano-sized Al_2O_3 and Y_2O_3. Fully dense (99.5%) bulk YAM ceramics were consolidated by spark plasma sintering (SPS) at 1800 ℃. We demonstrated the excellent damage tolerance and good machinability of YAM ceramics. Such properties are attributed to the easy slipping along the weakly bonded crystallographic planes, resulting in multiple energy dissipation mechanisms such as transgranular fracture, shear slipping and localized grain crushing.

Keywords: ceramic; synthesis; microstructure; property; damage tolerance

1 Introduction

In the Al_2O_3–Y_2O_3 phase diagram, three intermediate compounds exist: (i) $Y_3Al_5O_{12}$ (YAG) with cubic yttrium aluminum garnet structure, (ii) $YAlO_3$ (YAP) with orthorhombic yttrium aluminum perovskite or hexagonal structure, and (iii) $Y_4Al_2O_9$ (YAM) with monoclinic structure [1]. YAG has been widely investigated for a wide range of applications, including rare-earth doped phosphors [2–6] (where color and

efficiency are controlled by host lattice effects), and transparent ceramics for solid-state lasers [7–9]. In addition, more structural applications have been proposed for YAG as reinforcement fiber in ceramics and intermetallic composites [10] besides functional YAG thermal barrier coatings (TBC) [11] and thin films [12,13].

Compared to YAG, YAP shows complementary properties, and it is currently employed in laser system as an excellent gain medium material for scintillators, acousto-optics [14,15] and Q-switches [16,17].

Up to now, the luminescence properties of rare-earth doped YAM have been widely investigated. For example, Yadav *et al.* [18] studied the emitting properties of Eu^{3+}-doped YAM exposed to vacuum

* Corresponding authors.
E-mail: Chunfeng Hu, hucf@nimte.ac.cn;
　　　Shu Cai, caishu@tju.edu.cn;
　　　Qing Huang, huangqing@nimte.ac.cn

ultraviolet (VUV) and ultraviolet (UV). Wang and Wang [19] reported that YAM:Re (Re = Tb^{3+}, Eu^{3+}) phosphors are promising candidates for plasma display applications. The thermal stability and phase transformation of YAM at high temperature attract considerable research attention as well [20–24]. Furthermore, YAM might have some potential applications even as high-temperature oxidation/ thermal barrier coatings [25]. In fact, Zhan et al. [26] have shown that the lattice thermal conductivity of YAM is as low as 1.10 W/(m·K) on the basis of the first-principles calculation, and experimental results confirm a very low thermal conductivity of 1.56 W/(m·K) at 1000 ℃.

However, despite the significant number of scientific publications about YAM, its intrinsic mechanical properties are seldom described in literature. The present work aims to (i) synthesize high-purity single-phase YAM powders via a solid-state reaction, (ii) achieve fully dense bulk ceramics using spark plasma sintering (SPS), and (iii) characterize the microstructure and describe the mechanisms involved in its damage tolerance behavior and machinability.

2 Experimental procedure

Nano Al$_2$O$_3$ powders (140 nm, 99.99%, Taimei Chemcials Co Ltd., Osaka, Japan) and nano Y$_2$O$_3$ powders (50 nm, 99.99%, Aladdin Reagent Inc., Shanghai, China) with the molar ratio of 1 : 2 were wet mixed in a planetary ball milling machine. The slurry was freeze-dried in order to minimize particle agglomeration. Subsequently, the freeze-dried powders were sieved and heated in an alumina crucible at 1650 ℃ for 30 h by using a muffle furnace (Nabertherm P310, Bremen, Germany). The as-synthesized powder mixture was poured in a 20-mm diameter graphite mould and densified by an SPS furnace (FCT Systeme GmbH DH25, Frankenblick, Germany). The densification process was carried out via a multi-step heating method; the initial heating rate from room temperature (RT) to 1300 ℃ was 50 ℃/min, and at higher temperature it was reduced down to 25 ℃/min. The dwell time and uniaxial pressure were 15 min and 30 MPa, respectively. The sintered bulks were examined by X-ray diffraction (XRD, Bruker AXS Inc. D8 Discover, Madison, WI) and scanning electron microscope (SEM, Hitachi S-4800, Tokyo, Japan) equipped with energy dispersive spectrum

(EDS). The densities of sintering polished bodies were determined by Archimedes method. Vickers hardness (H_v) was measured by a Vickers hardness tester (Wilson Wolpert 432 SVD, Norwood, MA, USA) under various loads. For each load, six indents were made. The morphologies of indents were examined by SEM. The polished and fractured surfaces were also examined with SEM. The flexure strength was determined via a three-point bending test. The samples were cut into the dimensions of 3 mm × 3 mm × 18 mm with diamond wire saw. The crosshead in the test was 0.5 mm/min. The brittleness index (B) and CNC (computer numerical control) machining were used to quantitatively estimate the machinability of YAM ceramics. The thermal conductivity of disk YAM specimen (Φ 12.7 mm × 1 mm) was measured by the laser flash technique in the temperature range of 400–980 ℃. At various temperatures, the thermal conductivity (D_{th}) and constant pressure molar heat capacity (C_p) were measured by a laserflash thermal analyzer (NETZSCH LFA457, Selb, Germany).

3 Results and discussion

The synthesis of YAM powders was carried out via a solid-state reaction using nano-sized Al$_2$O$_3$ and Y$_2$O$_3$ powders as starting materials. The relatively high reaction temperature (1650 ℃) and prolonged dwelling time (30 h) contributed to the synthesis of pure YAM powders as confirmed by XRD.

The starting Al$_2$O$_3$ is α phase, which is the only allotrope thermodynamically stable at high temperature [27]. In addition, Al^{3+} and Y^{3+} show identical valence. Thus, no electron migration occurred in the reaction process; the reaction was mainly driven by the mutual diffusion of Al^{3+} and Y^{3+} ions. However, the ionic radius of Y^{3+} ion is larger than that of Al^{3+} ion, making the solid-state reaction largely depending on the diffusion rate of Y^{3+} ion. Therefore, it is reasonable to assume that the reaction kinetics was mainly controlled by the diffusion rate of Y^{3+} ions into Al^{3+}-rich side.

During the progress of the reaction, YAM, YAP and YAG phases appeared (detected by XRD, not shown here). The intermediated formed phases behaved as barrier inhibiting the mutual diffusion of Al^{3+} ions and Y^{3+} ions. Accordingly, in comparison with the phase-diagram temperatures, the formation of intermediate layers required higher reaction temperature and prolonged soaking time to complete the synthesis of YAM.

Due to the thermal stability even at 1650 ℃ for tens of hours (experimental synthesis conditions), in combination with the extremely low thermal conductivity [26], YAM might be considered as a promising candidate for thermal insulation material. In order to investigate the mechanical properties of YAM ceramics, the as-prepared powders were consolidated by SPS method. Figure 1 shows the XRD pattern of YAM bulk sintered at 1800 ℃ for 15 min. No phase transformation or decomposition is observed in comparison with the starting powders, and all the diffraction peaks belong to YAM phase.

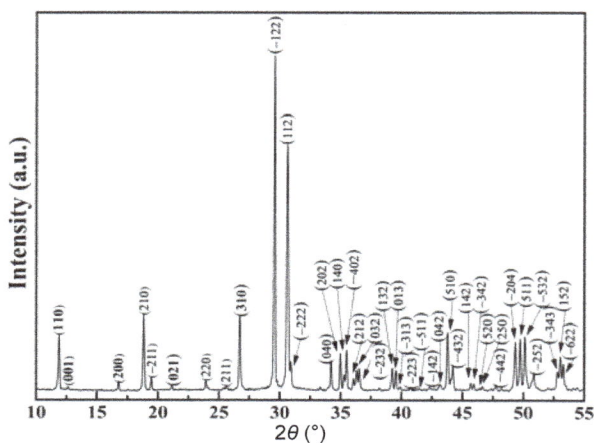

Fig. 1   XRD pattern of YAM bulk sintered at 1800 ℃ for 15 min. No detectable modification on the pattern is observed in comparison with the synthesized powders.

Figure 2 shows the absolute and relative densities of sintered YAM bulk as a function of SPS temperatures. The densities increase with the increment of sintering temperature. Full dense YAM, 99.5% (4.497 g/cm^3), is obtained when the sintering temperature is 1800 ℃. No phase decomposition is observed in the YAM, thus confirming the excellent temperature stability even up to 1800 ℃.

Figure 3 shows the SEM micrographs of polished and fracture surfaces of YAM ceramics sintered at 1750 ℃ and 1800 ℃. As confirmed in Figs. 3(a) and 3(b), no obvious porosity is observed on the polished surfaces, and this is consistent with the density measurements shown in Fig. 2. As observed in Figs. 3(c) and 3(d), the main fracture mode is transgranular. In addition, typical cleavage zone pointed by an arrow

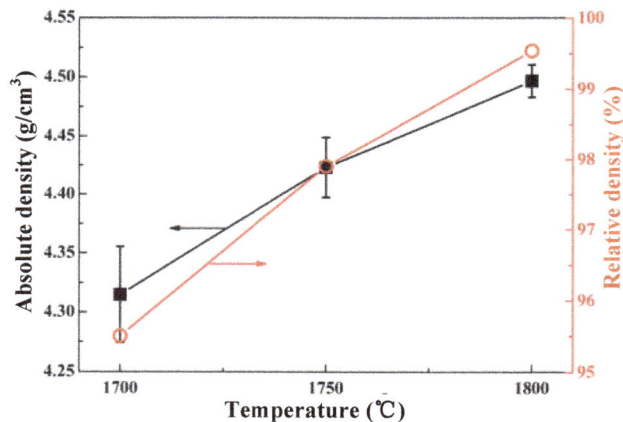

Fig. 2   Absolute and relative densities of sintered YAM bulk as a function of sintering temperatures.

in Fig. 3(c) is found widely distributed on the fracture surfaces. The specific fracture surface also indicates the presence of weak interface. Sun et al. [28] reported the similar fracture characteristic in high damage tolerant ceramics γ-$Y_2Si_2O_7$, where the fracture energy was believed to be consumed by crack deflection and cleavage. Similarly, the YAM fracture by cleavage is a more preferred means for energy dissipation since it requires less energy than crack propagation through the grain boundaries [29].

The insets of Fig. 4 show the Vickers hardness imprints of YAM performed under different indentation loads. Due to the elastic recovery of the sample during unloading, the so-called indentation size effect (ISE) [30] occurs and the Vickers hardness decrease with the load. The Vickers hardness of dense YAM ceramics under a load of 100 N approaches 4.3 GPa.

From the insets of Fig. 4, no cracks are generated at the corner of the indents for loads between 10 N and 100 N. The mechanical feedback behaviors of YAM to external loads imply that some special energy dissipation mechanisms might exist there. Consequently, the morphology of Vickers indentation imprints was investigated with SEM. It is found at a low load (10 N), the indentation energy is dissipated by grain crushing. When the load is increased up to 30 N, apart from being absorbed by grain crushing, the mechanical energy is directly dissipated at the contact area by the shearing slipping (Fig. 4 Inset (b)). When the loads are increased up to 50 N and 100 N, the contact pressure zones peel off the material from the surfaces, and this might be associated with pulling out and sliding effects induced by the high shearing stress.

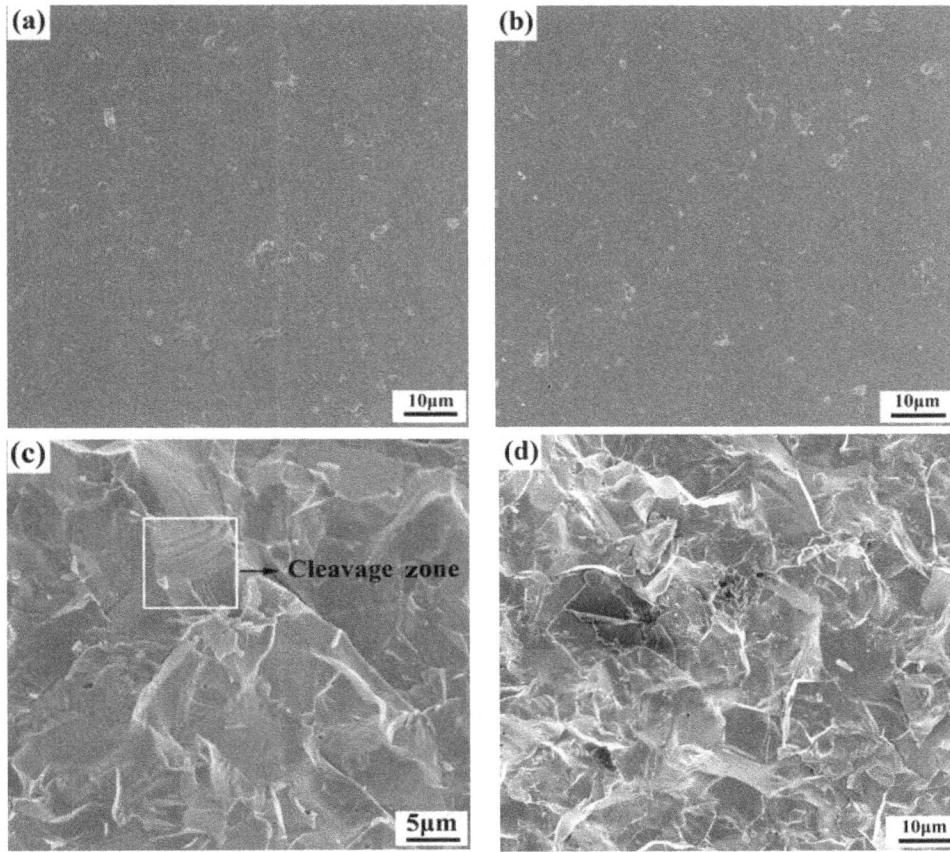

Fig. 3 SEM micrographs of polished surfaces (sintered at (a) 1750 ℃, (b) 1800 ℃) and fracture morphologies (sintered at (c) 1750 ℃, (d) 1800 ℃) of YAM bulks.

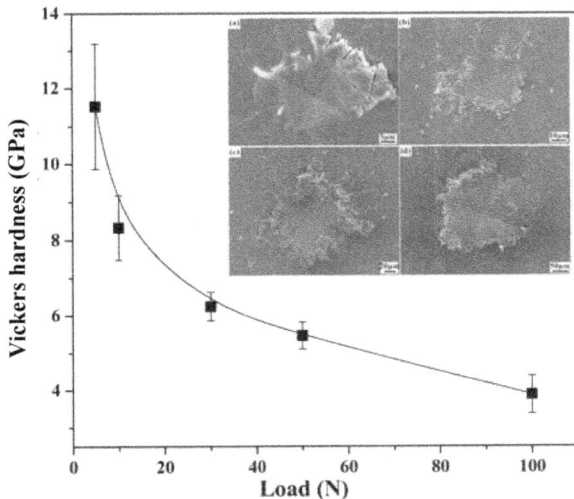

Fig. 4 Vickers hardness profile as a function of indentation load. The insets show the SEM micrographs of Vickers indents induced by different loads: (a) 10 N, (b) 30 N, (c) 50 N, and (d) 100 N.

The obvious protuberance observed along the edges of the imprints as evidenced in Fig. 4 Insets (c) and (d), suggests that the weakly bonded planes easily slip under the shear stress of the indentations [26]. The damage mechanisms consist of intergranular and transgranular fractures, grain crushing and pulling-out effect. It is then reasonable to conclude that YAM ceramics exhibit typical damage tolerant features like γ-$Y_2Si_2O_7$ and nano-layered MAX phases [28,31].

In order to investigate in more details the damage mechanisms involved, the cross section of indentation imprint made by a load of 50 N was analyzed. As indicated in Fig. 5(a), three layers distribute vertically downward to the damage zone, being surface layer, subsurface layer and substrate, respectively. It is observed that the surface layer is crushed into fine powders. Grain crushing is believed to be the dominating mechanism for energy dissipation. Together with grain crushing (Fig. 5(b), area B), it is possible to observe shear slipping (Fig. 5(c), area C) in the subsurface layer. As also evidenced in Fig. 5, the substrate remains intact, while most of the indentation energy is locally dissipated nearby the indentation by crushing and shearing of grains. This localized dissipating energy feature is comparable with contact damage tolerance of quasi-plastic ceramics such as MAX phases [31].

Fig. 5   (a) Magnification of protruded zone, evidencing (b) grain crushing (area B in (a)), and (c) shear zone (area C in (a)) obtained under an indentation load of 100 N of YAM sintered at 1800 ℃.

It is also found that dense bulk YAM ceramics can be easily machined by conventional cemented carbide (WC-Co) drill. Drilling test was performed using a 3-mm diameter drill with a rotating speed of 1250 rpm. No drilling fluid or coolant was used during the drilling. YAM ceramics show almost the same material removal rate compared to MAX phases. A macrograph of the drilled sample with 3 closely aligned holes is shown in Fig. 6(a). The holes couple well with the drill,

no visible cracking or chipping phenomena are observed around these holes. A high-magnification SEM image to the inner wall of the drilled hole is shown in Fig. 6(b). The inner wall of the hole is complanate and partially covered with a layer of smeared debris. The apparent roughness (R_a) is about 4 μm. Consequently, it is then reasonable to consider YAM as machinable ceramics. Although porosity, to some degree, determines the machinability of ceramics greatly [32], it is just not the case for YAM ceramics when considering the extremely high relative density of sintered bulks. The cleavage planes that behave as weak interfaces in fact partially contribute to the good machinability, as demonstrated in the case of mica-containing machinable glass ceramics [33]. Analogously, the successful fabrication of machinable Si_3N_4/h-BN composite is also benefitting from the self-possessing cleavage planes in h-BN [34,35]. In addition, the aforementioned specific energy dissipation mechanisms such as grain crushing, shear slipping or deflection and branching effects might as well contribute to the machinalibity of YAM ceramics.

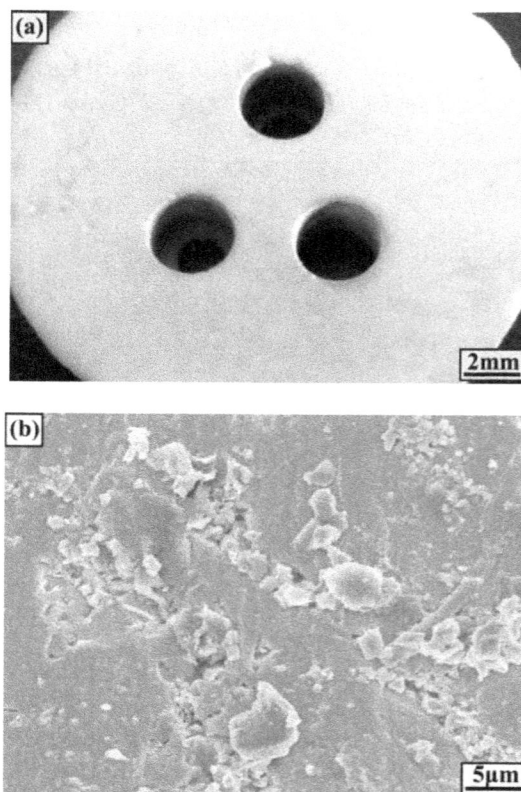

Fig. 6   (a) Optical photograph of the drilled YAM ceramics and (b) SEM image of the surface of the drilled hole.

First-principles calculations anticipate that YAM exhibits relative low shear modulus (74 GPa) [28]. The flexure strength of dense YAM ceramics prepared by SPS is 110 MPa. In agreement with the previous investigations on machinable oxide ceramics such as LaPO$_4$ [36] and CaWO$_4$ [37], both the low shear modulus and shear strength are crucial factors to obtain good machinability. Thus, the relative low shear modulus together with the low flexure strength of YAM might also have led to the good machinability.

Furthermore, brittleness index (B), originally introduced by Boccaccini [38], is used to quantitatively assess the machinability of YAM. The brittleness index (B) is defined as follows:

$$B = \frac{H_v}{K_{IC}} \qquad (1)$$

where H_v represents the intrinsic Vickers hardness; K_{IC} stands for the fracture toughness. Lower value of B corresponds to the superior machinability. For comparison, some typical machinable ceramics such as LaPO$_4$ [36], Macor [39], Ti$_3$SiC$_2$ [40], Nb$_4$AlC$_3$ [41], and γ-Y$_2$Si$_2$O$_7$ [28] are listed in Table 1. As indicated in Table 1, the brittleness index of YAM is 3.58 $\mu m^{-1/2}$, while for highly machinable MAX phases of Ti$_3$SiC$_2$ and Nb$_4$AlC$_3$, the B values are 0.67 $\mu m^{-1/2}$ and 0.37 $\mu m^{-1/2}$, respectively. Even if YAM does not show machinability as good as that of MAX phases, the machinability of YAM is comparable to γ-Y$_2$Si$_2$O$_7$ ceramics and it is slightly better than that of LaPO$_4$ ceramics.

Table 1   The calculated brittleness indexes of typical ceramics

	Vickers hardness H_v (GPa)	Fracture toughness K_{IC} (MPa·m$^{1/2}$)	Brittleness index B ($\mu m^{-1/2}$)
LaPO$_4$ [36]	4.86	1.0	4.86
Macor [39]	3.00	1.4	2.14
Ti$_3$SiC$_2$ [40]	4.00	6.0	0.67
Nb$_4$AlC$_3$ [41]	2.60	7.1	0.37
γ-Y$_2$Si$_2$O$_7$ [28]	6.20	2.1	2.95
YAM	4.30	1.2 [26]	3.58

As shown in Fig. 7, due to the potential application as TBC at high temperature, we measured the heat capacity and thermal conductivity of YAM up to 980 ℃. The thermal conductivity of YAM ceramics slightly increases between 400 ℃ and 600 ℃. In this temperature range, the dominant thermal conduction mechanism is phonon thermal conduction. In the temperature range between 600 ℃ and 980 ℃, the contribution of photon thermal conduction becomes significant. The thermal conductivity of dense YAM is 2.83 W/(m·K) at 980 ℃. The higher thermal conductivity compared with that ever reported by Zhan et al. [26] might be attributed to different morphology in local and the way of material preparation. The relatively low thermal conductivity together with excellent high temperature stability make YAM ceramics a promising candidate for TBC materials. As a comparison, the thermal conductivity of YAM is comparable to that of the widely-known TBC material—yttria-stabilized zirconia (YSZ, 2.0 W/(m·K) at 1127 ℃ [11]), while thermal stability of the former is definitely much better. As for YAG, another stable phase in the Y$_2$O$_3$–Al$_2$O$_3$ system with a thermal conductivity of 3.2 W/(m·K) at 1000 ℃, has been proposed as potential TBC materials [11]. Consequently, from this point of view, YAM might even have slight superiority over YAG in TBC material field.

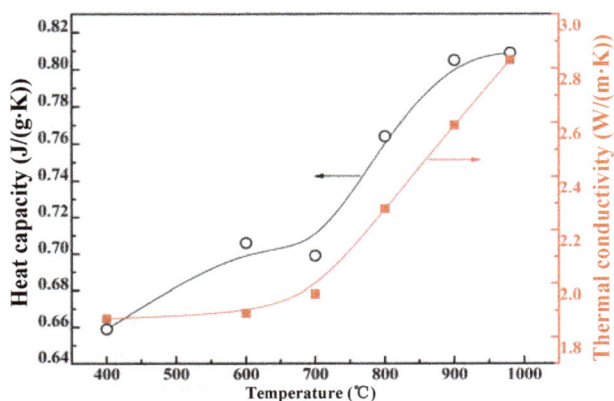

Fig. 7   Experimental measurements of heat capacity and thermal conductivity of the as-prepared YAM ceramics sintered at 1800 ℃.

4   Conclusions

Single-phase Y$_4$Al$_2$O$_9$ (YAM) powders were successfully synthesized by a solid-state reaction carried out at 1650 ℃ for 30 h starting from nano-sized Al_2O_3 and Y_2O_3 particles. Full dense YAM ceramics were obtained in a subsequent SPS consolidation step. When the SPS sintering temperature was 1800 ℃, the relative density of sintered body was as high as 99.5%. Fracture morphologies revealed the predominance of transgranular fracture mode. The intrinsic Vickers hardness of YAM was 4.3 GPa. The flexure strength of dense YAM ceramics was 110 MPa. It was confirmed that YAM ceramics were damage tolerant and machinable. The intergranular and transgranular energy dissipation mechanisms for crack propagation involved grain crushing and grain shear slipping. The thermal conductivity of YAM was 2.83 W/(m·K) at 980 ℃.

Acknowledgements

The present work was supported by "Chunlei Program" in Ningbo, "Hundred Talents Program" of the Chinese Academy of Sciences (No. KJCX2-EW-H06), National Natural Science Foundation of China (No. 51172248/E020301), and National Natural Science Foundation of China (Nos. 50772072 and 51072129).

References

[1] Abell JL, Harris IR, Cockayne B, et al. An investigation of phase stability in the Y_2O_3–Al_2O_3 system. J Mater Sci 1974, 9: 527–537.

[2] Xia GD, Zhou SM, Zhang JJ, et al. Structural and optical properties of YAG:Ce^{3+} phosphors by sol–gel combustion method. J Cryst Growth 2005, 279: 357–362.

[3] Won CW, Nersisyan HH, Won HI, et al. Efficient solid-state route for the preparation of spherical YAG:Ce phosphor particles. J Alloys Compd 2011, 509: 2621–2626.

[4] Rai P, Song M-K, Song H-M, et al. Synthesis, growth mechanism and photoluminescence of monodispersed cubic shape Ce doped YAG nanophosphor. Ceram Int 2012, 38: 235–242.

[5] Zhu QQ, Hu WW, Ju LC, et al. Synthesis of $Y_3Al_5O_{12}$:Eu^{2+} phosphor by a facile hydrogen iodide-assisted sol–gel method. J Am Ceram Soc 2013, 96: 701–703.

[6] Song HJ, Noh JH, Roh H-S, et al. Preparation and characterization of nano-sized $Y_3Al_5O_{12}$:Ce^{3+} phosphor by high-energy milling process. Curr Appl Phys 2013, DOI: 10.1016/j.cap.2013.01.033.

[7] Dong J, Shirakawa A, Ueda K, et al. Efficient Yb^{3+}:$Y_3Al_5O_{12}$ ceramic microchip lasers. Appl Phys Lett 2006, 89: 091114.

[8] Zhang WX, Zhou J, Liu WB, et al. Fabrication, properties and laser performance of Ho:YAG transparent ceramic. J Alloys Compd 2010, 506: 745–748.

[9] Qin XP, Yang H, Zhou GH, et al. Fabrication and properties of highly transparent Er:YAG ceramics. Opt Mater 2012, 34: 973–976.

[10] Fabrichnaya O, Seifert HJ, Ludwig T, et al. The assessment of thermodynamic parameters in the Al_2O_3–Y_2O_3 system and phase relations in the Y–Al–O system. Scand J Metall 2001, 30: 175–183.

[11] Su YJ, Trice RW, Faber KT, et al. Thermal conductivity, phase stability, and oxidation resistance of $Y_3Al_5O_{12}$ (YAG)/Y_2O_3–ZrO_2 (YSZ) thermal-barrier coatings. Oxid Met 2004, 61: 253–271.

[12] Chao W-H, Wu R-J, Wu T-B. Structural and luminescent properties of YAG:Ce thin film phosphor. J Alloys Compd 2010, 506: 98–102.

[13] Wu Y-C, Parola S, Marty O, et al. Structural characterizations and waveguiding properties of YAG thin films obtained by different sol–gel processes. Opt Mater 2005, 27: 1471–1479.

[14] Shim JB, Yoshikawa A, Nikl M, et al. Scintillation properties of the Yb-doped $YAlO_3$ crystals. Radiat Meas 2004, 38: 493–496.

[15] Medraj M, Hammond R, Parvez MA, et al. High temperature neutron diffraction study of the Al_2O_3–Y_2O_3 system. J Eur Ceram Soc 2006, 26: 3515–3524.

[16] Su H, Shen HY, Lin WX, et al. Computational model of Q-switch Nd:$YAlO_3$ dual-wavelength laser. J Appl Phys 1998, 84: 6519.

[17] Sullivan AC, Wagner GJ, Gwin D, et al. High power Q-switched Tm:$YAlO_3$ lasers. Advanced Solid-State Photonics, Santa Fe, New Mexico, February 1, 2004: WA7.

[18] Yadav R, Khan AF, Yadav A, et al. Intense red-emitting $Y_4Al_2O_9$:Eu^{3+} phosphor with short decay time and high color purity for advanced

plasma display panel. *Opt Express* 2009, **17**: 22023–22030.

[19] Wang DY, Wang YH. Photoluminescence of $Y_4Al_2O_9$:Re (Re = Tb^{3+}, Eu^{3+}) under VUV excitation. *J Alloys Compd* 2006, **425**: L5–L7.

[20] Toropov NA, Bondar IA, Galadhov FY, *et al.* Phase equilibria in the yttrium oxide–alumina system. *Russ Chem B+* 1964, **13**: 1076–1081.

[21] Yamane H, Omori M, Okubo A, *et al.* High-temperature phase transition of $Y_4Al_2O_9$. *J Am Ceram Soc* 1993, **76**: 2382–2384.

[22] Yamane H, Omori M, Hirai T. Twin structure of $Y_4Al_2O_9$. *J Mater Sci Lett* 1995, **14**: 561–563.

[23] Yamane H, Shimada M, Hunter BA. High-temperature neutron diffraction study of $Y_4Al_2O_9$. *J Solid State Chem* 1998, **141**: 466–474.

[24] Mah T-I, Keller KA, Sambasivan S, *et al.* High-temperature environmental stability of the compounds in the Al_2O_3–Y_2O_3 system. *J Am Ceram Soc* 1997, **80**: 874–878.

[25] Yang S, Lan X, Huang N. Role of $Y_4Al_2O_9$ in high temperature oxidation resistance of NiCoCrAlY–ZrO_2·Y_2O_3 coatings. *J Mater Sci Technol* 2007, **23**: 568–570.

[26] Zhan X, Li Z, Liu B, *et al.* Theoretical prediction of elastic stiffness and minimum lattice thermal conductivity of $Y_3Al_5O_{12}$, $YAlO_3$ and $Y_4Al_2O_9$. *J Am Ceram Soc* 2012, **95**: 1429–1434.

[27] López-Delgado A, Fillali L, Jiménez JA, *et al.* Synthesis of α-alumina from a less common raw material. *J Sol–Gel Sci Technol* 2012, **64**: 162–169.

[28] Sun ZQ, Zhou YC, Wang JY, *et al.* γ-$Y_2Si_2O_7$, a machinable silicate ceramic: Mechanical properties and machinability. *J Am Ceram Soc* 2007, **90**: 2535–2541.

[29] Mecholsky Jr. JJ, Freimam SW, Rice RW. Fracture surface analysis of ceramics. *J Mater Sci* 1976, **11**: 1310–1319.

[30] Gao H, Huang Y, Nix WD, *et al.* Mechanism-based strain gradient plasticity—I. Theory. *J Mech Phys Solids* 1999, **47**: 1239–1263.

[31] Bao YW, Hu CF, Zhou YC. Damage tolerance of nanolayer grained ceramics and quantitative estimation. *Mater Sci Tech-Long* 2006, **22**: 227–230.

[32] Padture NP, Evans CJ, Xu HHK, *et al.* Enhenced machinalibity of silicon carbide via microstructural design. *J Am Ceram Soc* 1995, **78**: 215–217.

[33] Baik DS, No KS, Chun JSS, *et al.* Mechanical properties of mica glass-ceramics. *J Am Ceram Soc* 1995, **78**: 1217–1222.

[34] Wang RG, Wei P, Jiang MN, *et al.* Investigation of the physical and mechanical properties of hot-pressed machinable Si_3N_4/h-BN composites and FGM. *Mat Sci Eng B* 2002, **90**: 261–268.

[35] Wang RG, Wei P, Chen J, *et al.* Fabrication and characterization of machinable Si_3N_4/h-BN functionally graded materials. *Mater Res Bull* 2002, **37**: 1269–1277.

[36] Morgan PED, Marshall DB. Ceramic composites of monazite and alumina. *J Am Ceram Soc* 1995, **78**: 1553–1563.

[37] Mogilevsky P, Parthasarathy TA, Petry MD. Anisotropy in room temperature microhardness and fracture of $CaWO_4$ scheelite. *Acta Mater* 2004, **52**: 5529–5537.

[38] Boccaccini AR. Machinability and brittleness of glass-ceramics. *J Mater Process Tech* 1997, **65**: 302–304.

[39] Davis JB, Marshall DB, Housley RM, *et al.* Machinable ceramics containing rare-earth phosphates. *J Am Ceram Soc* 1998, **81**: 2169–2175.

[40] Wang XH, Zhou YC. Microstructure and properties of Ti_3AlC_2 prepared by the solid–liquid reaction synthesis and simultaneous in-situ hot pressing process. *Acta Mater* 2002, **50**: 3143–3151.

[41] Hu CF, Li FZ, He LF, *et al.* In situ reaction synthesis, electrical and thermal, and mechanical properties of Nb_4AlC_3. *J Am Ceram Soc* 2008, **91**: 2258–2263.

Permissions

The contributors of this book come from diverse backgrounds, making this book a truly international effort. This book will bring forth new frontiers with its revolutionizing research information and detailed analysis of the nascent developments around the world.

We would like to thank all the contributing authors for lending their expertise to make the book truly unique. They have played a crucial role in the development of this book. Without their invaluable contributions this book wouldn't have been possible. They have made vital efforts to compile up to date information on the varied aspects of this subject to make this book a valuable addition to the collection of many professionals and students.

This book was conceptualized with the vision of imparting up-to-date information and advanced data in this field. To ensure the same, a matchless editorial board was set up. Every individual on the board went through rigorous rounds of assessment to prove their worth. After which they invested a large part of their time researching and compiling the most relevant data for our readers.

The editorial board has been involved in producing this book since its inception. They have spent rigorous hours researching and exploring the diverse topics which have resulted in the successful publishing of this book. They have passed on their knowledge of decades through this book. To expedite this challenging task, the publisher supported the team at every step. A small team of assistant editors was also appointed to further simplify the editing procedure and attain best results for the readers.

Apart from the editorial board, the designing team has also invested a significant amount of their time in understanding the subject and creating the most relevant covers. They scrutinized every image to scout for the most suitable representation of the subject and create an appropriate cover for the book.

The publishing team has been an ardent support to the editorial, designing and production team. Their endless efforts to recruit the best for this project, has resulted in the accomplishment of this book. They are a veteran in the field of academics and their pool of knowledge is as vast as their experience in printing. Their expertise and guidance has proved useful at every step. Their uncompromising quality standards have made this book an exceptional effort. Their encouragement from time to time has been an inspiration for everyone.

The publisher and the editorial board hope that this book will prove to be a valuable piece of knowledge for researchers, students, practitioners and scholars across the globe.

List of Contributors

Piyush R. DAS
Department of Physics, Institute of Technical Education & Research, Siksha 'O' Anusandahan University, Khandagiri, Bhubaneswar 751030, Odisha, India

B. N. PARIDA
Department of Physics, Institute of Technical Education & Research, Siksha 'O' Anusandahan University, Khandagiri, Bhubaneswar 751030, Odisha, India

R. PADHEE
Department of Physics, Institute of Technical Education & Research, Siksha 'O' Anusandahan University, Khandagiri, Bhubaneswar 751030, Odisha, India

R. N. P. CHOUDHARY
Department of Physics, Institute of Technical Education & Research, Siksha 'O' Anusandahan University, Khandagiri, Bhubaneswar 751030, Odisha, India

Valdirlei Fernandes FREITAS
UEM — Universidade Estadual de Maringá, GDDM — Grupo de Desenvolvimento de Dispositivos Multifuncionais, Departamento de Física, Maringá, Brazil

Gustavo Sanguino DIAS
UEM — Universidade Estadual de Maringá, GDDM — Grupo de Desenvolvimento de Dispositivos Multifuncionais, Departamento de Física, Maringá, Brazil

Otávio Algusto PROTZEK
UEM — Universidade Estadual de Maringá, GDDM — Grupo de Desenvolvimento de Dispositivos Multifuncionais, Departamento de Física, Maringá, Brazil

Diogo Zampieri MONTANHER
UEM — Universidade Estadual de Maringá, GDDM — Grupo de Desenvolvimento de Dispositivos Multifuncionais, Departamento de Física, Maringá, Brazil

Igor Barbosa CATELLANI
UEM — Universidade Estadual de Maringá, GDDM — Grupo de Desenvolvimento de Dispositivos Multifuncionais, Departamento de Física, Maringá, Brazil

Daniel Matos SILVA
UEM — Universidade Estadual de Maringá, GDDM — Grupo de Desenvolvimento de Dispositivos Multifuncionais, Departamento de Física, Maringá, Brazil

Luiz Fernando CÓTICA
UEM — Universidade Estadual de Maringá, GDDM — Grupo de Desenvolvimento de Dispositivos Multifuncionais, Departamento de Física, Maringá, Brazil

Ivair Aparecido dos SANTOS
UEM — Universidade Estadual de Maringá, GDDM — Grupo de Desenvolvimento de Dispositivos Multifuncionais, Departamento de Física, Maringá, Brazil

Lianfeng DUAN
Key Laboratory of Advanced Structural Materials, Ministry of Education, and Department of Materials Science and Engineering, Changchun University of Technology, Changchun 130012, China

Fenghui GAO
Key Laboratory of Advanced Structural Materials, Ministry of Education, and Department of Materials Science and Engineering, Changchun University of Technology, Changchun 130012, China

Limin WANG
State Key Laboratory of Rare Earth Resource Utilization, Changchun Institute of Applied Chemistry, Chinese Academy of Sciences, Changchun 130022, China

Songzhe JIN
Key Laboratory of Advanced Structural Materials, Ministry of Education, and Department of Materials Science and Engineering, Changchun University of Technology, Changchun 130012, China

Hua WU
Key Laboratory of Advanced Structural Materials, Ministry of Education, and Department of Materials Science and Engineering, Changchun University of Technology, Changchun 130012, China

Yogesh A. CHAUDHARI
Department of Physics, School of Physical Sciences, North Maharashtra University, Jalgaon 425001, India
Department of Engineering Sciences and Humanities (DESH), SRTTC-FOE, Kamshet, Pune 410405, India

Chandrashekhar M. MAHAJAN
Department of Engineering Sciences and Humanities (DESH), Vishwakarma Institute of Technology (VIT), Pune 411037, India

Prashant P. JAGTAP
Department of Physics, School of Physical Sciences, North Maharashtra University, Jalgaon 425001, India

Subhash T. BENDRE
Department of Physics, School of Physical Sciences, North Maharashtra University, Jalgaon 425001, India

Xuming PANG
Department of Mechanical Engineering, Nanjing Tech University, Nanjing 210009, China

Jinhao QIU
State Key Laboratory of Mechanics and Control of Mechanical Structures, Nanjing University of Aeronautics and Astronautics, Nanjing 210016, China

Kongjun ZHU
State Key Laboratory of Mechanics and Control of Mechanical Structures, Nanjing University of Aeronautics and Astronautics, Nanjing 210016, China

Awadesh Kumar MALLIK
CSIR -Central Glass & Ceramic Research Institute, Kolkata 700032, West Bengal, India

Sandip BYSAKH
CSIR -Central Glass & Ceramic Research Institute, Kolkata 700032, West Bengal, India

Monjoy SREEMANY
CSIR -Central Glass & Ceramic Research Institute, Kolkata 700032, West Bengal, India

Sudakshina ROY
CSIR -Central Glass & Ceramic Research Institute, Kolkata 700032, West Bengal, India

Jiten GHOSH
CSIR -Central Glass & Ceramic Research Institute, Kolkata 700032, West Bengal, India

Soumyendu ROY
Department of Physics, Indian Institute of Technology Bombay, Powai, Mumbai 400076, Maharashtra, India

Joana Catarina MENDES
Nanotechnology Research Division, Centre for Mechanical Technology and Automation, University of Aveiro, 3810-193, Portugal

Jose GRACIO
Instituto de Telecomunicações, Campus Universitário de Santiago, 3810-193, Portugal

Someswar DATTA
CSIR -Central Glass & Ceramic Research Institute, Kolkata 700032, West Bengal, India

Adhimoolam Bakthavachalam KOUSAALYA
aMaterials Processing Section, Department of Metallurgical and Materials Engineering, Indian Institute of Technology Madras, Chennai 600036, Tamil Nadu, India

Ravi KUMAR
Materials Processing Section, Department of Metallurgical and Materials Engineering, Indian Institute of Technology Madras, Chennai 600036, Tamil Nadu, India

Shanmugam PACKIRISAMY
Analytical Spectroscopy and Ceramics Group, PCM Entity, Vikram Sarabhai Space Centre, Thiruvananthapuram 695022, India

Mahmoud Aly HAMAD
Physics Department, Faculty of Science, Tanta University, Tanta, Egypt

Huiqin LI
State Key Laboratory Cultivation Base for Nonmetal Composite and Functional Materials, Southwest University of Science and Technology, Mianyang, Sichuan, China

Jingsong LIU
State Key Laboratory Cultivation Base for Nonmetal Composite and Functional Materials, Southwest University of Science and Technology, Mianyang, Sichuan, China

Hongtao YU
State Key Laboratory Cultivation Base for Nonmetal Composite and Functional Materials, Southwest University of Science and Technology, Mianyang, Sichuan, China

Shuren ZHANG
State Key Laboratory of Electronic Thin Films and Integrated Devices, University of Electronic Science and Technology of China, Chengdu, Sichuan, China

Umasankar DASH
School of Applied Sciences, KIIT University, Bhubaneswar-751 024, India

Subhanarayan SAHOO
Department of Electrical and Electronics Engineering, Trident Academy of Technology, Bhubaneswar-751 024, India

Paritosh CHAUDHURI
Institute for Plasma Research, Bhat, Gandhinagar-382 428, India

S. K. S. PARASHAR
School of Applied Sciences, KIIT University, Bhubaneswar-751 024, India

Kajal PARASHAR
School of Applied Sciences, KIIT University, Bhubaneswar-751 024, India

Qiming HANG
National Laboratory of Solid State Microstructures, School of Physics, Nanjing University, Nanjing 210093, China

Wenke ZHOU
National Laboratory of Solid State Microstructures, School of Physics, Nanjing University, Nanjing 210093, China

Xinhua ZHU
National Laboratory of Solid State Microstructures, School of Physics, Nanjing University, Nanjing 210093, China

Jianmin ZHU
National Laboratory of Solid State Microstructures, School of Physics, Nanjing University, Nanjing 210093, China

Zhiguo LIU
National Laboratory of Solid State Microstructures, Department of Materials Science and Engineering, Nanjing University, Nanjing 210093, China

Talaat AL-KASSAB
King Abdullah University of Science & Technology (KAUST), Physical Sci. and Eng., Thuwal 23955-6900, Kingdom of Saudi Arabia

Rafidah HASSAN
Physics Department, Faculty of Science, Universiti Putra Malaysia, 43400 UPM Serdang, Selangor, Malaysia

Jumiah HASSAN
Physics Department, Faculty of Science, Universiti Putra Malaysia, 43400 UPM Serdang, Selangor, Malaysia
Institute of Advanced Technology, University Putra Malaysia, 43400 UPM Serdang, Selangor, Malaysia

Mansor HASHIM
Physics Department, Faculty of Science, Universiti Putra Malaysia, 43400 UPM Serdang, Selangor, Malaysia
Institute of Advanced Technology, University Putra Malaysia, 43400 UPM Serdang, Selangor, Malaysia

Suriati PAIMAN
Physics Department, Faculty of Science, Universiti Putra Malaysia, 43400 UPM Serdang, Selangor, Malaysia

Raba'ah Syahidah AZIS
Physics Department, Faculty of Science, Universiti Putra Malaysia, 43400 UPM Serdang, Selangor, Malaysia
Institute of Advanced Technology, University Putra Malaysia, 43400 UPM Serdang, Selangor, Malaysia

Biplab Kumar PAUL
Physics Department, Jadavpur University, Kolkata-700 032, India

Kumaresh HALDAR
Physics Department, Jadavpur University, Kolkata-700 032, India

Debasis ROY
Physics Department, Jadavpur University, Kolkata-700 032, India

Biswajoy BAGCHI
Physics Department, Jadavpur University, Kolkata-700 032, India

Alakananda BHATTACHARYA
Physics Department, Jadavpur University, Kolkata-700 032, India

Sukhen DAS
Physics Department, Jadavpur University, Kolkata-700 032, India

Oleg SMORYGO
Powder Metallurgy Institute, 41, Platonov Str., 220005, Minsk, Belarus

Alexander MARUKOVICH
Powder Metallurgy Institute, 41, Platonov Str., 220005, Minsk, Belarus

Vitali MIKUTSKI
Powder Metallurgy Institute, 41, Platonov Str., 220005, Minsk, Belarus

Vladislav SADYKOV
Boreskov Institute of Catalysis, 5, Lavrentiev Ave., 630090, Novosibirsk, Russia

Abdel-Mageed H. KHAFAGY
Physics Department, Faculty of Science, Menufiya University, Shebin El-Koom 32511, Egypt

Sanaa M. EL-RABAIE
Physics and Engineering Mathematics Department, Faculty of Electronic Engineering, Menufiya University, Menouf 32952, Egypt

Mohamed T. DAWOUD
Physics and Engineering Mathematics Department, Faculty of Electronic Engineering, Menufiya University, Menouf 32952, Egypt

M. T. ATTIA
Physics and Engineering Mathematics Department, Faculty of Electronic Engineering, Menufiya University, Menouf 32952, Egypt

Laxman SINGH
Department of Chemistry, Centre of Advanced Study, Faculty of Science, Banaras Hindu University, Varanasi 221005, U.P., India

U. S. RAI
Department of Chemistry, Centre of Advanced Study, Faculty of Science, Banaras Hindu University, Varanasi 221005, U.P., India

Alok Kumar RAI
Department of Materials Science and Engineering, Chonnam National University, 300 Yongbong-dong, Bukgu, Gwangju 500-757, Republic of Korea

K. D. MANDAL
Department of Applied Chemistry, Indian Institute of Technology, Banaras Hindu University, Varanasi 221005, U.P., India

Paweł J. RUTKOWSKI
Department of Ceramics and Refractories, Faculty of Materials Science and Ceramics, AGH University of Science and Technology, al. Mickiewicza 30, 30-059 Krakow, Poland

Dariusz KATA
Department of Ceramics and Refractories, Faculty of Materials Science and Ceramics, AGH University of Science and Technology, al. Mickiewicza 30, 30-059 Krakow, Poland

R. V. BARDE
Department of Engineering Physics, Shri Hanuman Vyayam Prasarak Mandal's College of Engineering and Technology, Amravati 444 605, India

S. A. WAGHULEY
Department of Physics, Sant Gadge Baba Amravati University, Amravati 444 602, India

Hana NACEUR
Laboratory of Applied Mineral Chemistry, Department of Chemistry, Faculty of Sciences, University Tunis ElManar, Campus 2092, Tunis, Tunisia

Adel MEGRICHE
Laboratory of Applied Mineral Chemistry, Department of Chemistry, Faculty of Sciences, University Tunis ElManar, Campus 2092, Tunis, Tunisia

Mohamed EL MAAOUI
Laboratory of Applied Mineral Chemistry, Department of Chemistry, Faculty of Sciences, University Tunis ElManar, Campus 2092, Tunis, Tunisia

Samita PATTANAYAK
Department of Physics, Institute of Technical Education & Research, Siksha 'O' Anusandhan University, Bhubaneswar 751030, India

Ashwasa PRIYADARSHAN
Department of Physics, Institute of Technical Education & Research, Siksha 'O' Anusandhan University, Bhubaneswar 751030, India

Ritesh SUBUDHI
Department of Physics, Institute of Technical Education & Research, Siksha 'O' Anusandhan University, Bhubaneswar 751030, India

Ranjan Kumar NAYAK
Department of Physics, Institute of Technical Education & Research, Siksha 'O' Anusandhan University, Bhubaneswar 751030, India

Rajib PADHEE
Department of Physics, Institute of Technical Education & Research, Siksha 'O' Anusandhan University, Bhubaneswar 751030, India

P. S. DAS
Department of Physics, Midnapore College, Midnapore 721 101, India

P. K. CHAKRABORTY
Department of Physics, Midnapore College, Midnapore 721 101, India

Banarji BEHERA
School of Physics, Sambalpur University, Burla 768 019, Sambalpur, Odisha, India

N. K. MOHANTY
School of Physics, Sambalpur University, Burla 768 019, Sambalpur, Odisha, India

R. N. P. CHOUDHARY
Department of Physics, ITER, Siksha "O" Anusandhan University, Bhubaneswar 751 030, Odisha, India

Arvind YOGI
Materials Science Laboratory, School of Physics, Vigyan Bhawan, Devi Ahilya University, Khandwa Road Campus, Indore 452001, India
School of Physics, Indian Institute of Science Education and Research, Thiruvananthapuram 695016, India

Dinesh VARSHNEY
Materials Science Laboratory, School of Physics, Vigyan Bhawan, Devi Ahilya University, Khandwa Road Campus, Indore 452001, India

George A. GOGOTSI
Pisarenko Institute for Problems of Strength, 2, Timiryazevskaya Str., 01014, Kiev, Ukraine

Qian WANG
Ningbo Institute of Materials Technology and Engineering (NIMTE), Chinese Academy of Sciences (CAS), Ningbo 315201, China

Salvatore GRASSO
School of Engineering & Materials Science and Nanoforce Technology Ltd., Queen Mary University of London, Mile End Road, London E1 4NS, UK

Chunfeng HU
Ningbo Institute of Materials Technology and Engineering (NIMTE), Chinese Academy of Sciences (CAS), Ningbo 315201, China

Haibin ZHANG
China Academy of Engineering Physics, Mianyang 621900, China

Shu CAI
Key Laboratory for Advanced Ceramics and Machining Technology of Ministry of Education, Tianjin University, Tianjin 300072, China

Yoshio SAKKA
National Institute for Materials Science (NIMS), Tsukuba, Ibaraki 305-0047, Japan

Qing HUANG
Ningbo Institute of Materials Technology and Engineering (NIMTE), Chinese Academy of Sciences (CAS), Ningbo 315201, China